Electrical Power Systems Technology

Third Edition

Electrical Power Systems Technology

Third Edition

Stephen W. Fardo
Dale R. Patrick

THE FAIRMONT PRESS, INC.

CRC Press
Taylor & Francis Group

Library of Congress Cataloging-in-Publication Data

Fardo, Stephen W.
  Electrical power systems technology / Stephen W. Fardo, Dale R. Patrick. --3rd ed.
    p. cm.
  Includes index.
    ISBN-10: 0-88173-585-X (alk. paper) -- ISBN-10: 0-88173-586-8 (electronic) -- ISBN-13: 978-1-4398-0027-0 (Taylor & Francis : alk. paper)
    1. Electric power systems. 2. Electric machinery. I. Patrick, Dale R. II. Title.

TK1001.F28 2008
621.31--dc22

2008033611

Electrical Power Systems Technology / Stephen W. Fardo, Dale R. Patrick.
©2009 by The Fairmont Press. All rights reserved. No part of this publication may be reproduced or transmitted in any form or by any means, electronic or mechanical, including photocopy, recording, or any information storage and retrieval system, without permission in writing from the publisher.

Published by The Fairmont Press, Inc.
700 Indian Trail
Lilburn, GA 30047
tel: 770-925-9388; fax: 770-381-9865
http://www.fairmontpress.com

Distributed by Taylor & Francis Ltd.
6000 Broken Sound Parkway NW, Suite 300
Boca Raton, FL 33487, USA
E-mail: orders@crcpress.com

Distributed by Taylor & Francis Ltd.
23-25 Blades Court
Deodar Road
London SW15 2NU, UK
E-mail: uk.tandf@thomsonpublishingservices.co.uk

Printed in the United States of America
10 9 8 7 6 5 4 3 2 1

10: 0-88173-585-X (The Fairmont Press, Inc.)
13: 978-1-4398-0027-0 (Taylor & Francis Ltd.)

While every effort is made to provide dependable information, the publisher, authors, and editors cannot be held responsible for any errors or omissions.

Contents

Preface ix

UNIT I	POWER MEASUREMENT SYSTEMS AND FUNDAMENTALS 1
Chapter 1	Power Measurement Fundamentals 5
	Units of Measurement—Conversion of SI Units—Scientific Notation
Chapter 2	Power System Fundamentals 15
	The System Concept—Basic System Functions—A Simple Electrical System Example—Energy, Work, and Power—Types of Electrical Circuits—Power in DC Electrical Circuits—Maximum Power Transfer—Overview of Alternating Current (AC) Circuits—Vector and Phasor Diagrams—Impedance in AC Circuits—Power Relationships in AC Circuits—Power Relationships in Three—Phase Circuits
Chapter 3	Power Measurement Equipment 59
	Measurement Systems—Measuring Electrical Power—Measuring Electrical Energy—Measuring Three-Phase Electrical Energy—Frequency Measurement—Synchroscopes—Ground-Fault Indicators—Megohmeters—Clamp-On Meters Telemetering Systems
UNIT II	ELECTRICAL POWER PRODUCTION SYSTEMS 79
Chapter 4	Modern Power Systems 83
	Electrical Power Plants—Fossil Fuel Systems—Steam Turbines—Boilers—Hydroelectric Systems—Nuclear Fission Systems—Operational Aspects of Modem Power Systems
Chapter 5	Alternative Power Systems 117
	Potential Power Sources—Solar Energy Systems—Geothermal Power Systems—Wind Systems—Magnetohydrodynamic (MHD) Systems—Nuclear-Fusion Power Systems—Nuclear-Fusion Methods—Future of Nuclear Fusion—Fuel-Cell Systems—Tidal Power Systems—Coal-Gasification Fuel Systems—Oil-Shale Fuel-Production Systems—Alternative Nuclear Power Plants—Biomass Systems

Chapter 6   Alternating Current Power Systems 137
            Electromagnetic Induction—Basic Generator Operation—
            Single-Phase AC Power Systems—Single-Phase AC Generators—
            Three-Phase AC Generators—High-Speed and Low-Speed
            Generators—Generator Frequency—Generator Voltage
            Regulation—Generator Efficiency

Chapter 7   Direct Current Power Systems 157
            DC Production Using Chemical Cells—Characteristics of Primary
            Cells—Characteristics of Secondary Cells—DC Generating Sys-
            tems—DC Conversion Systems—DC Filtering Methods—DC Reg-
            ulation Methods

UNIT III   ELECTRICAL POWER DISTRIBUTION SYSTEMS 203

Chapter 8   Power Distribution Fundamentals 207
            Overview of Electrical Power Distribution—Power Transmission
            and Distribution—Radial, Ring, and Network Distribution
            Systems—Use of Transformers for Power Distribution—
            Conductors in Power Distribution Systems—Conductor Area—
            Resistance of Conductors—Conductor Sizes and Types—
            Ampacity of Conductors—Ampacity Tables—Use of Insulation in
            Power Distribution Systems

Chapter 9   Power Distribution Equipment... 239
            Equipment Used at Substations—Power System Protective
            Equipment—Power Distribution Inside Industrial and
            Commercial Buildings—The Electrical Service Entrance—Service
            Entrance Terminology

Chapter 10 Single-Phase and Three-Phase
            Distribution Systems 255
            Single-Phase Systems— Three-Phase Systems—Grounding
            of Distribution Systems—System Grounding—Ground-Fault
            Protection—Wiring Design Considerations for Distribution
            Systems—Branch Circuit Design Considerations—Feeder Circuit
            Design Considerations—Determining Grounding Conductor
            Size—Parts of Interior Electrical Wiring Systems

UNIT IV   ELECTRICAL POWER CONVERSION SYSTEMS 289

Chapter 11 Fundamentals of Electrical Loads.. 293
            Load Characteristics—Three-Phase Load Characteristics

| Chapter 12 | Heating Systems .... 307 |

Basic Heating Loads—Electrical Welding Loads—Power Considerations for Electric Welders—Electric Heating and Air Conditioning Systems

| Chapter 13 | Lighting Systems .... 327 |

Characteristics of Light—Electrical Lighting Circuits—Branch Circuit Design—Lighting Fixture Design—Factors in Determining Light Output

| Chapter 14 | Mechanical Systems .... 349 |

Basic Motor Principles—DC Motors—Specialized DC Motors—Single-Phase AC Motors—Three-Phase AC Motors—Specialized Mechanical Power Systems—Electric Motor Applications

| UNIT V | ELECTRICAL POWER CONTROL SYSTEMS .... 401 |

| Chapter 15 | Power Control Devices .... 405 |

Power Control Standards, Symbols, and Definitions—Power Control Using Switches—Control Equipment for Electric Motors—other Electromechanical Power Control Equipment—Electronic Power Control

| Chapter 16 | Operational Power Control Systems .... 427 |

Basic Control Systems—Motor—Starting Systems—Specialized Control Systems—Frequency—Conversion Systems—Programming the PLC

| Chapter 17 | Control Devices .... 453 |

Silicon Controlled Rectifiers—SCR Construction—SCR I-V Characteristics—DC Power Control with SCRs—AC Power Control with SCRs—Triac Power Control—Triac Construction—Triac Operation—Triac I-V Characteristics—Triac Applications—Static Switching—Start-Stop Triac Control—Triac Variable Power Control—Diac Power Control—Electronic Control Considerations

| Appendix A | Trigonometric Functions .... 471 |

| Appendix B | The Elements .... 473 |

| Appendix C | Metric Conversions .... 475 |

Index 481

Preface

Electrical Power Systems Technology (Third Edition) provides a broad overview of the production, distribution, control, conversion, and measurement of electrical power. The presentation method used in this book will allow the reader to develop an understanding of electrical power systems. The units of the book are organized in a systematic manner, beginning with electrical power production methods. The fundamentals of each major unit of the book are discussed at the beginning of the unit. These fundamentals provide a framework for the information that follows in each unit. The last unit has been expanded to include control devices.

This book deals with many important aspects of electrical power, not just with one or two areas. In this way, it will give the reader a better understanding of the *total* electrical power system—from the production of electricity to its conversion to other forms of energy. Each unit deals with a specific system, such as production, distribution, control, conversion, or measurement. Each system is broken down into subsystems. The subsystems are then explored in greater detail in the chapters that make up each unit.

In order to understand the contents of this book in depth, the reader should have a knowledge of basic electrical fundamentals. The mathematical presentations given are very simple and are used only to show the practical relationships that are important in electrical power system operation. This book is recommended as a textbook for an "electrical power" or "electrical generators and motors" course. It would be a suitable text for vocational-technical schools, community colleges, universities, and, possibly, some technical high school programs. Many illustrations are shown, to make the presentations that are given easier to understand. The content is presented in such a way that any reader should be able to learn a great deal about the operation of electrical power systems.

Stephen W. Fardo
Dale R. Patrick
Eastern Kentucky University
Richmond, KY 40475

UNIT I

Power Measurement Systems and Fundamentals

In order to understand *electrical power measurement systems,* we must first study the fundamentals of measurement. These fundamentals deal mainly with the characteristics and types of measurement systems. Measurement systems are discussed in Chapter 1.

Chapter 2 provides an overview of the *fundamentals* that are important in the study of electrical power systems.

Chapter 3 deals with *measurement equipment and methods* associated with electrical power systems. These measurement systems include single-phase and three-phase wattmeters, power factor meters, ground-fault indicators, and many other types of equipment used in the analysis of electrical power system operation.

Figure I shows a block diagram of the *electrical power systems model* used in this textbook. This model is used to divide electrical power systems into five important systems: (1) *Power Measurement,* (2) *Power Production,* (3) *Power Distribution,* (4) *Power Conversion,* and (5) *Power Control.*

UNIT OBJECTIVES

Upon completion of Unit I, Power Measurement Systems and Fundamentals, you should be able to:

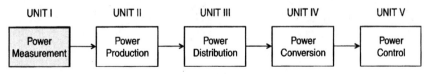

Power Measurement Fundamentals (Chapter 1)
Power System Fundamentals (Chapter 2)
Power Measurement Equipment (Chapter 3)

Figure I. Electrical power systems model

1. Compare the basic systems used for measurement.
2. Convert quantities from small units to large units of measurement.
3. Convert quantities from large units to small units of measurement.
4. Convert quantities from English to metric units.
5. Convert quantities from metric to English units.
6. Explain the parts of an electrical system.
7. Calculate power using the proper power formulas.
8. Draw diagrams illustrating the phase relationship between current and voltage in a capacitive circuit or inductive circuit.
9. Define capacitive reactance and inductive reactance.
10. Solve problems using the capacitive reactance formula and inductive reactance formula.
11. Define impedance.
12. Calculate impedance of series and parallel AC circuits.
13. Determine current in AC circuits.
14. Explain the relationship between AC voltages and current in resistive circuits.
15. Describe the effect of capacitors and inductors in series and in parallel.
16. Explain the characteristics of series and parallel AC circuits.
17. Solve Ohm's law problems for AC circuits.
18. Solve problems involving true power, apparent power, power factor, and reactive power in AC circuits.
19. Explain the difference between AC and DC.
20. Define the process of electromagnetic induction.
21. Describe factors affecting induced voltage.
22. Draw a simple AC generator and explain AC voltage generation.
23. Convert peak, peak-peak, average, and RMS/effective values from one to the other.
24. Describe voltage, current, and power relationships in three-phase AC circuits for wye and delta configurations.
25. Describe the following basic types of measurement systems:
    Analog Instruments
    Comparative Instruments
    CRT Display Instruments
    Numerical Readout Instruments
    Chart Recording Instruments

Unit Objectives

26. Explain the operation of an analog meter movement.
27. Describe the function of a Wheatstone bridge.
28. Explain the use of the dynamometer movement of a wattmeter to measure electrical power.
29. Describe the use of a watt-hour meter to measure electrical energy.
30. Interpret numerical readings taken by a watt-hour meter.
31. Explain the use of a power analyzer to monitor three-phase power.
32. Describe the measurement of power factor with a power factor meter.
33. Calculate power demand.
34. Explain the monitoring of power demand.
35. Explain the methods of measuring frequency.
36. Explain the use of a synchroscope.
37. Describe the use of a ground fault indicator.
38. Describe the use of a megohmmeter to measure high resistance values.
39. Describe the operation of a clamp-on current meter.
40. Describe a telemetering system.

Chapter 1

Power Measurement Fundamentals

Electrical power measurements are important quantities, which must be measured precisely. Electrical power systems are dependent upon accurate measurements for everyday operation. Thus, many types of measurements and measuring equipment are associated with electrical power systems. Measurement fundamentals will be discussed in the following sections.

Today, most nations of the world use the metric system of measurement. In the United States, the National Bureau of Standards began a study in 1968 to determine the feasibility and costs of converting the nation to the metric measurement system. Today, this conversion is incomplete.

The units of the metric system are decimal measures based on the kilogram and the meter. Although the metric system is very simple, several countries have been slow to adopt it. The United States has been one of these reluctant countries, because of the complexity of actions required by a complete changeover of measurement systems.

IMPORTANT TERMS

Chapter 1 deals with power measurement fundamentals. After studying this chapter, you should have an understanding of the following terms:

Units of Measurement
Measurement Standards
English System of Units
International System of Units (SI)
Unit Conversion Tables
Base Units

Derived Units
Small Unit Prefixes
Large Unit Prefixes
Conversion Scale
Scientific Notation
Powers of 10
Electrical Power Units

UNITS OF MEASUREMENT

Units of measurement have a significant effect on our lives, but we often take them for granted. Almost everything we deal with daily is measured by using some unit of measurement. For example, such units allow us to measure the distance traveled in an automobile, the time of day, and the amount of food we eat during a meal. Units of measurement have been in existence for many years; however, they are now more precisely defined than they were centuries ago. Most units of measurement are based on the laws of physical science. For example, distance is measured in reference to the speed of light, and time is measured according to the duration of certain atomic vibrations.

The *standards* we use for measurement have an important effect on modern technology. Units of measurement must be recognized by all countries of the world. There must be ways to compare common units of measurement among different countries. Standard units of length, mass, and time are critical to international marketing and to business, industry, and science in general.

The *English system of units*, which uses such units as the inch, foot, and pound, has been used in the United States for many years. However, many other countries use the metric system, which has units such as kilometers, centimeters, and grams. The metric system is also called the *International System of Units*, and is abbreviated SI. Although the English and SI systems of measurement have direct numerical relationships, it is difficult for individuals to change from one to the other. People form habits of using either the English or the SI system.

Since both systems of measurement are used, this chapter will familiarize you with both systems, and with the conversion of units from one to the other. The *conversion tables* of Appendix C should be helpful. The SI system, which was introduced in 1960, has several advantages over

Power Measurement Fundamentals

the English system of measurement. It is a decimal system that uses units commonly used in business and industry, such as volts, watts, and grams. The SI system can also be universally used with ease. However, the use of other units is sometimes more convenient.

The SI system of units is *based on seven units,* which are shown in Table 1-1. Other units are derived from the base units and are shown in Table 1-2.

Table 1-1. Base Units of the SE System

Measurement Quantity	Unit	Symbol
Length	meter	m
Mass	kilogram	kg
Time	second	s
Electric current	ampere	A
Temperature	kelvin	K
Luminous Intensity	candela	cd
Amount of substance	mole	mol

Table 1-2. Derived Units of the SI System

Measurement Quantity	Unit	Symbol
Electric capacitance	farad	F
Electric charge	coulomb	C
Electric conductance	siemen	S
Electric potential	volt	V
Electric resistance	ohm	Ω
Energy	joule	J
Force	newton	N
Frequency	hertz	Hz
Illumination	lux	lx
Inductance	henry	H
Luminous intensity	lumen	lm
Magnetic flux	weber	Wb
Magnetic flux density	tesla	T
Power	watt	W
Pressure	pascal	Pa

Some definitions of *base units* are included below:

1. Unit of length: METER (m)—the length of the path that light travels in a vacuum during the time of 1/29,792,458 second (the speed of light).
2. Unit of mass: KILOGRAM (kg)—the mass of the international prototype, which is a cylinder of platinum-iridium alloy material stored in a vault at Sevres, France, and preserved by the International Bureau of Weights and Measures.
3. Unit of time: SECOND (s)—the duration of 9,192,631,770 periods of radiation corresponding to the transition between two levels of a Cesium-133 atom. (This is extremely stable and accurate.)
4. Unit of electric current: AMPERE (A)—the current that, if maintained in two straight parallel conductors of infinite length, placed 1 meter apart in a vacuum, will produce a force of 2×10^{-7} newtons per meter between the two conductors.
5. Unit of temperature: KELVIN (K)—an amount of 1/273.16 of the temperature of the triple point of water. (This is where ice begins to form, and ice, water, and water vapor exist at the same time.) Thus, 0 degrees Centigrade = 273.16 Kelvins.
6. Unit of luminous intensity: CANDELA (cd)—the intensity of a source that produces radiation of a frequency of 540×10^{12} Hertz.
7. Unit of amount of substance: MOLE (mol)—an amount that contains as many atoms, molecules, or other specified particles as there are atoms in 0.012 kilograms of Carbon-12.

As you can see, these are highly precise units of measurement. The definitions are included to illustrate that point. Below, a few examples of *derived units* are also listed:

1. Unit of energy: JOULE (J)—the work done when one newton is applied at a point and displaced a distance of one meter in the direction of the force; 1 joule = 1 newton meter.
2. Unit of power: WATT (W)—the amount of power that causes the production of energy at a rate of 1 joule per second; 1 watt = 1 joule per second.
3. Unit of capacitance: FARAD (F)—the capacitance of a capacitor in which a difference of potential of 1 volt appears between its plates when it is charged to 1 coulomb; 1 farad = 1 coulomb per volt.

Power Measurement Fundamentals 9

4. Unit of electrical charge: COULOMB (C)—the amount of electrical charge transferred in 1 second by a current of 1 ampere; 1 coulomb = 1 ampere per second.

CONVERSION OF SI UNITS

Sometimes it is necessary to make conversions of SI units, so that very large or very small numerals may be avoided. For this reason, decimal *multiples* and *submultiples* of the base units have been developed, by using standard prefixes. These standard prefixes are shown in Table 1-3. Multiples and submultiples of SI units are produced by adding prefixes to the base unit. Simply multiply the value of the unit by the factors listed in Table 1-3. For example:

1 kilowatt = 1000 watts
1 microampere = 10^{-6} ampere
1 megohm = 1,000,000 ohms

Table 1-3. SI Standard Prefixes

Prefix	Symbol	Factor by Which the Unit is Multiplied
exa	E	1,000,000,000,000,000,000 = 10^{18}
peta	P	1,000,000,000,000,000 = 10^{15}
tera	T	1,000,000,000,000 = 10^{12}
giga	G	1,000,000,000 = 10^{9}
mega	M	1,000,000 = 10^{6}
kilo	k	1,000 = 10^{3}
hecto	h	100 = 10^{2}
deka	da	10 = 10^{1}
deci	d	0.1 = 10^{-1}
centi	c	0.01 = 10^{-2}
milli	m	0.001 = 10^{-3}
micro	μ	0.000001 = 10^{-6}
nano	n	0.000000001 = 10^{-9}
pico	p	0.000000000001 = 10^{-12}
femto	f	0.000000000000001 = 10^{-15}
atto	a	0.000000000000000001 = 10^{-18}

Small Units

The measurement of a value is often less than a whole unit, for example 0.6 V. 0.025 A, and 0.0550 W. Some of the *prefixes* used in such measurements are shown in Table 1-4.

For example, a millivolt (mV) is 0.001 V, and a microampere (μA) is 0.000001 A. The prefixes of Table 1-4 may be used with any electrical unit of measurement. The unit is divided by the fractional part of the unit. For example, to change 0.6 V to millivolts, divide by the fractional part indicated by the prefix. Thus, 0.6 V equals 600 mV, or 0.6 V ÷ 0.001 = 600 mV. To change 0.0005 A to microamperes, divide by 0.000001. Thus, 0.0005 A = 500 μA. When changing a base electrical unit to a unit with a prefix, move the decimal point of the unit to the right by the same number of places in the fractional prefix. To change 0.8 V to millivolts, the decimal point of 0.8 V is moved three places to the right (8.↲0↲0), since the prefix milli has three decimal places. So 0.8 V equals 800 mV. A similar method is used for converting any electrical unit to a unit with a smaller prefix.

Table 1-4. Prefixes of Units Smaller Than 1

Prefix	Abbreviation	Fractional Part of a Whole Unit
milli	m	1/1000 or 0.001 (3 decimal places)
micro	μ	1/1,000,000 or 0.000001 (6 decimal places)
nano	n	n 1/1,000,000,000 or 0.000000001 (9 decimal places)
pico	p	1/1,000,000,000,000 or 0.000000000001 (12 decimal places)

When a unit with a *prefix* is converted back to a *base unit*, the prefix must be multiplied by the fractional value of the prefix. For example, 68 mV is equal to 0.068 V. When 68 mV is multiplied by the fractional value of the prefix (0.01 for the prefix milli), this gives 68 mV × 0.001 = 0.0068 V. That is, to change a unit with a prefix into a base electrical unit, move the decimal in the prefix unit to the left by the same number of places as the value of the prefix. To change 225 mV to volts, move the decimal point in 225 three places to the left (2 2 5), since the value of the prefix milli has three decimal places. Thus, 225 mV equals 0.225 V.

Power Measurement Fundamentals

Table 1-5. Prefixes of Large Units

Prefix	Abbreviation	Number of Times Larger than 1
Kilo	k	1000
Mega	M	1,000,000
Giga	G	1,000,000,000

Large Units

Sometimes electrical measurements are very large, such as 20,000,000 W, 50,000, or 38,000 V. When this occurs, prefixes are used to make these numbers more manageable. Some prefixes used for large electrical values are shown in Table 1-5. To change a large value to a smaller unit, divide the large value by the value of the prefix. For example, 48,000,000 Ω is changed to 48 megohms (MΩ) by dividing by one million: 48,000,000 Ω ÷ 1,000,000 48 MΩ. To convert 7000 V to 7 kilovolts (kV), divide by one thousand: 7000 V ÷ 1000 = 7kv. To change a large value to a unit with a prefix, move the decimal point in the large value to the left by the number of zeros represented by the prefix. Thus 3600 V equals 3.6 kV (3 6 0 0). To convert a unit with a prefix back to a standard unit, the decimal point is moved to the right by the same number of places in the unit, or, the number may be multiplied by the value of the prefix. To convert 90 MΩ to ohms, the decimal point is moved six places to the right (90,000,000). The 90 MΩ value may also be multiplied by the value of the prefix, which is 1,000,000. Thus 90 MΩ × 1,000,000 = 90,000,000 Ω.

The simple *conversion scale* shown in Figure 1-1 is useful when converting standard units to units of measurement with prefixes. This scale uses either powers of 10 or decimals to express the units.

SCIENTIFIC NOTATION

Using *scientific notation* greatly simplifies arithmetic operations. Any number written as a multiple of a power of 10 and a number between 1 and 10 is said to be expressed in scientific notation. For example:

$$81{,}000{,}000 = 8.1 \times 10{,}000{,}000, \text{ or } 8.1 \times 10^7$$
$$500{,}000{,}000 = 5 \times 100{,}000{,}000, \text{ or } 5 \times 10^8$$
$$0.0000000004 = 4 \times 0.0000000001, \text{ or } 4 \times 10^{-10}$$

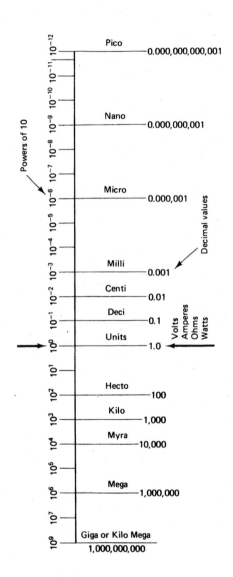

Figure 1-1. Simple conversion for large or small numbers

Power Measurement Fundamentals

Table 1-6 lists some of the *powers of 10*. In a whole-number *power of 10*, the power to which 10 is raised is positive and equals the number of zeros following the 1. In decimals, the power of 10 is negative and equals the number of places the decimal point is moved to the left of the 1.

Table 1-6. Power of 10

	Number	Power of 10
	1,000,000	10^6
	100,000	10^5
Whole	10,000	10^4
numbers	1000	10^3
	100	10^2
	10	10^1
	1.0	10^0
	0.1	10^{-1}
	0.01	10^{-2}
	0.001	10^{-3}
Decimals	0.0001	10^{-4}
	0.00001	10^{-5}
	0.000001	10^{-6}

Scientific notation simplifies multiplying and dividing large numbers or small decimals. For example:

$4800 \times 0.000045 \times 800 \times 0.0058$
$= (4.8 \times 10^3) \times (4.5 \times 10^{-5}) \times (8 \times 10^2) \times 5.8 \times (10^{-3})$
$= (4.8 \times 4.5 \times 8 \times 5.8) \, ´ \, (10^{3-5+2-3})$
$= 1002.24 \times 10^{-3}$
$= 1.00224$
$= 9.5 \times 10^4$
8
$= 1.1875 \times 10^8$
$= 118,750,000$

Other Electrical Power Units

Table 1-7 shows some common units used in the study of electrical power systems. These units will be introduced as they are utilized. You should review this figure and the sample problems included in Appendix A.

Table 1-7. Common Units

Quantity	SI Unit	Symbol
Angle	radian (1 rad ≈ 57.3°)	rad
Area	square meter	m²
Capacitance	farad	F
Conductance	siemens (mhos)	S
Electric charge	coulomb	C
Electric current	ampere	A
Energy (work)	joule	J
Force	newton	N
Frequency	hertz	Hz
Heat	joule	J
Inductance	henry	H
Length	meter	m
Magnetic field strength	ampere per meter	A/m
Magnetic flux	weber	Wb
Magnetic flux density	tesla (1 T = 1 Wb/m²)	T
Magnetomotive force	ampere	A
Mass	kilogram	kg
Potential difference	volt	V
Power	watt	W
Pressure	pascal (1 Pa = 1 N/m2)	Pa
Resistance	ohm	Ω
Resistivity	ohm-meter	Ωm
Specific heat	joule per kilogram-kelvin	J/kg K or J/kg = °C
Speed	meter per second	m/s
Speed of rotation	radian per second (1 rad/sec = 9.55 r/min)	rad/s
Temperature	kelvin	K
Temperature difference	kelvin or degree Celsius	K or °C
Thermal conductivity	watt per meter-kelvin	W/m K/or W/m= °C
Thermal power	watt	W
Torque	newton-meter	N-m
Volume	cubic meter	m³
Volume	liter	L

Chapter 2

Power System Fundamentals

One of the most important areas of electrical knowledge is the study of electrical power. Complex systems supply the vast need of our country for electrical power. Because of our tremendous power requirement, we must constantly be concerned with the efficient operation of our power production and power conversion systems. This textbook deals with the characteristics of electrical power production systems, power distribution systems, power conversion systems, and power control systems. In addition, an overview of electrical power measurement systems is included in this unit.

IMPORTANT TERMS

Systems Concept
Electrical System
    Source
    Path
    Control
    Load
    Indicator
Energy
    Kinetic Energy
    Potential Energy
Work
Power
Force
Electrical Power Systems Model
    Electrical Power Measurement
    Electrical Power Production
    Electrical Power Distribution

Electrical Power Conversion
Electrical Power Control
Electrical Circuits
    Resistive
    Inductive
    Capacitive
DC Power Calculation
Maximum Power Transfer
Purely Resistive AC Circuit
Counter-Electromotive Force (CEMF)
Magnetic Flux
Actual Power
Phase Angle (θ)
Resistive-Inductive (R-L) Circuit
Purely Inductive Circuit (R=0)
Negative Power
Inductance (L)
Inductive Reactance (X_L)
Capacitance (C)
Farad (F)/Microfarad (uF) Units
Electrostatic Field
Capacitive Reactance (X_C)
Vector (phasor) Diagram
Series AC Circuit
    Voltage Values: V_A, V_R, V_L, V_C, V_X
Parallel AC Circuit
    Current Values: I_T, I_R, I_L, I_C, I_X
Impedance (Z)
Total Reactance (X_T)
Impedance Triangle
Admittance (Y) Triangle
Conductance (G)
Inductive Susceptance (B_L)
Capacitive Susceptance (B_C)
Apparent Power (VA)
True Power (W)
Power Factor (pf)
Unity Power Factor (1.0)
Active Power

Reactive Power (VARs)
Three-phase System
Wye Configuration
Delta Configuration
Line Voltage (V_L)
Phase Voltage (V_P)
Line Current (I_L)
Phase Current (I_P)
Power per Phase (P_P)
Total Three-Phase Power (P_T)

THE SYSTEM CONCEPT

For a number of years, people have worked with jigsaw puzzles as a source of recreation. A jigsaw puzzle contains a number of discrete parts that must be placed together properly to produce a picture. Each part then plays a specific role in the finished product. When a puzzle is first started, it is difficult to imagine the finished product without seeing a representative picture.

Understanding a complex field such as electrical power poses a problem that is somewhat similar to the jigsaw puzzle, if it is studied by its discrete parts. In this case, too, it is difficult to determine the role that a discrete part plays in the operation of a complex system. A picture of the whole system, divided into its essential parts, therefore becomes an extremely important aid in understanding its operation.

The *system concept* will serve as the "big picture" in the study of electrical power. In this approach, a system will first be divided into a number of essential blocks. This will clarify the role played by each block in the operation of the overall system. After the location of each block has been established, the discrete component operation related to each block becomes more relevant. Through this approach, the way in which some of the "pieces" of electronic systems fit together should be made more apparent.

BASIC SYSTEM FUNCTIONS

The word *system* is commonly defined as an organization of parts that are connected together to form a complete unit. A wide variety of electrical systems is in use today. Each system has a number of unique features, or characteristics, that distinguish it from other systems. More importantly, however, there is a common set of parts found in each system. These parts play the same basic role in all systems. The terms energy *source*, transmission *path, control, load,* and *indicator* are used to describe the various system parts. A block diagram of these basic parts of the system is shown in Figure 2-1.

Each block of a basic system has a specific role to play in the overall operation of the system. This role becomes extremely important when a detailed analysis of the system is to take place. Hundreds and even thousands of discrete components are sometimes needed to achieve a specific block function. Regardless of the complexity of the system, each block must achieve its function in order for the system to be operational. Being familiar with these functions and being able to locate them within a complete system is a big step toward understanding the operation of the system.

The *energy source* of a system converts energy of one form into something more useful. Heat, light, sound, and chemical, nuclear, and mechan-

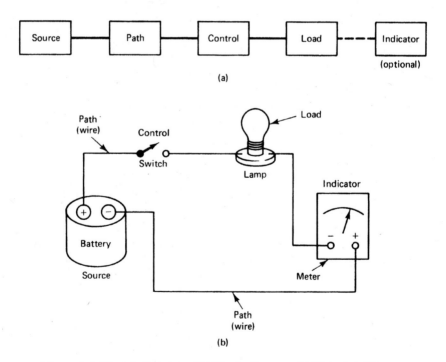

Figure 2-1. Electrical system: (A) Block diagram; (B) Pictorial diagram

ical energy are considered as primary sources of energy. A primary energy source usually goes through an energy change before it can be used in an operating system.

The *transmission path* of a system is somewhat simpler than other system functions. This part of the system simply provides a path for the transfer of energy (see Figure 2-2). It starts with the energy source and continues through the system to the load. In some cases, this path may be a single electrical conductor, light beam, or other medium between the source and the load. In other systems, there may be a supply line between the source and the load. In still other systems, there may be a supply line between the source and the load, and also a return line from the load to the source. There may also be a number of alternate or auxiliary paths within a complete system. These paths may be series connected to a number of small load devices, or parallel connected to many independent devices.

The *control* section of a system is by far the most complex part of the entire system. In its simplest form, control is achieved when a system is turned on or off. Control of this type can take place anywhere between the source and the load device. The term "full control" is commonly used to describe this operation. In addition to this type of control, a system may also employ some type of partial control. Partial control usually causes some type of an operational change in the system, other than an on or off condition. Changes in electric current or light intensity are examples of alterations achieved by partial control.

The *load* of a system refers to a specific part, or a number of parts, designed to produce some form of work (see Figure 2-2). Work, in this case, occurs when energy goes through a transformation or change. Heat, light, chemical action, sound, and mechanical motion are some of the common forms of work produced by a load device. As a general rule, a very large portion of all energy produced by the source is consumed by the load device during its operation. The load is typically the most prominent part of the entire system because of its obvious work function.

The *indicator* of a system is primarily designed to display certain operating conditions at various points throughout the system. In some systems the indicator is an optional part, while in others it is an essential part in the operation of the system. In the latter case, system operations and adjustments are usually critical and are dependent upon specific indicator readings. The term "operational indicator" is used to describe this application. Test indicators are also needed to determine different operating values. In this role, the indicator is only temporarily attached to the sys-

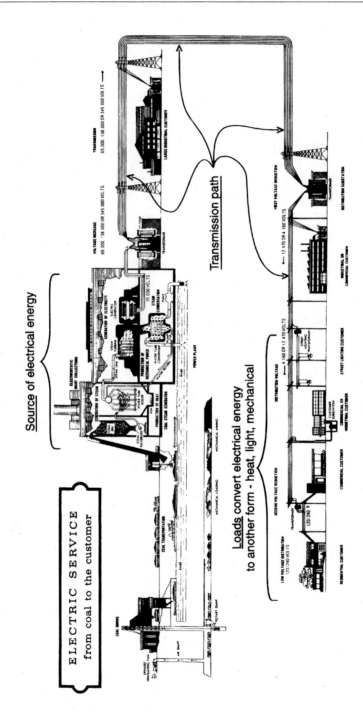

Figure 2-2. Distribution path for electrical power from its source to where it is used (*Courtesy Kentucky Utilities Co.*)

Power System Fundamentals   21

tem, in order to make measurements. Test lights, meters, oscilloscopes, chart recorders, and digital display instruments are some of the common indicators used in this capacity.

A SIMPLE ELECTRICAL SYSTEM EXAMPLE

A flashlight is a device designed to serve as a light source in an emergency, or as a portable light source. In a strict sense, flashlights can be classified as portable *electrical systems*. They contain the four essential parts needed to make this classification. Figure 2-3 is a cutaway drawing of a flashlight, with each component part shown in association with its appropriate system block.

The battery of a flashlight serves as the primary *energy source* of the system. The chemical energy of the battery must be changed into electrical energy before the system becomes operational. The flashlight is a synthesized system because it utilizes two distinct forms of energy in its operation. The energy source of a flashlight is a expendable item. It must be replaced periodically when it loses its ability to produce electrical energy.

The transmission path of a flashlight is commonly through a metal casing or a conductor strip. Copper, brass, and plated steel are frequently used to achieve the transmission function.

The control of electrical energy in a flashlight is achieved by a slide switch or a push-button switch. This type of control simply interrupts the transmission path between the source and the load device. Flashlights are primarily designed to have full control capabilities. This type of control is achieved manually by the person operating the system.

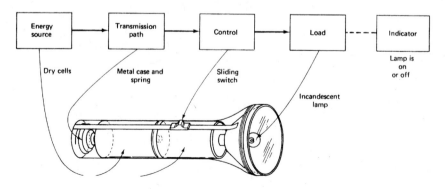

Figure 2-3. Cutaway drawing of a flashlight

The load of a flashlight is a small incandescent lamp. When electrical energy from the source is forced to pass through the filament of the lamp, the lamp produces a bright glow. Electrical energy is first changed into heat energy and then into light energy. A certain amount of the work is achieved by the lamp when this energy change takes place.

The energy transformation process of a flashlight is irreversible. It starts at the battery when chemical energy is changed into electrical energy. Electrical energy is then changed into heat energy and eventually into light energy by the load device. This flow of energy is in a single direction. When light is eventually produced, it consumes a large portion of the electrical energy coming from the source. When this energy is exhausted, the system becomes inoperative. The battery cells of a flashlight require periodic replacement in order to maintain a satisfactory operating condition.

Flashlights do not ordinarily employ a specific indicator as part of the system. Operation is indicated when the lamp produces light. In a strict sense, we could say that the load of this system also serves as an indicator. In some electrical systems the indicator is an optional system part.

ENERGY, WORK, AND POWER

An understanding of the terms "energy," "work," and "power" is necessary in the study of electrical power systems. The first term, "energy," means the capacity to do work. For example, the capacity to light a light bulb, to heat a home, or to move something requires energy. Energy exists in may forms, such as electrical, mechanical, chemical, and heat. If energy exists because of the movement of some item, such as a ball rolling down a hill, it is called kinetic energy. If energy exists because of the position of something, such as a ball that is at the top of the hill but not yet rolling, it is called potential energy. Energy is one of the most important factors in our society.

A second important term is "work." Work is the transferring or transforming of energy. Work is done when a force is exerted to move something over a distance against opposition, such as when a chair is moved from one side of a room to the other. An electrical motor used to drive a machine performs work. Work is performed when motion is accomplished against the action of a force that tends to oppose the motion. Work is also done each time energy changes from one form into another.

Power System Fundamentals

Sample Problem: Work

Work is done whenever a force (F) is moved a distance (d), or:

W = F × d, where
W = work in joules
F = force in newtons
d = distance the force moves in meters

Given: An object with a mass of 22Kg is moved 55 meters.
Find: The amount of work done when the object is moved.
Solution: The force of gravity acting on the object is equal to 9.8 (a constant that applies to objects on earth) multiplied by the mass of the object, or:

F = 9.8 × 22 Kg = 215.6 newtons
W = F × d
 = 215.6 × 55
W = 11,858 joules

A third important term is "power." Power is the rate at *which* work is done. It concerns not only the work that is performed but the amount of time in *which* the work is done. For instance, electrical power is the rate at which work is done as electrical current flows through a wire. Mechanical power is the rate at which work is done as an object is moved against opposition over a certain distance. Power is either the rate of production of energy or the rate of use of energy. The watt is the unit of measurement of power.

Sample Problem: Power

Power is the time rate of doing work, which is expressed as:

$$P = \frac{W}{t}, \text{ where}$$

P = power in watts
W = work done in joules
t = time taken to do the work in seconds

Given: An electric motor is used to move an object along a conveyor line. The object has a mass of 150 kg and is moved 28 meters in 8 seconds.

Find: The power developed by the motor in watts and horsepower units.

Solution:

$$\text{Force (F)} = 9.8 \times \text{mass}$$
$$= 9.8 \times 150 \text{ kg}$$
$$F = 1470 \text{ newtons}$$

$$\text{Work (W)} = F \times d$$
$$= 1470 \times 28 \text{ m}$$
$$W = 41,160 \text{ joules}$$

$$\text{Power (P)} = W/t$$
$$= 41,160/8$$
$$P = 5,145 \text{ watts}$$

$$\text{Horsepower} = \frac{P}{746}, \text{ since}$$

1 horsepower = 746 W.

hp = 5,145/746 = 6.9 hp

The Electrical Power System

A block diagram of the electrical power systems model used in *this* textbook is shown in Figure 2-4. Beginning on the left, the first block is Electrical Power Measurement. Power measurement is critical to the efficient operation of electrical power systems. Measurement fundamentals and power measurement equipment are discussed in Unit I of this textbook. The second block is Electrical Power Production. Unit II presents the electrical power production systems used in our country. Once electrical power has been produced, it must be distributed to the location where it is used. Electrical Power Distribution Systems are discussed in Unit III. Power distribution systems transfer electrical power from one location to another. Electrical Power Conversion Systems (Unit IV), also called electrical loads, convert electrical power into some other form, such as light, heat, or mechanical energy. Thus, power conversion systems are an extremely important part of the electrical power system. The last block, Electrical Power Control (Unit V), is probably the most complex of all the parts of the electrical power system. There are almost unlimited types of devices, circuits, and equipment used to control electrical power systems.

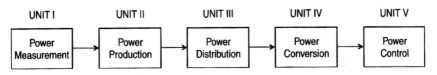

Figure 2-4. Electrical Power Systems Model

Each of the blocks shown in Figure 2-4 represents one important part of the electrical power system. Thus, we should be concerned with each one as part of the electrical power system, rather than in isolation. In this way, we can develop a more complete understanding of how electrical power systems operate. This type of understanding is needed to help us solve problems that are related to electrical power. We cannot consider only the production aspect of electrical power systems. We must understand and consider all parts of the system.

TYPES OF ELECTRICAL CIRCUITS

There are several basic fundamentals of electrical power systems. Therefore, the basics must be understood before attempting an in-depth study of electrical power systems. The types of electrical circuits associated with electrical power production or power conversion systems are (1) resistive, (2) inductive, and (3) capacitive. Most systems have some combination of each of these three circuit types. These circuit elements are also called loads. A load is a part of a circuit that converts one type of energy into another type. A resistive load converts electrical energy into heat energy.

In our discussions of electrical circuits, we will primarily consider alternating current (AC) systems at this time, as the vast majority of the electrical power that is produced is alternating current. Direct current (DC) systems will be discussed in greater detail in Chapter 7.

POWER IN DC ELECTRICAL CIRCUITS

In terms of voltage and current, power (P) in watts (W) is equal to voltage (in volts) multiplied by current (in amperes). The formula is $P = V \times I$. For example, a 120-V electrical outlet with 4 A of current flowing from it has a power value of

$$P = V \times I = 120 \text{ V} \times 4 \text{ A} = 480 \text{ W}.$$

The unit of electrical power is the watt. In the example, 480 W of power are converted by the load portion of the circuit. Another way to find power is:

$$P = \frac{V^2}{R}$$

This formula is used when voltage and resistance are known, but current is not known. The formula $P = F \times R$ is used when current and resistance are known. DC circuit formulas are summarized in Figure 2-5. The quantity in the center of the circle may be found by any of the three formulas along the outer part of the circle in the same part of the circle. This circle is handy to use for making electrical calculations for voltage, current, resistance, or power in DC circuits.

It is easy to find the amount of power converted by each of the resistors in a series circuit, such as the one shown in Figure 2-6. In the circuit shown, the amount of power converted by each of the resistors, and the total power, are found as follows:

1. Power converted by resistor R_1:
$P_1 = I^2 \times R_1 = 2^2 \times 20 \text{ }\Omega = 80 \text{ W}$

2. Power converted by resistor R_2:
$P_2 = I^2 \times R_2 = 2^2 \times 30 \text{ }\Omega = 120 \text{ W}$

3. Power converted by resistor R_3:
$P_3 = I^2 \times R_3 = 2^2 \times 50 \text{ }\Omega = 200 \text{ W}$

4. Power converted by the circuit:
$P_T = P_1 + P_2 + P_3 = 80 \text{ W} + 120 \text{ W} + 200 \text{ W}$
$ = 400 \text{ W, or}$

$P_T = V_T \times I = 200 \text{ V} \times 2 \text{ A} = 400 \text{ W}$

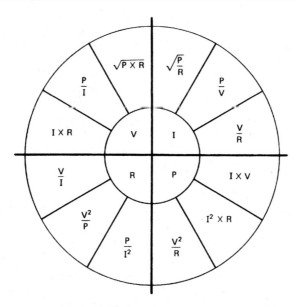

Figure 2-5. Formulas for finding voltage, current, resistance, or power

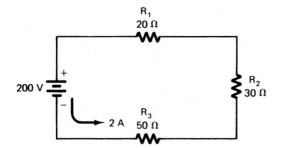

Figure 2-6. Finding power values in a series circuit

When working with electrical circuits, you can check your results by using other formulas.

Power in parallel circuits is found in the same way as power in series circuits. In the example shown in Figure 2-7, the power converted by each of the resistors, and the total power of the parallel circuit, are found as follows:

1. Power converted by resistor R_1:

$$P_1 = \frac{V^2}{R_1} = \frac{30^2}{5} = \frac{900}{5} = 180 \text{ W}$$

2. Power converted by resistor R_2:

$$P_2 = \frac{V^2}{R_2} = \frac{30^2}{10} = \frac{900}{10} = 90 \text{ W}$$

3. Power converted by resistor R_3:

$$P_3 = \frac{V^2}{R_3} = \frac{30^2}{20} = \frac{900}{20} = 45 \text{ W}$$

4. Total power converted by the circuit:

$$P_T = P_1 = P_2 + P_3 = 180 \text{ W} + 90 \text{ W} + 45 \text{ W} = 315 \text{ W}$$

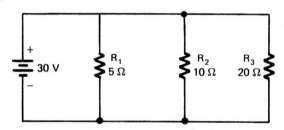

Figure 2-7. Finding power values in a parallel circuit.

The watt is the basic unit of electrical power. To determine an actual quantity of electrical energy, one must use a factor that indicates how long a given power value continued. Such a unit of electrical energy is called a watt-second. It is the product of watts (W) and time (in seconds). The watt-second is a very small quantity of energy. It is more common to measure electrical energy in kilowatt-hours (kWh). It is the kWh quantity of electrical energy that is used to determine the amount of electrical utility bills. A kilowatt-hour is 1000 W in 1 h of time, or 3,600,000 W per second.

As an example, if an electrical heater operates on 120 V, and has a resistance of 200, what is the cost to use the heater for 200 h at a cost of 5 cents per kWh?

1.
$$P = \frac{V^2}{R} = \frac{120^2}{20\Omega} = \frac{14{,}400}{20\Omega} = 720 \text{ W} = 0.72 \text{ kW}.$$

2. There are 1000 W in a kilowatt (1000 W = 1 kW).

3. Multiply the kW that the heater has used by the hours of use:

   kW × 200 h = kilowatt-hours (kWh)
   0.72 × 200 h = 144 kWh

4. Multiply the kWh by the cost:
   kWh × cost = 1.44 KWh × 0.05 = $7.20

Some simple electrical circuit examples have been discussed in this chapter. They become easy to understand after practice with each type of circuit. It is very important to understand the characteristics of series, parallel, and combination circuits.

MAXIMUM POWER TRANSFER

An important consideration in relation to electrical circuits is *maximum power transfer*. Maximum power is transferred from a voltage source to a load when the load resistance (RL) is equal to the internal resistance of the source (RS). The source resistance limits the amount of power that can be applied to a load. Electrical sources and loads may be considered as diagrammed in Figure 2-8.

For example, as a flashlight battery gets older, its internal resistance increases. This increase in the internal resistance causes the battery to supply less power to the lamp load. Thus, the light output of the flashlight is reduced.

Figure 2-9 shows an example that illustrates maximum power transfer. The source is a 100 V battery with an internal resistance of 5 Ω. The values of I_L, V_{out}, and power output (P_{out}) are calculated as follows:

$$I_L = \frac{V_T}{R_L + R_S}; \quad V_{out} = I_L \times R_L; \quad P_{out} = I_L \times V_{out}$$

Notice the graph shown in Figure 2-9. This graph shows that maximum power is transferred from the source to the load when $R_L = R_S$. This is an important circuit design consideration for power sources.

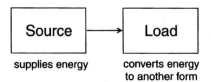

(A) Circuits have a source and a load.

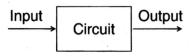

(B) Circuits have an input and an output.

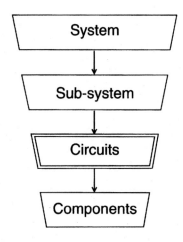

(C) Circuits are a part of any electrical system.

Figure 2-8. Electrical circuits and systems: (A) Circuits have a source and a load; (B) Circuits have an input and an output; (C) Circuits are a part of any electrical system.

OVERVIEW OF ALTERNATING CURRENT (AC) CIRCUITS

The following discussion provides an overview of the three basic types of alternating current (AC) circuits. These basic circuits are: (1) resistive, (2) inductive, and (3) capacitive. The basic characteristics of each of these circuits should be examined to gain a fundamental understanding of electrical power systems.

Power System Fundamentals

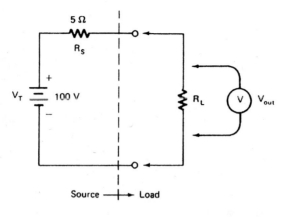

R_L	I_L	V_{out}	Power Output (W)
0	$\frac{100\ V}{5\ \Omega} = 20\ A$	20 A × 0 Ω = 0 V	20 A × 0 V = 0 W
2.5 Ω	$\frac{100\ V}{7.5\ \Omega} = 13.3\ A$	13.3 A × 2.5 Ω = 33.3 V	13.3 A × 33.3 V = 444 W
5 Ω	$\frac{100\ V}{10\ \Omega} = 10\ A$	10 A × 5 Ω = 50 V	10 A × 50 V = 500 W
7.5 Ω	$\frac{100\ V}{12.5\ \Omega} = 8\ A$	8 A × 12.5 Ω = 60 V	8 A × 60 V = 480 W
10 Ω	$\frac{100\ V}{15\ \Omega} = 6.7\ A$	6.7 A × 10 Ω = 67 V	6.7 A × 67 V = 444 W

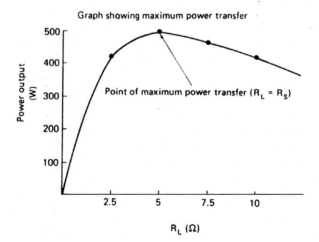

Figure 2-9. Problem that shows maximum power transfer

Resistive AC Circuits

The simplest type of AC electrical circuit is a resistive circuit, such as the one shown in Figure 2-10A. The purely resistive circuit offers the same type of opposition to AC power sources as it does to pure DC power sources. In DC circuits,

$$\text{Voltage (V)} = \text{Current (I)} \times \text{Resistance (R)}$$

$$\text{Current (I)} = \frac{\text{Voltage (V)}}{\text{Resistance (R)}}$$

$$\text{Resistance (R)} = \frac{\text{Voltage (V)}}{\text{Current (I)}}$$

$$\text{Power (P)} = \text{Voltage (V)} \times \text{Current (I)}$$

These basic electrical relationships show that when voltage is increased, the current in the circuit increases proportionally. Also, as resistance is increased, the current in the circuit decreases. By looking at the waveforms of Figure 2-10B, we can see that the voltage and current in a purely resistive circuit, with AC applied, are in phase. An in-phase relationship exists when the minimum and maximum values of both voltage and current occur at the same time interval. Also, the power converted by the circuit is a product of voltage times current (P = V × I). The power curve is also shown in Figure 2-10B. Thus, when an AC circuit contains only resistance, its behavior is very similar to that of a DC circuit. Purely resistive circuits are seldom encountered in electrical power systems designs, although some devices are primarily resistive in nature.

Inductive AC Circuits

The property of inductance (L) is very commonly encountered in electrical power systems. This circuit property, shown in Figure 2-11A, adds more complexity to the relationship between voltage and current in an AC circuit. All motors, generators, and transformers exhibit the property of inductance. The occurrence of this property is due to a counter electromotive force (cemf), which is produced when a magnetic field is developed around a coil of wire. The magnetic flux produced around the coils affects circuit action. Thus, the inductive property (cemf) produced by a magnetic field offers an opposition to change in the current flow in a circuit.

Power System Fundamentals

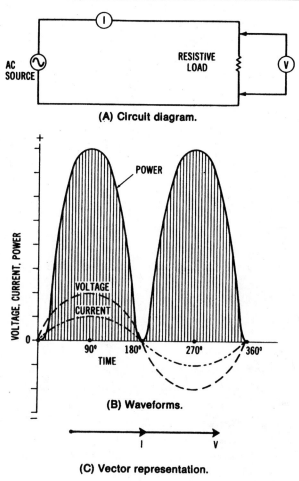

(A) Circuit diagram.

(B) Waveforms.

(C) Vector representation.

Figure 2-10. Resistive circuit: (A) Circuit diagram; (B) Waveforms; (C) Vector representation

Sample Problem: Energy Stored in an Inductor

A coil stores energy in its magnetic field because of current flow through it. The amount of energy is defined by the equation:

$$W + \frac{1}{2} \times L \times I^2, \text{ where}$$

W = energy stored in joules
L = coil inductance in henries
I = current flow through the coil in amperes

Given: a 10-henry coil has 5.8 amperes of current flowing through it.
Find: the amount of energy stored in the coil.
Solution:

$$W = \frac{1}{2} \times L \times I^2$$

$$= \frac{1}{2} \times 10 \times (5.8)^2$$

$$W = 168.2 \text{ joules}$$

The opposition to change of current is evident in the diagram of Figure 2-11B. In an inductive circuit, we can say that voltage leads current or that current lags voltage. If the circuit were purely inductive (containing no resistance), the voltage would lead the current by 90° (Figure 2-11B), and no actual power would be converted in the circuit. However, since all actual circuits have resistance, the inductive characteristic of a circuit typically causes the condition shown in Figure 2-12 to exist. Here, the voltage is leading the current by 30°. The angular separation between voltage and current is called the "phase angle." The phase angle increases as the inductance of the circuit increases. This type of circuit is called a "resistive-inductive (RL) circuit."

In terms of power conversion, a purely inductive circuit would not convert any actual power in a circuit. All AC power would be delivered back to the power source. Refer back to Figure 2-11B, and note points A and B on the waveforms. These points show that the value at the peak of each waveform is zero. The power curves shown are equal and opposite in value and will cancel each other out. Where both voltage and current are positive, the power is also positive, since the product of two positive values is positive. When one value is positive and the other is negative, the product of the two values is negative; therefore, the power converted is negative. Negative power means that electrical energy is being returned from the load device to the power source without being converted into another form of energy. Therefore, the power converted in a purely inductive circuit (900 phase angle) would be equal to zero.

Compare the purely inductive circuit waveforms (Figure 2-11B) to those of Figure 2-12B. In the practical resistive-inductive (RL) circuit, part of the power supplied from the source is converted in the circuit. Only

Power System Fundamentals 35

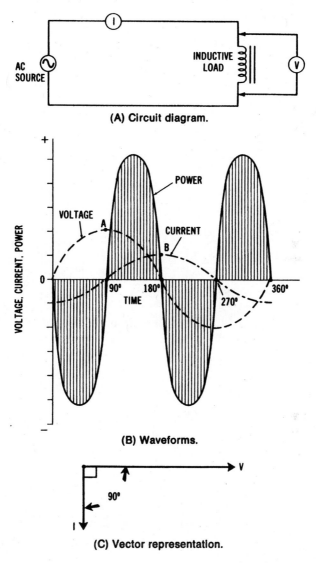

Figure 2-11. Inductive circuit: (A) Circuit diagram; (6) Waveforms; (C) Vector representation

during the intervals from 0° to 30° and from 180° to 210° does negative power result. The remainder of the cycle produces positive power; therefore, most of the electrical energy supplied by the source is converted into another form of energy.

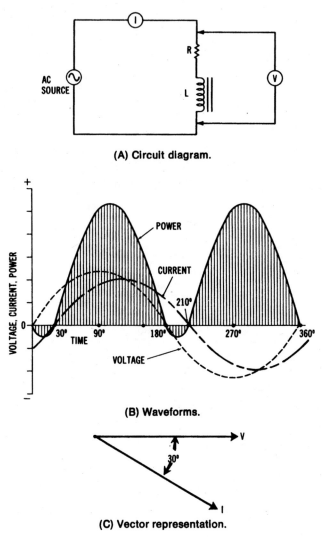

Figure 2-12. Resistive-inductive (R-L) circuit: (A) Circuit diagram; (B) Waveforms; (C) Vector representation

Any inductive circuit exhibits the property of inductance (L), which is the opposition to a change in current flow in a circuit. This property is found in coils of wire (which are called inductors) and in rotating machinery and transformer windings. Inductance is also present in electrical power transmission and distribution lines to some extent. The unit of

Power System Fundamentals

measurement for inductance is the *henry (H)*. A circuit has a 1-henry inductance if a current changing at a rate of 1 ampere per second produces an induced counter electromotive force (cemf) of 1 volt.

In an *inductive circuit* with AC applied, an opposition to current flow is created by the inductance. This type of opposition is known as *inductive reactance* (X_L). The inductive reactance of an AC circuit depends upon the inductance (L) of the circuit and the rate of change of current. The frequency of the applied AC establishes the rate of change of the current. Inductive reactance (X_L) may be expressed as:

$$X_L = 2\pi fL$$

where:

X_L is the inductive reactance in ohms,

2π is 6.28, the mathematical expression for one sine wave of AC (0° to 360°),

f is the frequency of the AC source in hertz, and

L is the inductance of the circuit in henries.

Sample Problem:

Given: frequency = 60 Hz, and inductance = 20 henries.
Find: inductive reactance.
Solution:

$$X_L = 2\pi \times 60 \text{ Hz} \times 20 \text{ H}$$
$$X_L = 7536 \ \Omega$$

Capacitive AC Circuits

Figure 2-13A shows a *capacitive* device connected to an AC source. We know that whenever two conductive materials (plates) are separated by an insulating (dielectric) material, the property of *capacitance* is exhibited. Capacitors have the capability of storing an electrical charge. They have many applications in electrical power systems.

The operation of a capacitor in a circuit is dependent upon its ability to charge and discharge. When a capacitor charges, an excess of electrons (negative charge) is accumulated on one plate, and a deficiency of electrons (positive charge) is created on the other plate. *Capacitance* (C) is determined by the size of the conductive material (plates) and by their separation (determined by the thickness of the dielectric or insulating material). The type of insulating material is also a factor in determining capacitance. Capacitance is directly proportional to the plate size, and inversely

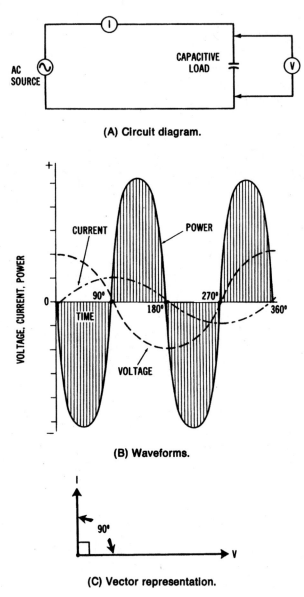

Figure 2-13. Capacitive circuit: (A) Circuit diagram; (B) Waveforms; (C) Vector representation

proportional to the distance between the plates. The unit of capacitance is the *farad (F)*. A capacitance of 1 farad results when a potential of 1 volt causes an electrical charge of 1 coulomb (a specific mass of electrons) to accumulate on a capacitor. Since the farad is a very large unit, *microfarad* (μF) values are ordinarily assigned to capacitors.

Sample Problem: Energy Stored in a Capacitor

Energy is stored by a capacitor in its electrostatic field when voltage is applied to the capacitor. The amount of energy is defined by the equation:

$$W = \frac{1}{2} \times C \times V^2, \text{ where}$$

W = energy stored in a capacitor in joules,
C = capacitance of the capacitor in farads, and
V = applied voltage in volts.

Given: a 100 µF capacitor has 120 volts applied.
Find: the amount of energy stored in the capacitor.
Solution:

$$W = \frac{1}{2} \times C \times V^2$$

$$W = \frac{1}{2} \times 100^{-6} \times 120^2 = 72 \text{ joules}$$

If a direct current is applied to a capacitor, the capacitor will charge to the value of that DC voltage. After the capacitor is fully charged, it will block the flow of direct current. However, if AC is applied to a capacitor, the changing value of current will cause the capacitor to alternately charge and discharge. In a *purely capacitive circuit*, the situation shown in Figure 2-13B would exist. The greatest amount of current would flow in a capacitive circuit when the voltage changes most rapidly. The most rapid change in voltage occurs at the 0° and 180° positions where the polarity changes. At these positions, maximum current is developed in the circuit. When the rate of change of the voltage value is slow, such as near the 90° and 270° positions, a small amount of current flows. In examining Figure 2-13B, we can observe that current leads voltage by 90° in a purely capacitive circuit, or the voltage lags the current by 90°. Since a 90° phase angle exists, no power would be converted in this circuit, just as no power was developed in the purely inductive circuit. As shown in Figure 2-13B, the positive and negative power waveforms will cancel one another out.

Since all circuits contain some resistance, a more practical circuit is the *resistive-capacitive (RC)* circuit, shown in Figure 2-14A. In an RC circuit, the current leads the voltage by some phase angle between 0° and 90°. As capacitance increases with no corresponding increase in resistance, the phase angle becomes greater. The waveforms of Figure 2-14B show an RC circuit in which current leads voltage by 30°. This circuit is similar to the RL circuit in Figure 2-12. Power is converted in the circuit except during the 00 to 30° interval and the 180° to 210° interval. In the RC circuit shown, most of the electrical energy supplied by the source is converted into another form of energy in the load.

Due to the *electrostatic field* that is developed around a capacitor, an opposition to the flow of AC exists. This opposition is known as capacitive reactance (X_C). Capacitive reactance is expressed as:

$$X_C = \frac{1}{2\pi fC}$$

where:
X_C is the capacitive reactance in ohms,
2π is the mathematical expression of one sine wave (0° to 360°),
f is the frequency of the source in hertz, and
C is the capacitance in farads.

Sample Problem:
Given: frequency = 50 Hz, and capacitance = 200 µF.
Find: capacitive reactance.
Solution:

$$X_C = \frac{1}{2\pi \times 50 \times 200^{-6}}$$

$$X_C = 15.92 \Omega$$

VECTOR AND PHASOR DIAGRAMS FOR AC CIRCUITS

In Figures 2-10C, 2-11C, 2-12C, 2-13C and 2-14C, a *vector diagram* was shown for each circuit condition that was illustrated. *Vectors* are straight lines that have a specific direction and length. They may be used to rep-

Power System Fundamentals   41

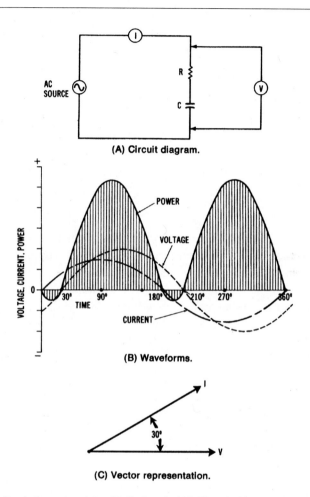

Figure 2-14. Resistive-capacitive (RC) circuit; (A) Circuit diagram; (6) Waveforms; (C) Vector representation

resent voltage or current values. An understanding of vector diagrams (sometimes caned *phasor diagrams*) is important when dealing with alternating current. Rather than using waveforms to show phase relationships, it is possible to use a vector or phasor representation.

Ordinarily, when beginning a *vector diagram*, a horizontal line is drawn with its left end as the reference point. Rotation in a counterclockwise direction from the reference point is then considered to be a positive direction. Note that in the preceding diagrams, the voltage vector was the reference. For the inductive circuits, the current vector was drawn in a clockwise direction, indicating a lagging condition. A leading condition is shown for the capacitive circuits by the use of a current vector drawn in a counterclockwise direction from the voltage vector.

Use of Vectors for Series AC Circuits

Vectors may be used to compare voltage drops across the components of a series circuit containing resistance, inductance, and capacitance (an RLC circuit), as shown in Figure 2-15. In a *series AC circuit,* the current is the same in all parts of the circuit, and the voltages must be added by using vectors. In the example shown, specific values have been assigned. The voltage across the resistor (V_R) is equal to 4 volts, while the voltage across the capacitor (V_C) equals 7 volts, and the voltage across the inductor (V_L) equals 10 volts. We diagram the capacitive voltage as leading the resistive voltage by 90° and the inductive voltage as lagging the resistive voltage by 90°. Since these two values are in direct opposition to one another, they may be subtracted to find the resultant reactive voltage (V_X). By drawing lines parallel to V_R and V_X, we can find the resultant voltage applied to the circuit. Since these vectors form a right triangle, the value of V_T can be expressed as:

$$V_T = \sqrt{V_R^2 + V_X^2}$$

where:
  V_T is the total voltage applied to the circuit,
  V_R is the voltage across the resistance, and
  V_X is the total reactive voltage ($V_L - V_C$ or $V_C - V_L$, depending on which is the larger).

Sample Problem:

  Given: resistive voltage = 25 volts, and reactive voltage = 18 volts.
  Find: total applied voltage.
  Solution:

$$V_T = \sqrt{V_R^2 + V_X^2}$$

$$V_T = \sqrt{25^2 + 18^2}$$

$$V_T = \sqrt{625 + 324}$$

$$V_T = 30.8 \text{ V}$$

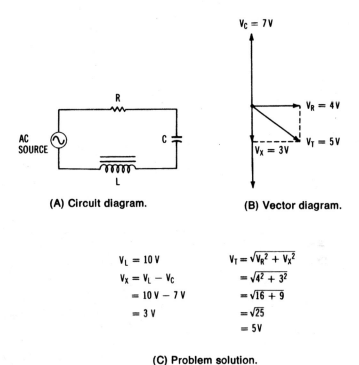

Figure 2-15. Voltage vector relationship in a series RLC circuit: (A) Circuit diagram; (B) vector diagram; (C) problem solution

Use of Vectors for Parallel AC Circuits

Vector representation is also useful for parallel AC circuit analysis. Voltage in a *parallel AC circuit* remains the same across all the components, and the currents through the components of the circuit can be shown using vectors. A *parallel RLC circuit* is shown in Figure 2-17.

The current through the capacitor (I_C) is shown leading the current through the resistor (I_R) by 90°. The current through the inductor (I_L) is shown lagging I_R by 90°. Since I_L and I_C are 180° out of phase, they are subtracted to find the total reactive current (I_X). By drawing lines parallel to I_R and I_X, we can find the total current of the circuit (I_T). These vectors form a right triangle; therefore, total current can be expressed as:

$$I_T = \sqrt{I_R^2 + I_X^2}$$

Sample Problem:
  Given: resistive current = 125 amperes, and reactive current = 65 amperes.
   Find: total current.
   Solution:

$$I_T = \sqrt{I_R^2 + I_X^2}$$

$$I_T = \sqrt{125^2 + 65a^2}$$

$$I_T = \sqrt{5625 + 4225}$$

$$I_T = 99.25 \text{ A}$$

A similar method of vector diagramming can be used for voltages in RL and RC series circuits. This method may also be used for currents in RL and RC parallel circuits. RLC circuits were used in the examples to illustrate the method used to find a resultant reactive voltage or current in a circuit.

IMPEDANCE IN AC CIRCUITS

Another application of the use of vectors is determining the total opposition of an AC circuit to the flow of current. This total opposition is called impedance (Z) and is measured in ohms.

Impedance in Series AC Circuit
Both resistances and reactances in AC circuits affect the opposition to current flow.
   Impedance (Z) of an AC circuit may be expressed as:

$$Z = \frac{V_T}{I}$$

or

$$Z = \sqrt{R_2 + (X_L - X_C)^2}$$

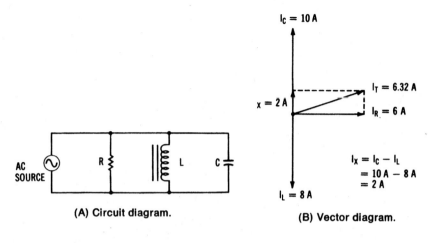

$I_T = \sqrt{I_R^2 + I_x^2}$
$= \sqrt{6^2 + 2^2}$
$= \sqrt{36 + 4}$
$= \sqrt{40}$
$= 6.32$ A

(C) Problem solution.

Figure 2-16. Current vector relationship in a parallel RLC circuit: (A) Circuit diagram; (B) Vector diagram; (C) problem solution

Sample Problem:

Given: an AC circuit has the following component values: R = 100, L = 1 H, C = 10µF, and 1 = 60 Hz.

Find: total impedance of the circuit.

Solution:

$$X_L = 2\pi \times f \times L$$
$$= 6.28 \times 60 \text{ Hz} \times 1 \text{ H}$$
$$X_L = 376.8 \, \Omega$$

$$X_C = \frac{1}{2\pi \times f \times C}$$

$$X_C = \frac{1}{6.28 \times 60 \times 0.1^{-6}} = 265.4 \, \Omega$$

$$Z = \sqrt{R_2 + (X_L - X_C)^2}$$

$$\sqrt{100^2 + (376.8 - 265.4)^2}$$

$$\sqrt{10{,}000 \ \Omega + 12{,}410 \ \Omega}$$

$$Z = 149.7 \ \Omega$$

This formula may be clarified by using the vector diagram shown in Figure 2-17. The total *reactance* (X_T) of an AC circuit may be found by subtracting the smallest reactance (X_L or X_C) from the largest reactance. The impedance of a series AC circuit is determined by using the preceding formula, since a right triangle (called an *impedance triangle*) is formed by the three quantities that oppose the flow of alternating current. A sample problem for finding the total impedance of a series AC circuit is shown in Figure 2-17.

Impedance in Parallel AC Circuits

When components are connected in parallel, the calculation of impedance becomes more complex. Figure 2-18 shows a simple parallel RLC circuit. Since the total impedance in the circuit is smaller than the resistance or reactance, an impedance triangle, such as the one shown in the series circuit of Figure 2-17, cannot be developed. A simple method used to find impedance in parallel circuits is the *admittance triangle*, shown in Figure 2-18B. The following quantities may be plotted on the triangle:

$$\text{admittance} = \frac{1}{Z}, \ \text{conductance} = \frac{1}{R}, \ \text{inductive susceptance} = \frac{1}{X_L}$$

$$\text{and capacitive susceptance} = \frac{1}{X_C}$$

Notice that these quantities are the reciprocals of each type of opposition to alternating current. Therefore, since total *impedance (Z)* is the smallest quantity in a parallel AC circuit, it becomes the largest value on the admittance triangle. The sample problem of Figure 2-18 shows the procedure used to find total impedance of a parallel RC circuit.

Power System Fundamentals

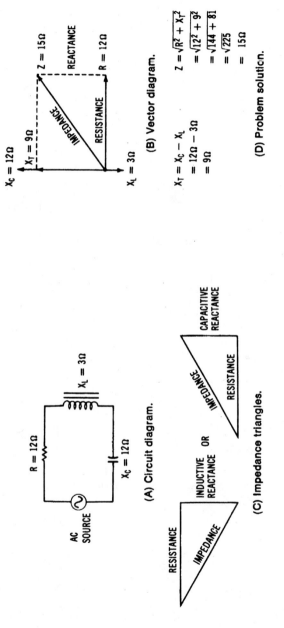

Figure 2-17. Impedance in series AC circuits: (A) Circuit diagram; (B) Vector diagram; (C) Impedance triangles; (D) problem solution

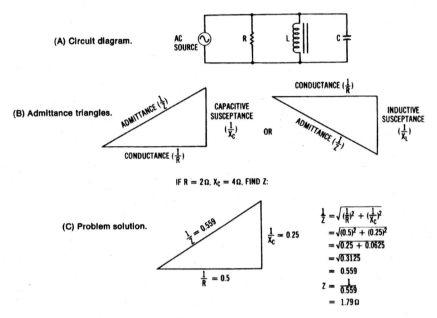

Figure 2-18. Impedance in parallel AC circuits: (A) Circuit diagram; (8) Vector diagram; (C Impedance triangles, (D) Problem solution

POWER RELATIONSHIPS IN AC CIRCUITS

An understanding of basic *power relationships* in AC circuits is very important when studying complex electrical power systems. In the previous sections, resistive, inductive, and capacitive circuits were discussed. Also, power converted in these circuits was discussed in terms of power waveforms, which were determined by the phase angle between voltage and current. In a DC circuit, power is equal to the product of voltage and current ($P = V \times I$). This formula is also true for purely resistive circuits. However, when a reactance (either inductive or capacitive) is present in an AC circuit, power is no longer a product of voltage and current.

Since reactive circuits cause changes in the method used to compute power, the following described techniques express the basic power relationships in AC circuits. The product of voltage and current is expressed in volt-amperes (VA) or kilovolt-amperes (kVA), and is known as *apparent power*. When meters are used to measure power in an AC circuit, apparent power is the voltage reading multiplied by the current reading. The actual power that is converted into another form of energy by the circuit is

Power System Fundamentals

measured with a wattmeter. This actual power is referred to as *true power*. Ordinarily, it is desirable to know the ratio of true power converted in a circuit into apparent power. This ratio is called the *power factor* and is expressed as:

$$pf = \frac{P}{VA}$$

or

$$\%pf = \frac{P}{VA} \times 100$$

where:
  pf is the power factor of the circuit,
  P is the true power in watts, and
  VA is the apparent power in volt-amperes.

Sample Problem:

Given: a 240 volt, 60 hertz, 30 ampere electric motor is rated at 6000 Watts.
Find: power factor at which the motor operates.
Solution:

$$\%pf = \frac{P}{VA} \times 100$$

$$\frac{6000 \text{ W}}{240V \times 30A} = 100$$

$$\% \text{ pf} = 83.3\%$$

The maximum value of the power factor is 1.0, or 100%, which would be obtained in a purely resistive circuit. This is referred to as the *unity power* factor.

The phase angle between voltage and current in an AC circuit determines the power factor. If a purely inductive or capacitive circuit existed, the 90° phase angle would cause a power factor of zero to result. In practical circuits, the power factor varies according to the relative values of resistance and reactance.

The power relationships we have discussed may be simplified by looking at the power triangle shown in Figure 2-19. There are two components that affect the power relationship in an AC circuit. The in-phase (resistive) component that results in power conversion in the circuit is called *active power*. Active power is the *true power* of the circuit and it is measured in watts. The second component is that which results from an inductive or capacitive reactance, and it is 900 out of phase with the active power. This component, called *reactive power*, does not produce an energy conversion in the circuit. Reactive power is measured in volt-amperes reactive *(vars)*.

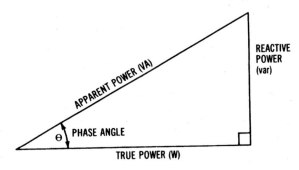

Figure 2-19. Power triangle

The power triangle of Figure 2-19 shows *true power (watts)* on the horizontal axis, *reactive power (var)* at a 90° angle from the true power, and volt-amperes (VA) as the longest side (hypotenuse) of the right triangle. Note the similarity among this right triangle, the voltage triangle for series AC circuits in Figure 2-15B, the current triangle for parallel AC circuits in Figure 2-16B, the impedance triangles in Figure 2-17C, and the admittance triangles in Figure 2-18B. Each of these right triangles has a horizontal axis that corresponds to the resistive component of the circuit, while the vertical axis corresponds to the reactive component. The hypotenuse represents the resultant, which is based on the relative values of resistance and reactance in the circuit. We can now see how important vector representation and an understanding of the right triangle are in analyzing AC circuits.

We can further examine the power relationships of the power triangle by expressing each value mathematically, based on the value of apparent power (VA) and the phase angle (θ). Remember that the *phase angle* is the amount of phase shift, in degrees, between voltage and current in the circuit. *Trigonometric ratios*, which are discussed in Appendix A, show that the sine of an angle of a right triangle is expressed as:

Power System Fundamentals

$$\sin \theta = \frac{\text{opposite side}}{\text{hypotenuse}}$$

Since this is true, the phase angle can be expressed as:

$$\sin \theta = \frac{\text{reactive power (var)}}{\text{apparent power (VA)}}$$

Therefore,

$$\text{var} = \text{VA} \times \sin \theta.$$

We can determine either the phase angle or the var value by using trigonometric functions. We also know that the cosine of an angle of a right triangle is expressed as:

$$\sin \theta = \frac{\text{adjacent side}}{\text{hypotenuse}}$$

Thus, in terms of the power triangle:

$$\sin \theta = \frac{\text{true power (W)}}{\text{apparent power (VA)}}$$

Therefore, true power can be expressed as:

$$W = \text{VA} \times \cos \theta.$$

Sample Problem:

Given: a circuit has the following values: applied voltage = 240, current = 12 amperes, power factor = 0.83.

Find: true power of the circuit.

Solution:
$$\begin{aligned} W &= \text{VA} \times \cos \theta \\ &= 240 \text{ V} \times 12 \text{ A} \times 0.83 \\ W &= 2390 \text{ watts} \end{aligned}$$

Note that the expression

$$\frac{\text{true power}}{\text{apparent power}}$$

is the *power factor* of a circuit; therefore, the power factor is equal to the cosine of the phase angle (pf = cosine θ).

Right triangle relationships can also be expressed as equations that determine the value of any of the sides of the power triangle when the other two values are known. These expressions are as follows:

$$W = \sqrt{VA^2 - var^2}$$

$$VA = \sqrt{W^2 - var^2}$$

$$var = \sqrt{VA^2 - W^2}$$

Sample Problem:

Given: total reactive power = 54 var, applied; voltage = 120 volts; current = 0.5 amperes.

Find: true power of the circuit.

Solution:

$$W = \sqrt{VA^2 - var^2}$$

$$\sqrt{(120 \times 0.5)^2 - 54^2}$$

$$\sqrt{3600 - 2916}$$

26.15 watts

Appendix B should be reviewed in order to gain a better understanding of the use of right triangles and trigonometric ratios for solving AC circuit problems.

POWER RELATIONSHIPS IN THREE-PHASE AC CIRCUITS

To illustrate the basic concepts of three-phase power systems, we will use the example of a three-phase AC generator. This type of generator will be discussed in more detail in Chapter 6. A simplified pictorial diagram of a three-phase generator is shown in Figure 2-20. A three-phase voltage diagram is shown in Figure 2-21.

In Figure 2-20, poles A', B', and C' represent the beginnings of each of the phase windings, while poles A, B, and C represent the ends of each of the windings. These windings may be connected in either of two ways. These methods of connection, called the wye configuration and the delta configuration, are the basic types of three-phase power systems. These three-phase connections are shown schematically in Figure 2-22. Keep in mind that these methods of connection apply not only to three-phase AC generators, but also to three-phase transformer windings and three-phase motor windings.

In the wye connection of Figure 2-22A, the beginnings or the ends of each winding are connected together. The other sides of the windings become the AC lines from the generator. The voltage across the AC lines (V_L) is equal to the square root of 3 (1.73) multiplied by the voltage across the phase windings (V_P), or:

$$V_L = V_P \times 1.73.$$

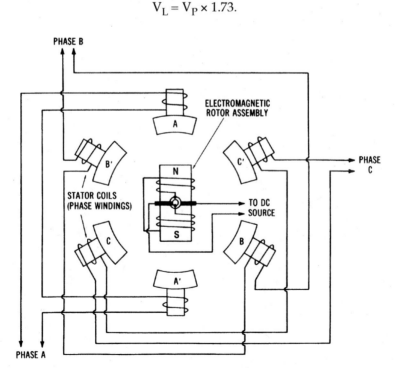

Figure 2-20. Simplified drawing showing the basic construction of a three-phase AC alternator

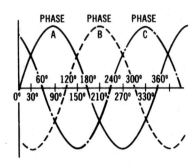

Figure 2-21. Three-phase output waveform developed by an alternator

Sample Problem:

Given: phase voltage = 120 volts for a three-phase wye system.
Find: line voltage.
Solution:

$$V_L = V_P \times 1.73$$
$$= 120 \text{ V} \times 1.73$$
$$V_L = 208 \text{ volts}$$

The line currents (I_L) are equal to the phase currents (I_P), or:

$$I_L = I_P$$

In the *delta* connection of Figure 2-22B, the end of one phase winding is connected to the beginning of the adjacent phase winding. The *line voltages* (V_L) are equal to the *phase voltages* (V_P). The *line currents* (I_L) are equal to the *phase current* (I_P) multiplied by 1.73.

The power developed in each phase (P_P) for either a wye or a delta circuit is expressed as:

$$P_P = V_P \times I_P \times pf$$

Sample Problem:

Given: a three-phase delta system has a phase voltage of 240 volts, a phase current of 20 amperes, and a power factor of 0.75.

Find: power per phase.
Solution:

$$P_P = V_P \times I_P \times pf$$
$$P_P = 240 \text{ V} \times 20 \text{ A} \times 0.75$$
$$P_P = 3600 \text{ watts}$$

where pf is the power factor of the load.

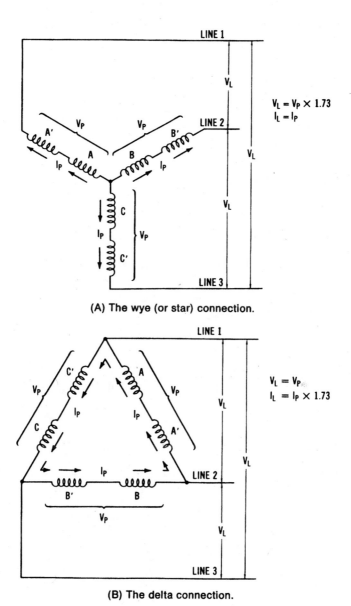

(A) The wye (or star) connection.

(B) The delta connection.

Figure 2-22. The two methods of connecting three-phase stator coils together: (A) The wye (or star) connection; (B) The delta connection

The *total power* (P_T) developed by all three phases of a three-phase system is expressed as:

$$P_T = 3 \times 277\ V \times 10\ A \times 0.85$$
$$= 3 \times V_p \times I_p \times pf$$
$$= 1.73 \times V_L \times I_L \times pf.$$

Sample Problem:

Given: a three-phase wye system has a phase voltage of 277 volts, a phase current of 10 amperes, and a power factor of 0.85.

Find: total three-phase power.

Solution:
$$P_T = 3 \times 277\ V \times 10\ A \times 0.85$$
$$P_T = 7063.5\ watts$$

We can summarize three-phase power relationships as follows:

$$\text{Volt-amperes per phase } (VA_P) = V_P I_P$$
$$\text{Total volt-amperes } (VA_T) = 3\ V_P I_P$$
$$= 1.73\ V_L I_L$$

Sample Problem:

Given: a three-phase delta system has a line voltage of 208 volts and a line current of 4.86 amperes.

Find: total three-phase volt-amperes.

Solution:
$$VA = 1.73 \times 208\ V \times 4.86\ A$$
$$VA = 1748.8\ \text{volt-amperes}$$

$$\text{power factor (pf)} = \frac{\text{true power (W)}}{1.73\ V_L I_L}$$

$$= \frac{W}{3\ V_P I_P}$$

$$\text{power per phase (Pp)} = 3\ V_P I_P \times pf$$
$$= 1.173\ V_L I_L \times pf$$

where:
    V_L is the line voltage in volts,
    I_L is the line current in amperes,
    V_P is the phase voltage in volts, and
    I_P is the phase current in amperes.

Sample Problem:

Given: a three-phase wye system has the following values: phase voltage = 120 V, phase current = 18.5 A, and power factor = 0.95.

Find: total three-phase power of the circuit.

Solution:

$$\begin{aligned} P_T &= 3 \times V_P \times I_P \times pf \\ &= 3 \times 120 \text{ V} \times 18.5 \text{ A} \times 0.95 \\ &= P_T = 6{,}327 \text{ W} = 6.327 \text{ kW} \end{aligned}$$

Calculations involving three-phase power are somewhat more complex than single-phase power calculations. We must keep in mind the difference between phase values and line values to avoid making mistakes.

SUMMARY

In this chapter, we have examined some of the fundamentals of electrical power systems. We need to have some understanding of the three basic types of circuits—*resistive, inductive,* and *capacitive*—in order to understand the operation of power-producing systems, such as generators, chemical cells, and other power-conversion systems (such as electric lights and electric motors).

Resistive circuits exhibit similar characteristics with either applied AC or DC. The power converted in a resistive circuit is expressed in all cases as:

$$P = V \times I$$

The property of *inductance* occurs in systems because of coils of wire or windings that exhibit electromagnetic characteristics. The current developed in an inductive AC circuit *lags* behind the applied AC voltage because of this electromagnetic effect or *counter electromotive force (cemf)*.

The power developed in an inductive circuit is dependent upon the *power factor (pf)* of the circuit. Power factor is a ratio of *apparent power* (volt-amperes) and *true power* (watts) of a circuit, an expressed as:

$$Pf = \frac{\text{true power}}{\text{apparent power}}$$

Capacitance causes current to *lead* voltage in an AC circuit. The *electrostatic field* produced by a capacitor is responsible for this effect. The power developed in a capacitive circuit is expressed as:

$$P = V \times I \times pf$$

just as in an inductive circuit.

The *power relationships*, as well as voltage and current relationships, in AC circuits, may be simplified by using *vector or phasor diagrams*. These diagrams allow us to examine, by a visual analysis, the effect of resistance, inductance, or capacitance on a circuit.

Both *single-phase* and *three-phase* AC power systems are used extensively. Electrical power systems involve:

1. Electrical power *sources*, such as generators.
2. *Distribution* of electrical power, mainly by specialized conductors.
3. *Control* of electrical power by various methods.
4. Electrical power *conversion* systems or loads, such as electrical lights and motors.
5. *Measurement* of electrical-power-related quantities with specialized equipment.

The units of this book discuss each of these five aspects of electrical-power systems.

Chapter 3

Power Measurement Equipment

Chapter 3 provides an overview of the types of equipment used to *measure* electrical power quantities. Specific applications of measurement equipment are discussed further in the chapters that follow. There are many different types of equipment used to *measure* quantities associated with electrical power.

IMPORTANT TERMS

Upon completion of this chapter, you should have an understanding of the following terms:

Measurement Systems
Analog Instruments
Comparative Instruments
CRT Display Instruments
Numerical Readout Instruments
Chart Recording Instruments
Volt-Ohm-Milliammeter (VOM)
Meter Movement
d'Arsonval Principle
Multifunction Meter
Single-Function Meter
Meter Scale
Wheatstone Bridge
Unknown Resistance (R_x)
Standard Resistance (R_s)
Wattmeter
Watt-Hour Meter

True Power
Apparent Power
Power Factor
Three-Phase Power Analyzer
Power Factor Meter
Power Demand Meter
Average Power
Peak Power
Frequency Measurement
Synchroscope
Phase Sequence
On-Line Operation
Ground Fault Indicator
Megohmmeter
Clamp-on Meter
Telemetering

MEASUREMENT SYSTEMS

All *measurement systems* have certain basic characteristics. Usually, a specific quantity is monitored, either periodically or continuously. Therefore, some type of visual indication of the quantity being monitored must be available. For this purpose, several types of instruments for measuring electrical and physical quantities are available. The basic types of measurement systems can be classified as: (1) *analog instruments*, (2) *comparative instruments*, (3) *cathode ray tube (CRT) display instruments*, (4) *numerical-readout instruments*, and (5) *chart-recording instruments*.

Analog Instruments

Instruments that rely upon the motion of a hand or pointer are referred to as *analog instruments*. The *volt-ohm-milliammeter (vom)* is one type of analog instrument. The vom is a *multifunction, multirange* meter. Single-function analog meters can also be used to measure electrical or physical quantities. The basic part of an electrical analog meter is called the *meter movement* and is shown in Figure 3-1. The movement of the pointer along a calibrated *scale* is used to indicate an electrical or physical quantity. For example, physical quantities such as air flow or fluid pressure can be monitored by analog meters.

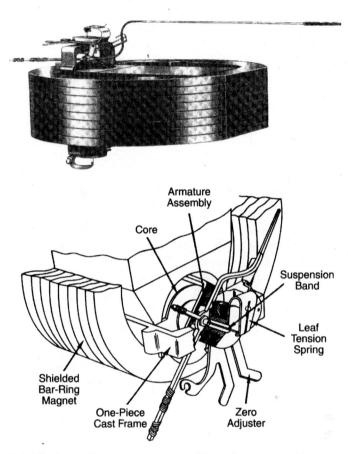

Figure 3-1. Moving-coil meter movement for analog meters (*Courtesy Triplett Corp.*)

Many meters employ an *analog movement* called the *d'Arsonval*, or *moving-coil*, type. The basic operational principle of this type of movement is shown in Figure 3-2. The pointer or needle of the movement remains stationary on the left portion of the calibrated scale until a current flows through the electromagnetic coil that is centrally located within a permanent magnetic field. When current flows through the coil, a reaction between the electromagnetic field of the coil and the stationary, permanent magnetic field develops. This reaction causes the hand (pointer) to deflect toward the right portion of the *scale*. This basic moving-coil meter movement operates on the same principle as an electric motor. It may be used for either *single-function meters*, which measure only one quantity, or for *multifunction meters.*

The basic meter movement may be modified so that it will measure almost any electrical or physical quantity.

Comparative Instruments

Another group of measuring instruments can be classified as *comparative instruments*. Usually, a comparative instrument is designed to compare a component of known value to one with an unknown value. The accuracy of comparative instruments is ordinarily much better than that of analog instruments, described above.

A *Wheatstone bridge* is a typical type of comparative instrument. The technique used for measurement with a Wheatstone bridge is illustrated in Figure 3-3. A voltage source is used in conjunction with a resistive bridge circuit and a sensitive zero-centered moving-coil meter movement. The bridge circuit is completed by adding the external *unknown resistance* (R_x) that is to be measured. When R_s is adjusted, so that the resistive path formed by R_x and R_s is equal to the path formed by R_1 and R_2 no current will flow through the meter. In this condition, the meter will indicate zero (referred to as a *null*), and the bridge is said to be balanced. The value of

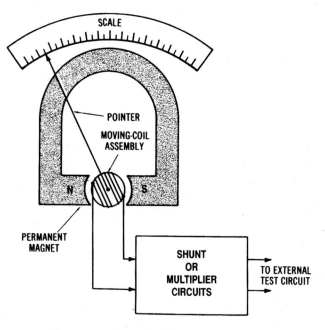

Figure 3-2. Operating principle of a meter movement

standard resistance (R_s) is marked on the meter, so that the value of the *unknown resistance* (R_x) can be determined. Resistors R_1 and R_2 form the "ratio arm" of the indicator. The value of an unknown resistance (R_x) may be mathematically expressed as:

$$R_x = \frac{R_1}{R_2} \times R_s$$

Sample Problem:

Given: in the bridge circuit of Figure 3-3, R_1 = 500 ohms, R_2 = 2000 ohms, and R_s = 1000 ohms.
Find: the value of unknown resistance (R_x).
Solution:

$$R_x = \frac{R_1}{R_2} \times R_s$$

$$\frac{5002}{20000} \times 1000 \; \Omega$$

$$R_x = 250 \text{ Ohms}$$

The *Wheatstone bridge* will measure most values of resistance with considerable accuracy. Several other types of comparative instruments also use the Wheatstone bridge principle.

Cathode Ray Tube (CRT) Instruments

A *cathode ray tube display instrument,* usually called an *oscilloscope,* is an important type of instrument. Using an oscilloscope makes it possible to visually monitor the voltages of a system. The basic operational part of the oscilloscope is the cathode ray tube. Figure 3-4 shows the construction and electron gun arrangement of a CRT.

Several different types of CRT display instruments are available. *General-purpose oscilloscopes* are used for electronic servicing and for displaying simple types of waveforms. *Triggered-sweep oscilloscopes* are used when it is preferable to apply an external voltage to the oscilloscope for comparative purposes. Other oscilloscopes, classified as laboratory types, have very high sensitivity and good frequency response over wide ranges. A CRT display instrument may be used to measure AC and DC voltages, frequency, and phase relationships, and for various timing and numerical-control applications as well. *Memory* and *storage type* instruments are available for use for more sophisticated measurement purposes.

Numerical-readout Instruments

Many modern instruments employ *numerical readouts*. These simplify the measurement processes and permit more accurate measurements to be made. Numerical-readout instruments rely upon the operation of *digital* circuitry in order to produce a numerical display of the measured quantity. They are very popular measuring instruments.

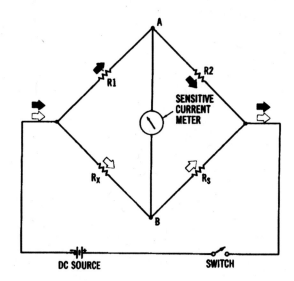

Figure 3-3. Schematic diagram of a Wheatstone bridge

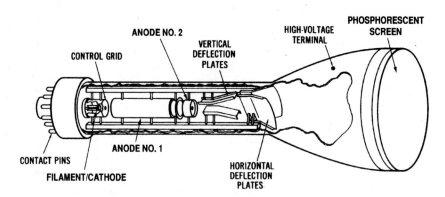

Figure 3-4. Construction of a cathode ray tube

Chart-recording Instruments

The types of instruments discussed previously are used when no permanent record of the measured quantity values is needed. However, instruments can be employed that provide a permanent record of the quantity values. Also, values measured over a specific time period can be recorded. A *chart-recording instrument* is such an instrument. The types of chart recorders include both *pen and ink* recorders and *inkless* recorders.

Pen and ink recorders have a pen attached to the instrument. This pen is controlled by either electrical or mechanical means, which cause it to touch a paper chart and leave a permanent record of the measured quantity on the chart. The charts utilized may be either roll charts (which revolve on rollers under the pen mechanism) or circular charts (which revolve on an axis under the pen). Chart recorders may use more than one pen to record several quantities simultaneously. In this case, each pen mechanism is connected so as to measure a specific quantity. The pen of a chart recorder is a capillary tube device that is actually an extension of the basic meter movement. The pen must be connected to a constant source of ink. The pen is moved by the torque that is exerted by the meter movement, just as the pointer of a hand-deflection type of meter is moved. The chart used for recording the measured quantity usually contains lines that correspond to the radius of the pen movement. Increments on the chart are marked according to time intervals. The chart must be moved under the pen at a constant speed. Either a spring-drive mechanism, a synchronous AC motor, or a DC servomotor can be used to drive the chart. Recorders are also available that use a single pen to make permanent records of measured quantities on a single chart. In this case, either coded lines or different colored ink can be used to record the quantities.

Inkless recorders may use a voltage applied to the pen point to produce an impression on a sensitive paper chart. In another process, the pen is heated to cause a trace to be melted along the chart paper. The obvious advantage of inkless recorders is that ink is not required.

Chart-recording instruments are commercially available for measuring almost any electrical or physical quantity. For many applications, the recording system is located a great distance from the device being measured. For accurate system monitoring, a central instrumentation system may be used. Power plants, for instance, ordinarily use chart recorders at a centralized location to monitor the various electrical and physical quantities involved in the power plant operation.

The operation of a typical *roll-chart* recording instrument involves several basic principles. The user should assure that the chart roll has enough paper to last throughout the duration of the time that it is to be used. The ink supply should be checked. If ink is needed, the well should be filled to the proper level. Also, the pen should be checked for proper pressure on the roll chart and for accurate adjustment along the incremental scale of the roll chart. The user should also assure that the meter is properly connected to the external circuit (which is needed for making the desired measurement).

MEASURING ELECTRICAL POWER

Electrical power is measured with a *wattmeter*. A meter movement called a dynamometer movement, shown in Figure 3-5, is used in most wattmeters. Note that this movement has two electromagnetic coils. One coil, called the *current coil*, is connected in series with the load to be measured. The other coil, called the *potential coil*, is connected in parallel with the load. Thus, the strength of each electromagnetic field affects the movement of the meter pointer. The operating principle of this movement is similar to that of the moving-coil type of movement, except that there is a fixed electromagnetic field rather than a permanent-magnetic field.

When we measure DC power, the total power is the product of voltage times current ($P = V \times I$). However, when measuring AC power, we must consider the *power factor* of the load, since $P = V \times I \times pf$). The *true power* of an AC circuit may be read directly with a wattmeter. When a load is either inductive or capacitive, the true power will be less than the *apparent power* ($V \times I$).

MEASURING ELECTRICAL ENERGY

The amount of electrical *energy* used over a certain period of time may be measured by using a watt-hour meter. A *watt-hour meter*, illustrated in Figure 3-6 relies upon the operation of a small motor mounted inside its enclosure. The speed of the motor is proportional to the power applied to it. The rotor is an aluminum disk that is connected to a numerical register, which usually indicates the number of kilowatt-hours of electrical energy used. Figure 3-7 shows the dial-type face plate that is frequently used on watt-hour meters. Other types of watt-hour meters have a direct

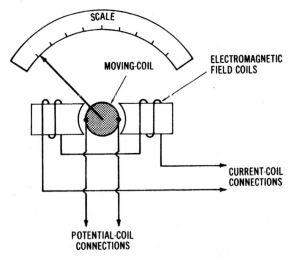

(A) Basic dynamometer movement.

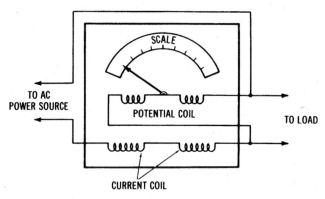

(B) Wattmeter movement using dynamometer movement.

Figure 3-5. Measuring electrical power: (A) Basic dynamometer movement, showing potential coil and current coil connections; (B) Wattmeter movement using dynamometer movement, showing source and load connections; (C) Thermal pens for a chart recorder (Courtesy Esterline Angus); (D) Programmable thermal array recorder (Courtesy Soltec Corp.)

numerical readout of the kilowatt-hours used. Study Figure 3-7 to learn to read a watt-hour meter.

The *watt-hour meter* is connected between the incoming power lines and the branch circuits of an electrical power system. In this way, all

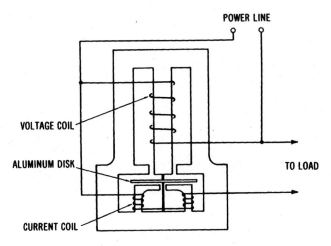

Figure 3-6. Construction diagram of a watt-hour meter

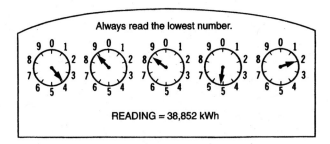

Figure 3-7. Reading the dials of a watt-hour meter

electrical energy that is used must pass through a watt-hour meter. The same type of system is used for home, industrial, or commercial service entrances.

The operation of a *watt-hour meter* is, in many ways, similar to the conventional wattmeter. A potential coil is connected across the incoming power lines to monitor voltage, while a current meter is placed in series with the line to measure current. Both meter sections are contained within the watt-hour meter enclosure. The voltage and current of the power system affect the movement of an aluminum-disk rotor that is part of the watt-hour meter assembly. The operation of the watt-hour meter may be considered as similar to that of an AC induction motor. The stator is an electromagnet that has two sets of windings—the voltage windings and the current windings. The field developed in the voltage windings causes a current to be induced into the aluminum disk. The torque produced is proportional to the voltage and the in-phase current of the system. Therefore, the watt-hour meter will monitor the *true power* converted in a system.

MEASURING THREE-PHASE ELECTRICAL ENERGY

For industrial and commercial applications, it is usually necessary to monitor the *three-phase energy* that is utilized. It is possible to use a combination of single-phase wattmeters to measure the total three-phase power, as shown in Figure 3-8. The methods shown are ordinarily not very practical, since the sum of the meter readings would have to be found in order to calculate the total power of a three-phase system. *Three-phase power analyzers* are designed to monitor the true power of a three-phase system.

Measuring Power Factor

Power factor is the ratio of the *true power* of a system to the *apparent power* (volts × amperes). To determine power factor, we could use a relationship of pf = W/VA. However, it would be more convenient to use a *power factor meter* in situations where the power factor must be monitored.

The principle of a *power factor meter is* shown in Figure 3-9. The power factor meter is similar to a wattmeter, except that it has two armature coils that rotate because of their electromagnetic field strengths. The armature coils are mounted on the same shaft so that their alignment is about 90° apart. One coil is connected across the AC line (in series with a resistance), while the other coil is connected across the line through an inductance. The resistive path through the coil reacts to produce a flux proportional to the *in-phase* component of the power. The inductive path reacts in proportion to the *out-of-phase* component of the power.

If a *unity (1.0) power factor* load is connected to the meter, the current in the resistive path through coil A will develop full torque. Since there is no out-of-phase component, no torque will be developed through the inductive path. The meter movement will now indicate full-scale or unity power factor. As the power factor decreases below 1.0, the *torque* developed by the inductive path through coil B becomes greater. This torque will be in opposition to the torque developed by the resistive path. There-

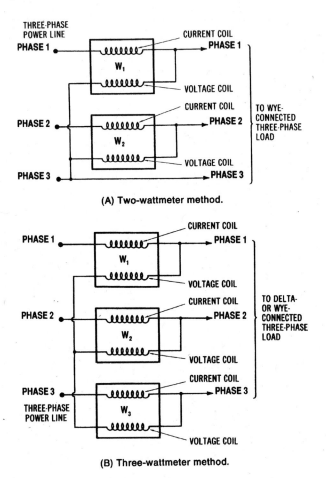

Figure 3-8. Using single-phase wattmeters to measure three-phase power: (A) Two-wattmeter method; (B) Three wattmeter method

fore, a power factor of less than 1.0 will be indicated. The scale must be calibrated to measure power factor ranges from zero to unity.

Power-demand Meters

Power-demand monitors are instruments used to perform an important industrial function. Power demand is expressed as:

$$\text{power demand} = \frac{\text{peak power (kW)}}{\text{average power used (kW)}}$$

Power Measurement Equipment

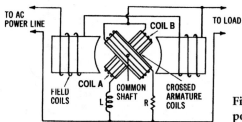

Figure 3-9. Schematic diagram of a power factor meter circuit

Sample Problem:

Given: an industry uses 5000 kW peak power and 3880 kW average power over a 24-hour period.

Find: the industry's demand factor over the 24-hour period.

Solution:

$$\text{Demand} = \frac{\text{Peak Power}}{\text{Avg. Power}}$$

$$= \frac{5000 \text{kW}}{3800 \text{kW}}$$

$$\text{Demand} = 1.31$$

This ratio is important since it indicates the amount above the *average power* consumption that a utility company must supply to an industry. Power demand is usually calculated over a 15-, 30-, or 60-minute interval, and then converted into figures that represent longer periods of time.

Industries may be penalized by a utility company if their *peak power demand* far exceeds their *average demand*. *Power-demand monitors* help industries to better utilize their electrical power. A high peak demand means that the equipment for an industrial power-distribution system must be rated higher. The closer that the peak demand approaches the value of the average demand, the more efficient the industrial power system is in terms of power utilization.

FREQUENCY MEASUREMENT

Another power measurement that is very important is frequency. The frequency of the power source must remain stable, or the operation of many types of equipment can be affected. Frequency refers to the number of cycles of voltage or current that occur in a given period of time. The unit of measurement for frequency is the hertz (Hz), which means cycles per second. A table of frequency bands is shown in Figure 3-10. The standard power frequency in the United States is 60 hertz. Some other countries use 50 hertz.

Frequency can be measured with several different types of meters. An electronic counter is one type of frequency indicator. *Vibrating-reed* instruments are also commonly used for measuring power frequencies. An *oscilloscope* can also be used to measure frequency. *Graphic recording instruments* may be used to provide a visual display of frequency over a period of time. The electrical power industry commonly uses this method to monitor the frequency output of its alternators.

SYNCHROSCOPES

The major application for a *synchroscope* is in electrical power plants. Most power plants have more than one alternator. In order to connect two or more alternators onto the same AC line, the following conditions must be met: (1) their voltage outputs must be equal, (2) their frequencies must be equal, (3) their voltages must be in phase, and (4) the phase sequence of the voltages must be the same.

Voltage output levels may be checked easily with a voltmeter of the transformer-type that is used to monitor high voltages. Frequency is adjusted by varying the speed of an alternator and is also easy to monitor. The phase sequence is established on each alternator when it is installed and connected to the electrical power system. In addition, before alternators are paralleled, the output voltages of both alternators must be monitored, to ensure that they are in phase. This is done by using a synchroscope (see Figure 3-11).

A *synchroscope* is used to measure the relationship between the phases of the system and the alternator that is to be put in parallel or "*online.*" A synchroscope also indicates whether the alternator is running faster or slower than the system to which it is being connected. The basic design of the indicator utilizes a phase-comparative network of two RLC circuits, which are connected between the operating system and the alternator to be paralleled. The meter scale shows whether the new alternator is running too slow or too fast (see Figure 3-11B). When an in-phase relationship exists, along with the other three factors previously

Power Measurement Equipment   73

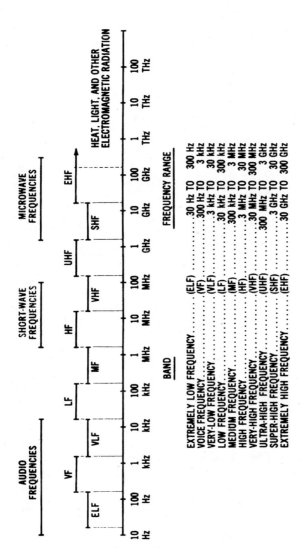

Figure 3-10. Classification of frequency bands

74   Electrical Power Systems Technology

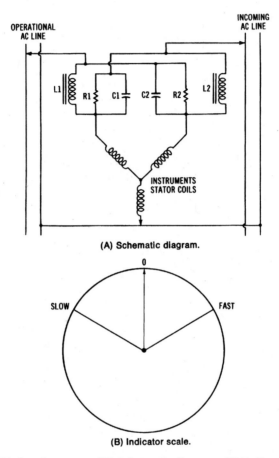

Figure 3-11. Synchroscope: (A) Schematic diagram; (B) Indicator scale

discussed, the alternator may be put "on-line" or connected in parallel with a system that is already operational. The addition of another alternator to the system will allow a higher power capacity to be produced by the electrical power system.

GROUND-FAULT INDICATORS

Ground-fault indicators are used to locate faulty system or equipment grounding conditions. The equipment of electrical power systems must be properly grounded. Proper grounding procedures, principles, and ground-fault interrupters are discussed in Chapter 10.

A *ground-fault indicator* may be used to check for faulty grounding at various points in an electrical power system. Several conditions can exist that might be hazardous. These faulty wiring conditions include: (1) hot and neutral wire reversed, (2) open equipment ground wire, (3) open neutral wire, (4) open hot wire, (5) hot and equipment ground wires reversed, or (6) hot wire on the neutral terminal and the neutral is unconnected. Each of these conditions would present a serious problem in the electrical power system. Although proper wiring eliminates most of these problems, a periodic check with a *groundfault indicator* will ensure that the electrical wiring is safe and efficient.

MEGOHMMETERS

Megohmmeters are used to measure high resistances that are beyond the range of a typical ohmmeter. These indicators are used primarily for checking the quality of insulation on electrical power equipment (mainly motors). The quality of equipment insulation varies with age, moisture content, and the applied voltage. A megohmmeter is similar to a typical ohmmeter, except that some types use a *hand-cranked* permanent-magnet DC generator as a voltage source, rather than a battery. The operator cranks the DC generator while making an insulation test. Figure 3-12 shows a diagram of a megohmmeter circuit. This circuit is essentially the

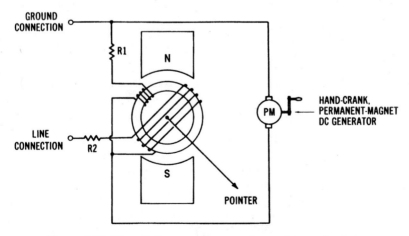

Figure 3-12. Circuit diagram of a megohmmeter, analog type

same as in any series ohmmeter, with the exception that a DC generator is used as the voltage source. *Digital* megohmmeters are also available.

Periodic *insulation tests* should be made on all large power equipment. As insulation breaks down with age, the equipment starts to malfunction. A good method is to develop a periodic schedule for checking and recording insulation resistance. Then, it can be predicted when a piece of equipment will need to be replaced or repaired. A resistance-versus-time graph (see Figure 3-13) can be made, and the trend shown on the graph can be noted. A downward trend (a decrease in insulation resistance) over a specific time period indicates that an insulation problem exists.

CLAMP-ON METERS

Clamp-on meters are popular for measuring current in power lines. This indicator may be used for periodic checks of the current by clamping it around a power line. It is an easy-to-use and convenient maintenance and testing instrument.

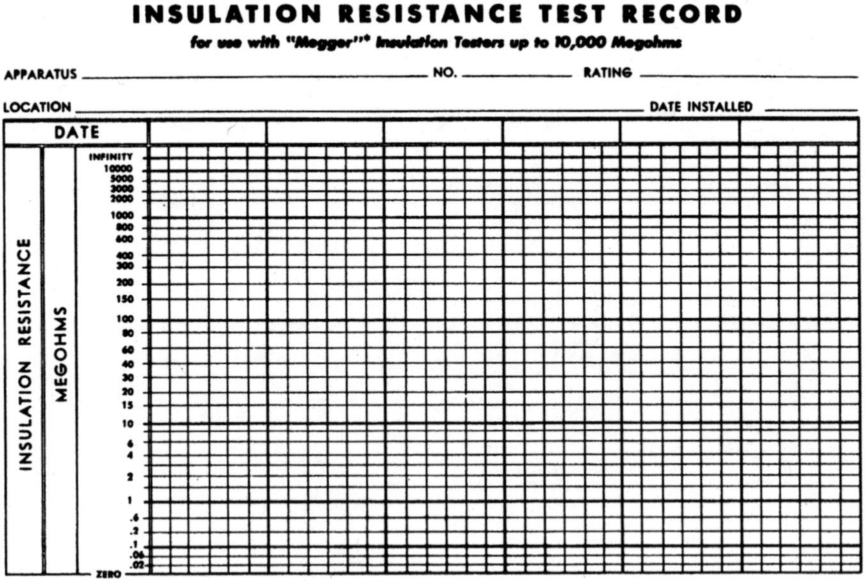

INSULATION RESISTANCE TEST RECORD
for use with "Megger"® Insulation Testers up to 10,000 Megohms

Figure 3-13. Resistance versus time chart to be used with a megohmmeter (Courtesy *Biddle Instruments*)

Power Measurement Equipment   77

The simplified circuit of a *clamp-on* analog current meter is shown in Figure 3-14. Current flow through a conductor creates a magnetic field around the conductor. The varying magnetic field induces a current into the iron core of the clamp portion of the meter. The meter scale is calibrated so that when a specific value of current flows in a power line, it will be indicated on the scale. Of course, the current flow in the power line is proportional to the current induced into the iron core of the clamp-on meter. The *clamp-on meter* may also have a voltage and resistance function that utilizes external test leads. Thus, the meter can be used to measure other quantities.

TELEMETERING SYSTEMS

When a quantity being measured is indicated at a location some distance from its transducer or sensing element, the measurement process is referred to as *telemetering*. Many types of metering systems fit this definition. However, telemetering systems are usually used for *long-distance* measurement, or for centralized measurement systems. For instance, many industries group their indicating systems together to facilitate process control. Another example of telemetering is the centralized monitoring (on a regional basis) of electrical power by utility companies. These systems are similar to other measuring systems, except that a transmitter/

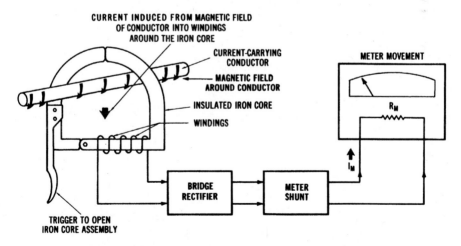

Figure 3-14. Circuit diagram of a damp-on current meter

receiver communication system is usually involved.

Many types of electrical and physical quantities can be monitored by using *telemetering* systems. The most common transmission media for telemetering systems are: (1) wire-such as telephone lines, (2) superimposed signals—which are 30- to 200-kHz signals carried on electrical power distribution lines, and (3) radio frequency signals—from AM, FM, and phase-modulation transmitters. The block diagram of one type of *telemetering* system is shown in Figure 3-15. In this type of system, a DC voltage from the transducer is used to modulate an AM or FM transmitter. The radio frequency (RF) signal is then received at another location and converted back into a DC voltage to activate some end device. The end device, which may be located at a considerable distance from the transducer, might be a chart recorder, a hand-deflection meter, or possibly a process controller. *Digital telemetering* is also used, since binary signals are well suited for data transmission. In this system, the transducer output is converted to a binary code for transmission.

Telemetering is the measurement of some quantity at an area that is distant from its origin. For instance, it is possible, by using telemetering systems, to monitor on one meter the power used at several different locations. Almost any quantity value, either electrical or non-electrical, can be transmitted by using some type of telemetering system. A basic telemetering system has: (1) a *transmitting* unit, (2) a *receiving* unit, and (3) an *interconnection* method. Electrical power systems frequently utilize telemetering systems for the monitoring of power.

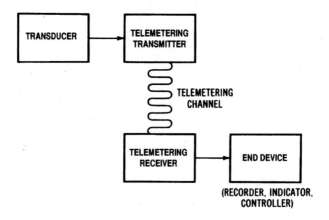

Figure 3-15. Block diagram of a basic telemetering system

UNIT II
Electrical Power Production Systems

Unit II of this book describes *power production systems*. The fundamentals of power production are examined in Chapter 4. *Electrical power production systems* include modern power systems such as *fossil fuel* (coal, oil, gas) *systems, hydroelectric systems,* and *nuclear fission systems.* These systems are discussed in Chapter 4. However, there are also many *alternative energy systems* that are being studied as potential sources of electrical power. These alternative energy systems, which include *nuclear fusion, geothermal, solar, wind, fuel-cell, Goal gasification, tidal,* and *magnetohydrodynamic IMHD) systems,* are discussed in Chapter 5.

Most of the electrical power that is produced is *alternating current (AC) power.* Chapter 6 deals with the *single-phase* and *three-phase* alternators that are used to produce AC power. *Direct current (DC) power* can be produced by chemical action or rotating machinery, or it can be converted from AC sources by the process of rectification. Chapter 7 deals with the systems used for production of direct current.

Figure II shows the *electrical power systems model* used in this book and the major topics of Unit II, Power Production.

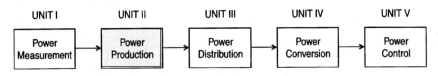

Figure II. Electrical power systems model

Modern Power Systems (Chapter 4)
Alternative Power Systems (Chapter 5)
Alternating Current (AC) Power Systems (Chapter 6)
Direct Current (DC) Power Systems (Chapter 7)

UNIT OBJECTIVES

Upon completing this unit, you should be able to:

1. Explain the basic operation of fossil fuel, hydroelectric, and nuclear fission electrical power systems.
2. Define the terms "load (demand) factor" and "capacity factor" for electrical power systems.
3. Describe coal, oil, and natural gas fossil fuel power production systems.
4. Describe the functions of steam turbines, boilers, and other auxiliary systems of power production systems.
5. Explain the difference between traditional hydroelectric systems and pumped storage hydroelectric systems.
6. Describe load-demand control of modem electrical power production systems.
7. List several alternative power production systems.
8. Explain the basic operation of the following alternative power production systems:
   Solar Energy Systems
   Geothermal Systems
   Wind Systems
   Magnetohydrodynamic (MHO) Systems
   Nuclear Fusion Systems
   Fuel-Cell Systems
   Tidal Power Systems
   Coal Gasification Systems
   Oil Shale Systems
   Biomass
   Fuel-Cell Systems
   Tidal Power Systems
9. State Faraday's law for electromagnetic induction.
10. Describe single-phase AC, three-phase AC, and DC generators.
11. Describe the effect of electrical generator speed, frequency calculation, voltage regulation, and efficiency.
12. Explain the difference between primary and secondary cells.
13. Calculate internal resistance of a DC cell.
14. Describe the following types of primary cells:
    Carbon-zinc
    Alkaline

Mercury
Nuclear
15. Describe the following types of secondary cells:
Lead-Acid
Nickel-Iron
Nickel-Cadmium
Silver-Oxide-Zinc
16. Explain the following types of DC generators:
Permanent Magnet
Separately Excited
Self-Excited
17. Explain the following types of DC rectification systems:
Single-Phase Half Wave
Single-Phase Full Wave
Single-Phase Bridge
Three-Phase Half Wave
Three-Phase Full Wave
18. Describe the effects of rotary converters, filtering methods, and regulation methods for the conversion of AC into DC.

Chapter 4

Modern Power Systems

There are many residential, commercial, and industrial customers of electrical power systems in the United States today. To meet this vast demand for electrical power, power companies work in combination to produce tremendous quantities of electrical power. This vast quantity of electrical power is supplied by *power generating plants*. Individual generating units that supply over 1000 megawatts of electrical power are now in operation at some power plants.

Electrical power can be produced in many ways, such as from chemical reactions, heat, light, or mechanical energy. The great majority of our electrical power is produced by *power plants* located throughout our country, which convert the energy produced by burning coal, oil, or natural gas, by falling water, or by nuclear reactions into electrical energy. Electrical *generators* at these power plants are driven by steam or gas turbines, or by hydraulic turbines in the case of hydroelectrical plants. This chapter will investigate the types of power systems that produce the great majority of the electrical power used today.

Various other methods, some of which are in the experimental stages, may become *future* power production methods. These include *solar cells, geothermal* systems, *wind-powered* systems, *magnetohydrodynamic (MHD)* systems, *nuclear fusion* systems, and *fuel cells*. These *alternative power systems* will be discussed in greater detail in Chapter 5.

IMPORTANT TERMS

This chapter deals with modern electrical power production systems. After studying this chapter, you should have an understanding of the following terms:

Fossil Fuel System
Electrical Power Plant
Steam Turbine

Load Factor (Demand)
Capacity Factor
Boiler
Superheater
Desuperheater
Feedwater
Reaction Turbine
Economizer
Feedwater Heater
Condenser
Feedwater Purifier
Coal Pulverizer
Hydroelectric System
Hydraulic Turbine
Pumped Storage Hydroelectric System
Nuclear Fission System
Nuclear Reactor
Moderator
Boiling-Water Reactor (BWR)
Pressurized-Water Reactor (PWR)

ELECTRICAL POWER PLANTS

Most electrical power in the United States is produced at *power plants* that are either *fossil fuel* steam plants, nuclear fission steam plants, or hydroelectric plants. Fossil fuel and nuclear fission plants utilize steam turbines to deliver the mechanical energy needed to rotate the large *three-phase alternators* that produce massive quantities of electrical power. *Hydroelectric plants* ordinarily use vertically mounted hydraulic turbines. These units convert the force of flowing water into mechanical energy to rotate three-phase alternators.

The *power plants* may be located near the energy sources, near cities, or near the large industries where great amounts of electrical power are consumed. The generating capacity of power plants in the United States is greater than the combined capacity of the next four leading countries of the world. Thus, we can see how dependent we are upon the efficient production of electrical power.

Supply and Demand

The *supply* and *demand* situation for electrical energy is much different from that of other products that are produced by an organization and later sold to consumers. Electrical energy must be supplied at the same time that it is demanded by consumers. There is no simple storage system that may be used to supply additional electrical energy at *peak demand* times. This situation is unique, and it necessitates the production of sufficient quantities of electrical energy to meet the demand of the consumers at any time. Accurate *forecasting* of load requirements at various given times must be maintained by utility companies in order that they may recommend the necessary power plant output for a particular time of the year, week, or day.

Plant Load and Capacity Factors

There is a significant variation in the load requirements that must be met at different times. Thus, the power plant generating capacity utilization is subject to continual change. For this reason, much of the generating capacity of a power plant may be idle during low demand times. This means that not all the generators at the plant will be in operation.

There are two mathematical ratios with which power plants are concerned. These ratios are called load factor *(demand factor)* and *capacity factor*. They are expressed as:

$$\text{Load (demand) factor} = \frac{\text{Average load for a time period}}{\text{Peak load for a time period}}$$

Sample Problem:

Given: a power plant has an average load of 220 MW and a peak load of 295 MW over a 24-hour period.

Find: the load factor of the power plant during the 24 hour period.
Solution:

$$\text{Load factor} = \frac{\text{Average Load}}{\text{Peak Load}}$$

$$= \frac{220 \text{ MW}}{295 \text{ MW}}$$

$$\text{Load factor} = 0.745$$

$$\text{Capacity factor} = \frac{\text{Average load for a time period}}{\text{Output capacity of a power plant}}$$

Sample Problem:
Given: a power plant has an average load of 112 MW and an output capacity of 166 MW.
Find: the capacity factor of the power plant.
Solution:

$$\text{Capacity factor} = \frac{\text{Average Load}}{\text{Capacity}}$$

$$= \frac{112 \text{MW}}{166 \text{ MW}}$$

$$\text{Capacity factor} = 0.675$$

It would be ideal, in terms of energy conservation, to keep these ratios as close to unity as possible.

FOSSIL FUEL SYSTEMS

Millions of years ago, large deposits of organic materials were formed under the surface of the earth. These deposits, which furnish our *coal, oil,* and *natural gas,* are known as *fossil fuels.* Of these, the most abundant fossil fuel is coal, and *coal-fired electrical power systems* produce about one-half of the electrical power used in the United States. Natural-gas-fired systems are used for about one-fourth of our electrical power, while oil-fired systems produce around one-tenth of the power at the present time. The relative contributions of all of these systems to the total electrical power produced in the United States are subject to change, as the result of the addition of new power generation facilities and fuel availability. At the present time, over 80 percent of our electrical energy is produced by fossil fuel systems. It is important to note that these percentages vary from year to year.

A basic *fossil fuel power system* is shown in Figure 4-1. In this type of system, a fossil fuel (coal, oil, or gas) is burned to produce heat energy. The heat from the combustion process is concentrated within a boiler where circulating water is converted to steam. The high-pressure steam is used to ro-

Electrical Power Production Systems

tate a turbine. The turbine shaft is connected directly to the electrical generator and provides the necessary mechanical energy to rotate the generator. The generator then converts the mechanical energy into *electrical energy.*

Fossil Fuels

Fossil fuels are used to supply heat, by means of chemical reactions, for many different purposes. Such fuels contain carbon materials that are burned as a result of their reaction with oxygen. These fossil fuels are used as a direct source of heat when burned in a furnace, and are used as a heat source for steam production when used in a power plant boiler system. The steam that is generated is used for rotating the steam turbines in the power plants.

Fossil fuels vary according to (1) their natural state (solid, liquid, or gas), (2) their ability to produce heat, and (3) the type of flame or heat that they produce. *Coal* and *coke* are solid fossil fuels, and coal is used extensively for producing heat to support electrical power production. *Oil, gas-*

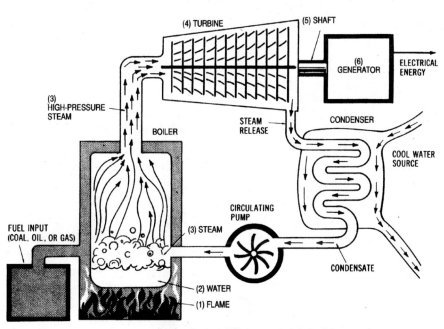

Heat from burning fuel (1) changes water in boiler (2) into steam (3), which spins turbine (4) connected by shaft (5) to generator (6), producing electrical energy.

Figure 4-1. A basic fossil fuel power system

oline, and diesel fuel, which are liquid fossil fuels derived by petroleum processing, are used mostly in conjunction with internal combustion engines. However, oil is used as a heat source for many power plants. *Natural gas* is the primary gaseous fuel used for electrical power production.

Coal-fired Systems

The use of *coal* as a fuel to supply the necessary heat energy at a power plant requires the use of specialized stokers or grating units. These units reduce the size of the lump coal. Most of these units are mechanical systems that agitate the coal to reduce it into smaller lumps. The coal used at a power plant usually is sent through a stoker or grating unit by conveyer belts. A large gravity-feed hopper is often used to route very small lumps of coal into a pulverizing unit.

The *pulverizer* looks very similar in construction to a large ball-bearing unit. The coal is routed into the pulverizer unit, where large, rotating steel balls crush the coal until it is in particles about the same size and consistency as face powder. These fine particles are routed into the furnace by air pressure produced by *force-draft fans*. The coal is held in suspension until it is ignited. It then releases a large amount of heat energy. The suspended powder-fine coal particles allow sustained combustion to take place in the furnace. The pulverized coal speeds up the combustion process.

Coal-fired Plant Operation

Since the majority of electrical power produced today is from *coal-fired systems*, we will discuss the basic operation of this type of system. In a steam plant that produces electrical power, most of the operations are used for rotating the *steam turbine*. Remember that in any steam plant, heat must be produced. This heat produces steam, which moves the steam turbine, which produces a rotary motion, which finally produces electrical power. Figure 4-2 shows the layout of a typical *coal-fired electrical power plant*. Notice that it is located near a river, so that cooling water can be easily provided. The water is also used to produce steam to operate the *steam turbine* that rotates the generator unit. A cross-section of a coal-fired power plant is shown in Figure 4-3.

The maximum *efficiency* of the coal-fired plant of today is approximately 40 percent when using a powdered-coal spraying process. Efficiency is calculated as:

Electrical Power Production Systems

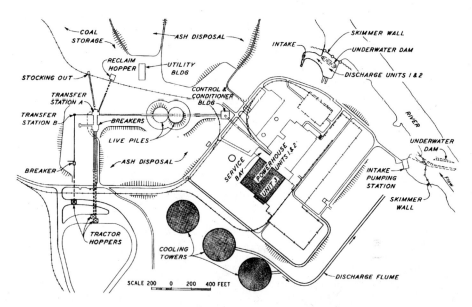

Figure 4-2. Layout of a typical coal-fired electrical power plant. (*Courtesy Tennessee Valley Authority*)

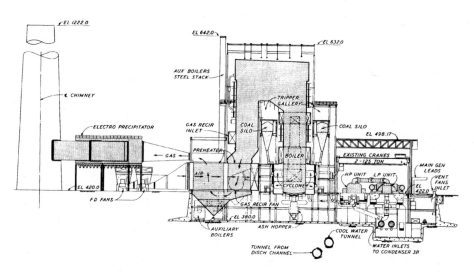

Figure 4-3. Cross-section view of a typical coal-fired power plant. (*Courtesy Tennessee Valley Authority*)

$$\text{Efficiency} = \frac{\text{Power Output}}{\text{Power Input}}$$

where power output is electrical power produced.

Coal requires extensive handling equipment. The coal itself must be handled, and then the ash and dust particles must be removed. At the power plant, the coal is moved to overhead hoppers by means of conveyor belts. These hoppers may typically be as large as eight stories high. The coal usually is fed into *pulverizing mills* by means of gravity. It is ground to a consistency similar to that of face powder, using the method discussed previously. The powdered coal is then dried using plant exhaust gases, and then blown into a *furnace*. The coal is ordinarily blown through a tangential or "T" burner into the furnace. These burners are placed in the four corners of a square furnace to create the turbulence needed for complete combustion.

Another method of firing the furnace is the fluidized bed. Advantages of the *fluidized bed* are that it produces less pollution and can burn a lower quality coal. In a fluidized bed, coal is crushed to form 1/8-inch to 1/4-inch diameter particles. When air is blown through a layer of this coal, the particles will float on a cushion of air. The pressure has to be adjusted very accurately, so that the particles are fluidized without being blown away from the bed. The fluidized bed is the basis of a direct-combustion process. If the bed is hot enough, the flow of air through the bed leads to almost total combustion, and can provide a greater efficiency with less ash and dust.

Power plant *boilers*, such as the one shown in Figure 4-4 incorporate several special units to improve their thermal efficiency and economy of operation. An *economizer* is placed at the exhaust exit to preheat the water coming into the boiler. The economizer also preheats the air blowing into the furnace. A *superheater* is a bank of tubes located at the hottest spot of the furnace. These tubes take up the steam after it leaves the boiler and before it enters the turbine. The purpose of the superheater is to raise the temperature of the *steam*. Increased superheat decreases the percent of water per unit volume in the steam, which increases turbine life. A desuperheater is the next part of the system. The *desuperheater* brings the steam down to a temperature at which it can be condensed. The *feedwater* in a power plant is used over and over, with water added only to account for losses. The feedwater must be very pure to ensure long life of the boiler tubes. Some common power plant terminology is summarized in Table 4-1.

Electrical Power Production Systems

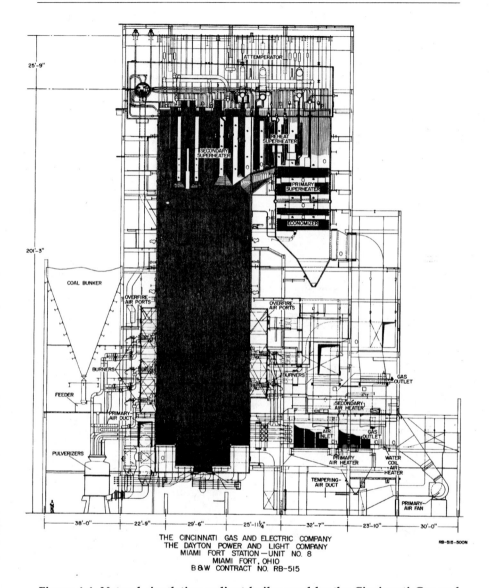

Figure 4-4. Natural circulation radiant boiler used by the Cincinnati Gas and Electric Company *(Courtesy Babcock and Wilcox)*

Table 4-1. Power Plant Terminology

Boiler (steam generator)	An enclosed unit that is used to produce steam by heat developed within the unit. Water that flows through it is converted to steam as a result of the heat of combustion.
Superheater	A grouping of tubes inside a boiler that absorbs heat from the combustion gases to raise the temperature of the steam to a very high temperature.
Feedwater	Water that is supplied to the boiler to produce steam.
Coal Pulverizer	A machine that reduces coal to a fineness suitable for burning in suspension. Coal is ground to a talcum powder consistency.
Forced Draft (FD) Fan	A fan to supply preheated combustion air under pressure to the furnace, for mixture with the fuel stream.
Economizer	A heat-absorbing section of the boiler that preheats incoming cold boiler feedwater by transferring heat from the outgoing combustion gases to the water.
Condenser	A unit that converts steam into water.

After steam has been produced, a rotary motion must be developed. This rotary motion is produced by a *steam turbine*. A steam turbine, shown in Figure 4-5 is typically made up of as many as 1500 blades. The rotor is usually divided into two parts—the *high pressure rotor* and the *low pressure rotor*. The low-pressure rotor is larger in diameter than the high-pressure rotor. Steam is channeled to the high-pressure rotor, and it is then routed to the low-pressure rotor.

Steam turbines ordinarily achieve a maximum efficiency of less than 30 percent, but only when run at very high speeds. Some turbines can produce as much as 160,000 horsepower. A speed of 3600 rpm is needed to develop a 60-Hz electrical power output. The standard power frequency in the U.S. is 60-Hz. Large three-phase AC generators are connected to DC *exciters*. Generators are often cooled with hydrogen because hydrogen has less than one-tenth the density of water. Therefore, much less energy is required to recirculate the hydrogen for cooling purposes.

The process just described summarizes the operation of a steam power plant. There are several variations to the basic process; however, most plants use similar methods. The individual parts of a steam-generating system will be discussed in more detail in the following sections.

Electrical Power Production Systems

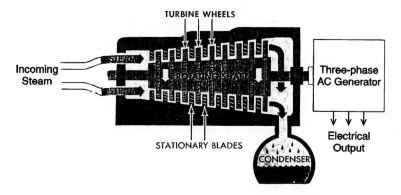

Steam from the boiler rushes through the turbine, rotating a series of bladed wheels mounted on a shaft. As the steam leaves the turbine, a condenser cools it and changes it into liquid water. This creates a vacuum that pulls steam through the turbine.

Expanded View of Turbine

Sets of stationary blades fastened to the turbine aim the steam at the wheels. These blades guide the steam so that it strikes the wheels at the correct angle.

Figure 4-5. Steam turbine principle of operation

STEAM TURBINES

Steam turbine systems are now used to produce over 80 percent of the electrical power used in the United States. The force of steam produces a rotary motion (mechanical energy) in a steam turbine. This mechanical energy is converted to electrical energy by three-phase generators connected by a common shaft. Both fossil fuel systems and nuclear fission systems utilize steam turbines as *prime movers* (rotary motion producers).

A *reaction turbine* channels high-velocity steam through a set of blades mounted on a rotary shaft. The *reaction turbine* usually has more than one set of blades, with each set having a different diameter. As the steam passes though the first section of blades, its pressure is reduced, and its volume is increased. Due to the increased volume, the additional sections of blades must have larger diameters and longer sets of blades. These combined sections of blades direct the high-velocity steam in such a way that a maximum rotational force is produced by the turbine.

The design of a *steam turbine* is very critical for the efficient production of electrical power. Several characteristics of steam turbines cause design problems. Steam turbines must be operated at high rotational speeds, so the blades must be designed to withstand a tremendous amount of centrifugal force. The rotor and blade assemblies for steam turbines are usually machined from a forged piece of chromium and steel alloy. This assembly must be very precisely balanced before the machine is put into operation. The leakage of steam from the enclosed rotor and blade assembly must be prevented. Solid seals cannot be used along the rotor shaft, so so-called "steam" seals are used to provide a minimum clearance between the seals and the shaft. The bearings of a steam turbine must be carefully designed to withstand both axial and end pressures of high magnitudes.

Steam turbines used in electrical power production must be rotated at a constant speed. If turbine *speed* changes, the *frequency* of the generator output voltage will be changed from the standard 60-Hz value. Therefore, a system of governors is used in a steam turbine to regulate its speed. The *governor system* adjusts the turbine speed by compensating for changes in generator power demand. As more load is placed on the generator (increased consumption of electrical power), the generator offers an increased resistance to rotation. Thus, power input to the turbine must be increased accordingly. The governor system of the turbine automatically adjusts the steam input to the turbine blades to compensate for increases and decreases in the load demand placed upon the generator that it drives.

BOILERS

Boilers (see Figures 4-4 and 4-6) are an important part of steam power production systems. The function of a boiler is to provide an enclosure in which pressurized water can be heated to a high temperature to produce

Electrical Power Production Systems

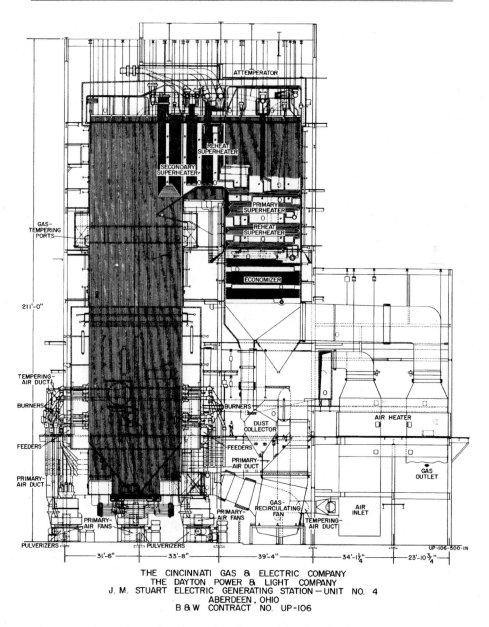

Figure 4-6. Once-through universal boiler used by the Cincinnati Gas and Electric Company (*Courtesy Babcock and Wilcox*)

steam. The heat from burning fossil or nuclear fuels is transferred to an area through which pressurized water flows, and the water is converted to steam through this procedure.

The transfer of heat within a boiler utilizes the three methods of *heat transfer: radiation, convection,* and *conduction.* The *radiation* method involves the movement of heat energy from a warm area to a cool area, and is dependent upon temperature difference and the ability of materials to absorb heat. The *conduction* method requires contact between the heat source and the heated area, and it relies upon the heat conductivity of the heated material. *Convection* is the movement of heat from a hot area to a cooler area by means of an intermediate substance, such as gas. Each of these three methods of heat transfer occurs in varying amounts, depending upon boiler design.

A *boiler* that functions properly is a very critical part of the power production system, as the boiler operation determines the quantity of steam available to produce the rotary motion of the turbine. When more power input for the generating process is required because of an increased load on the system, the boiler must deliver more steam to the turbine. Boilers must be able to provide effective water circulation, efficient fuel combustion, and maximum heat transfer to the circulating water. The boilers used in most steam power plants today are called water-tube boilers. Their design consists of banks of tubes, separated by heat insulation; water is circulated through the tubes under high temperature and high pressure. Boiler design is very important for an efficient steam power plant operation.

Boiler Auxiliary Systems

There are several auxiliary systems used in a steam power plant to increase the operation and efficiency of the boilers. Some of these systems were mentioned briefly in previous sections. One auxiliary system is called an *economizer.* Economizers utilize the hot exhaust gases from the fuel reactions within a boiler to preheat the cold feedwater that is pumped into the boiler. Thus, the economizer uses the waste gases, which would otherwise be emitted through the exhaust stack, for an important purpose. This improves the efficiency of the power plant.

In addition to the *economizer, feedwater heaters* and *preheaters* are used to increase the water temperature before its entry into the boiler. These systems heat the pumped feedwater by means of steam, which is circulated through the unit. In some plants, systems called *superheaters* are used.

These units consist of banks of tubes located at the hottest area of the boiler. Steam flows through these tubes before its entry into the steam turbine. The purpose of the superheater is to cause the steam to reach a higher temperature so as to produce more energy in the steam turbine. Each of these auxiliary systems helps to improve the efficiency of steam power plants.

Condensers and Purifiers

The *condensers* and *feedwater purifiers* used at steam power plants are also important in the production of energy. *Condensers* are used to cool the used steam that has passed through the steam turbine. The condensed water is continuously recirculated through the system. *Feedwater purifiers* are used to clean the impurities from the feedwater, which is obtained from a water source located adjacent to the power plant. The feed water purifiers play an important part in power plant operation. Without them, the metal used in the construction of the boiler would corrode, producing a slag build-up on the boiler walls, which eventually would destroy the boiler. Also, impurities in the steam can cause damage to the precision blades used in the steam turbines. In addition, the gases that are contained in the feed water must also be removed. These gases are removed by a unit called a *deaerator.*

The quantity of *feedwater* that reaches the steam turbine in the form of steam is dependent on the amount of evaporation that takes place in the system. In the power plant, a comparative analysis must be made of the quantity of water entering the boiler and the quantity of steam coming out of the steam turbine. Adjustments in feedwater flow are based on this comparison.

Future of Coal-fired Power Systems

Pulverized-coal systems have been used for many years to produce energy for conversion to electrical power. However, there are now more environmental restrictions on these systems. Major problems include sulfur dioxide and nitrogen oxide emission controls for the power plants, particularly sulfur-dioxide controls. More stringent environmental controls increase the capital cost of power system operations. Even though coal is the most abundant fossil fuel, it is also the dirtiest in terms of environmental factors. Thus, a significant problem of electrical power technology is how to utilize coal in an environmentally acceptable way.

Electrical power systems are the largest consumers of coal in the United States. With a decrease in the availability of oil and natural gas, coal must again be relied upon as the prime electrical power source of the future. Power production systems must be designed that will produce electrical energy in the most economical and environmentally responsible way.

Coal-fired System Simulator

A *coal-fired electrical power system simulator* can't be used to study this method of power production. The various subsystems of the coal-fired power plant are interconnected within the overall system. There are basically five subsystems for this type of power plant. These subsystems are the *feedwater system*, the *fuel and air system*, the *coal-pulverizing system*, the *boiler-water system*, and the *steam and turbine system*. Each of these five subsystems is shown in the diagrams of Figures 4-7 through 4-11.

Oil-fired System Simulator

An *oil-fired electrical power system simulator* is very similar to the coal-fired system. The subsystems are identical to those used in the coal-fired

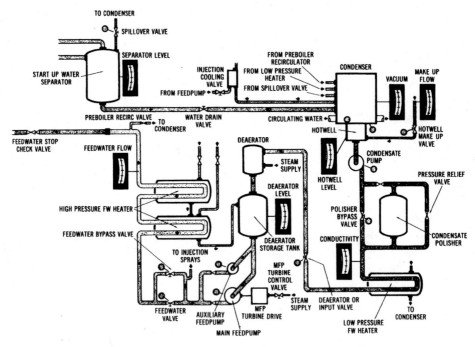

Figure 4-7. The feedwater subsystem (*Courtesy Omnidata, Inc.*)

Electrical Power Production Systems

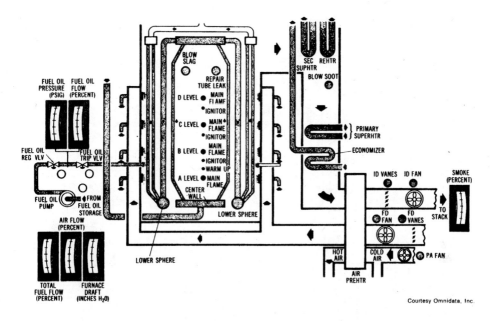

Figure 4-8. The fuel and air subsystem (*Courtesy Omnidata, Inc.*)

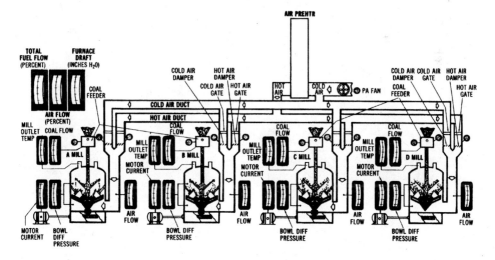

Figure 4-9. The coal pulverizer subsystem (*Courtesy Omnidata, Inc.*)

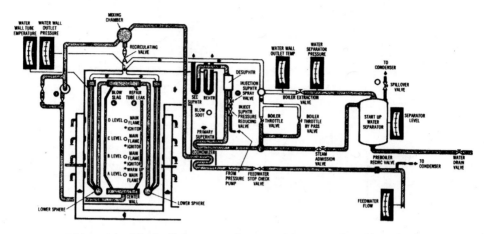

Figure 4-10. The boiler-water subsystem (*Courtesy Omnidata, Inc.*)

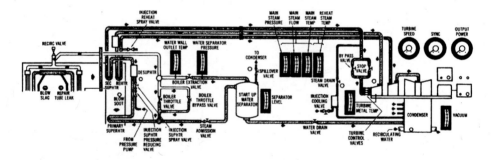

Figure 4-11. The steam and turbine subsystem (*Courtesy Omnidata, Inc.*)

system except, of course, that there is no coal-pulverizing system. These subsystems may be studied to learn the various parts of a power system in more detail.

HYDROELECTRIC SYSTEMS

The use of water power goes back to ancient times. It has been developed to a very high degree, but is now taking a secondary role due to the emphasis on other power sources that are being developed in our country today. Electrical power production systems using water power were de-

Electrical Power Production Systems

veloped in the early twentieth century.

The energy of flowing water may be used to generate electrical power. This method of power production is used in *hydroelectric power systems,* as shown by the simple system illustrated in the diagram in Figure 4-12. *Water,* which is confined in a large reservoir, is channeled through a control gate, which adjusts the flow rate. The flowing water passes through the blades and control vanes of a *hydraulic turbine,* which produces rotation. This mechanical energy is used to rotate a generator that is connected directly to the turbine shaft. Rotation of the alternator causes electrical power to be produced. However, hydroelectric systems are limited by the availability of large water supplies. Many hydroelectric systems are part of multipurpose facilities. For instance, a hydroelectric power system may be part of a project planned for flood control, recreation, or irrigation. Some hydroelectric power systems are shown in Figures 4-12 through 4-14.

Hydroelectric System Operation

The *turbines* used as the mechanical energy sources of hydroelectric systems are very efficient machines. They are ordinarily connected

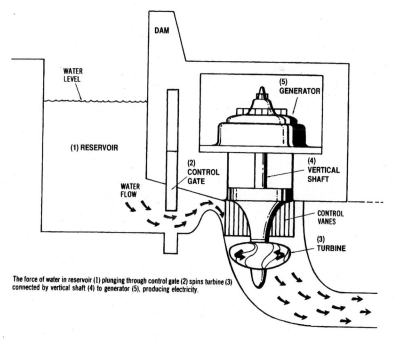

Figure 4-12. Drawing of a basic hydroelectric power system

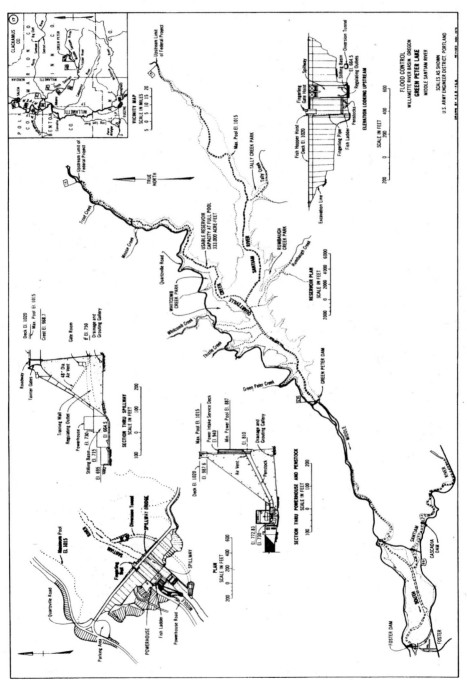

Figure 4-13. Site layout of a hydroelectric project (*Courtesy Portland District, U.S. Army Corp of Engineers*)

Electrical Power Production Systems 103

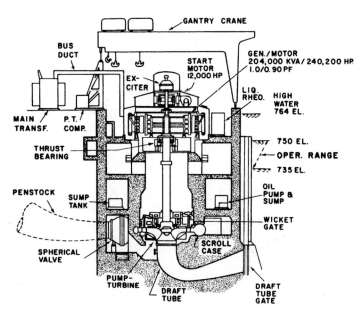

Figure 4-14. Cutaway drawing of a hydroelectric power section at a main unit showing equipment arrangement (*Courtesy Allis-Chalmers Co.*)

directly to the shaft of a *three-phase generator*, which produces the electrical power. *Water* is channeled in from a higher level down into a spiral set of blades on the turbine. The force of water flowing onto these blades causes a rotation of the turbine in the desired direction. The water that flows past the turbine blades is then channeled into a lower-level lake or *reservoir* area. The angle of the *turbine blades* can be adjusted to control the speed of rotation of the turbine. Since rotational speed must remain constant to produce a 60-Hz *frequency*, the blade-angle adjustment and the amount of water channeled onto the blades must be adjusted continuously. Also, varying amounts of force are required to turn the turbine so that different amounts of power are delivered by the turbine to rotate the generator. As the *load demand* delivered by the generator increases, the power input to the turbine must be increased accordingly. This control is accomplished by adjusting the angle of the blades and the amount of water channeled into the blades. The adjustments are automatically accomplished by servocontrol systems.

Hydraulic Turbines

The production of electrical energy by *hydroelectric* systems is dependent upon the operation of *hydraulic turbines*. Hydraulic turbines convert the energy produced by the force of moving water into mechanical energy. This type of turbine is connected to the shaft of a generator at a hydroelectric plant. Since AC generators at power plants must rotate at a constant speed, the hydraulic turbine must turn at a fixed rate of speed. The *efficiency* of hydraulic turbines is much greater (in excess of 85 percent) than that of most rotating machines.

The type of *hydraulic turbine* used with a hydroelectric power system determines whether the generators have horizontal or vertical shafts. Vertical shaft designs are the most common. Electrical power is produced by a *three-phase AC generator* connected directly to the shaft of the *hydraulic turbine*. Several hydroelectric systems are used as "reserve" systems for peak load times. They may be put into operation much faster than steam-driven power systems. It is also possible for the generators of a hydroelectric system to be operated as three-phase synchronous motors during low-demand periods. The motor can rotate the hydraulic turbine, which is then capable of pumping water. Sufficient water is pumped so that a higher external water level is achieved. The higher water elevation will then assist in the production of power during *peak load* intervals.

Future of Hydroelectric Systems

About 10 percent of the power produced in the United States is produced by *hydroelectric* power systems. After the initial cost of constructing a hydroelectric generating facility, the electrical power production cost is relatively inexpensive. Hydroelectric systems are easier to start up and stop than are other power production systems in use today. There are other advantages to hydroelectric systems that are not associated with the production of electrical power. These benefits, derived from the construction of multipurpose dams, include navigational control of waterways, flood control, irrigation, and development of recreational area. Another advantage is that hydroelectric systems do not cause a consumption of the energy source that produces the electrical power, as do other systems in use today.

Hydroelectric generating projects are considered to be low cost, and they produce little pollution. However, in the United States, we have already used the most desirable sites for installing hydroelectric systems. Since the cost of developing other alternative power systems, such as nuclear and geothermal, has become so great, the development of less desirable hydroelectric sites is now feasible. Future development of hydroelectric power systems may be inevitable.

A motivation for the use of water to produce electrical power is the fact that, if we used this natural resource to its full potential, it would give us other benefits. These were discussed earlier. Although the cost to produce electrical power with *hydroelectric* systems depends on a number of factors, it is generally considered to be a very cheap source of energy. The costs are primarily dependent upon the location of the power plant. The desirability of the site is dependent upon its natural characteristics, which affect the cost of development, and its regional characteristics, which, in turn, affect the market for the power.

In the late 1930s, water power supplied about 40 percent of the electrical power in the United States. Now, however, water power supplies only about 10 percent of the nation's electrical power. This is due to the massive development of other power production methods. It is estimated that, in the future, water power will account for an even smaller percentage of electrical power generation. Despite this projected decrease, *hydroelectric* plants are still being built, and the hydroelectric capacity of the United States is still substantial. Hydroelectric systems are not now being developed rapidly; however, with our ever-increasing energy problems and the shortages of our other natural resources, water systems may still have a useful potential.

Pumped-Storage Hydroelectric Systems

Several megawatts of electrical power are produced in the United States by *pumped-storage hydroelectric systems*. This type of system operates by pumping water to a higher elevation and storing it in a reservoir until it is released to drop to a lower elevation to drive the hydraulic turbines of a hydroelectric power-generating plant.

The variable nature of the electrical *load demand* makes *pumped-storage* systems desirable systems to operate. During low-load periods, the hydraulic turbines may be used as pumps to pump water to a storage reservoir of a higher elevation, from a water source of a lower elevation. The water in the upper reservoir can be stored for long periods of time, if necessary. When the electrical load demand on the power system increases, the water in the upper reservoir can be allowed to flow (by gravity feed) through the hydraulic turbines, which will then rotate the three-phase generators in the power plant. Thus, electrical power can be generated without any appreciable consumption of fuel. The pump-turbine and motor-generator units are constructed so that they will operate in two ways: (1) as a pump and motor, and (2) as a turbine and generator. In both cases, the two machines are connected by a common shaft and operate together. However, the multiple use of these machines, although economically very attractive, limits the amount of time that a pumped-storage system can generate electrical power.

Future of Pumped-storage Systems

The future of *pumped-storage* systems depends primarily on economic factors. If fuel and capital construction costs continue to rise, pumped-storage systems might be developed. The conversion of conventional hydroelectric systems to pumped-storage systems has been considered. Also, underground pumped-storage systems have been studied. The underground system would have an upper reservoir at ground level and the lower reservoir underground. The operating principle is the same as in a conventional pumped-storage system.

NUCLEAR FISSION SYSTEMS

Nuclear power plants in operation today employ reactors that utilize the *nuclear-fission* process. Nuclear fission is a complex reaction that results in the division of the nucleus of an atom into two or more nuclei. This splitting of the atom is brought about by the bombardment of the nucleus with neutrons, gamma rays, or other charged particles, and is referred to as induced fission. When an atom is split, it releases a great amount of heat.

In recent years, several *nuclear fission* power plants have been put into operation. A nuclear fission power system, shown in Figure 4-15, relies upon heat produced during a nuclear reaction process. Nuclear reactors "burn" nuclear material, whose atoms are split, causing the release of heat. This reaction is referred to as nuclear fission. The heat from the fission process is used to change circulating water into steam. The high-pressure steam rotates a turbine, which is connected to an electrical generator. This is shown in the diagram of Figure 4-16.

The *nuclear fission system* is very similar to fossil fuel systems in that heat is used to produce high-pressure steam that rotates a turbine. The source of heat in the nuclear fission system is a nuclear reaction; in the fos-

Electrical Power Production Systems

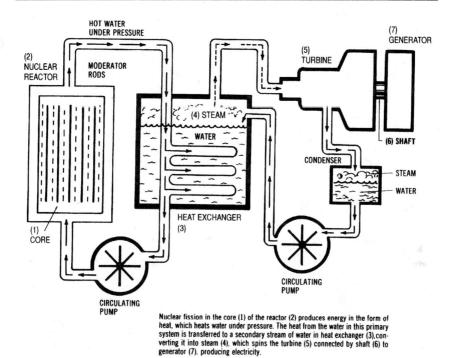

Figure 4-15. Drawing illustrating the principles of a nuclear fission power system.

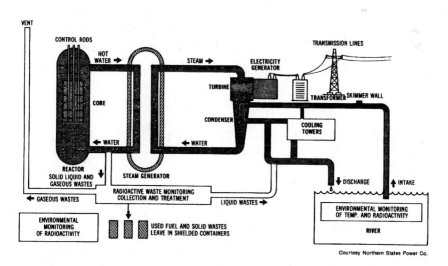

Figure 4-16. Diagram of the nuclear reaction process (*Courtesy Northern States Power Company*)

sil-fuel system, heat is developed by a burning fuel. At the present time, less than 10 percent of the electrical power produced in the United States comes from nuclear fission sources. However, this percentage is also subject to rapid change as new power facilities are put into operation. A typical nuclear power plant site layout is given in Figure 4-17.

Nuclear Power Fundamentals

In order to better understand the process involved in producing electrical power by *nuclear fission* plants, we should review some fundamentals. An atom is the smallest particle into which an element can be broken. The central part of an atom is called its *nucleus* (this is how the term "nuclear power" was derived). The nucleus of an atom is composed of

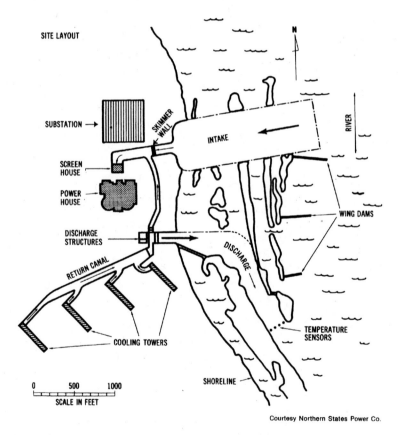

Courtesy Northern States Power Co.

Figure 4-17. Typical site layout of a nuclear power plant (*Courtesy Northern States Power Company*)

protons, which are positively charged particles, and neutrons, which have no electrical charge. Electrons, which are negatively charged particles, orbit around the nucleus. An atom of any element is electrically neutral in its natural state, since the number of protons in the nucleus is equal to the number of electrons that orbit around the nucleus.

The number of protons (+) and electrons (–) contained by an atom varies from one atom to another. (For further information concerning atomic number, mass, et cetera, refer to the table of Elements given in Appendix B.) The number of neutrons (0) in an atom is not always the same as the number of protons and electrons. *Atoms* that have additional neutrons are called isotopes. For instance, a hydrogen atom normally has one electron, one proton, and no neutrons. If one neutron is added to this atomic structure, heavy hydrogen, or deuterium, is formed. Deuterium is an isotope of hydrogen.

The element *uranium* has many different isotopes, each of which contains 92 protons. If the isotope has 143 neutrons in the nucleus, uranium-235 is formed. *Uranium-235* has proved to be a valuable nuclear fuel, but less than 1 percent of the uranium metal ore mined is of the uranium-235 type.

The *fission* or splitting reaction of uranium-235, or other nuclear fuels, is an interesting process. It requires separate controlled neutrons, traveling at high velocities, to penetrate the orbiting electrons around the nucleus of the U-235 isotope. Once a high-velocity neutron has struck the nucleus, the nucleus will split into smaller nuclei. This reaction causes a large quantity of heat to be released. When a nucleus splits, other neutrons from within it are released. These neutrons can cause additional fission reactions in other U-235 isotopes. Thus, the fission reaction occurs as a chain reaction, which causes massive amounts of heat energy to be given off.

Nuclear Fuels

A sustained *nuclear fission* reaction is dependent upon the use of the proper type of fuel. The most desirable fuels for nuclear fission reactions are uranium-233, uranium-235, and plutonium-239. These three nuclear materials are the only fissionable isotopes capable of producing sustained reactions. Of these nuclear fuels, the only one that occurs naturally is uranium-235. The other two isotopes are produced by artificial means. Ordinarily, nuclear reactors that use uranium-235 as a fuel are called *converter* reactors.

The possibility of a nuclear fission reaction producing as much or more fuel than is used has been investigated. Such reactors are called *breeder reactors* and use uranium-233 and plutonium-239 as fuels. During the nuclear reactions that take place in a breeder reactor, materials that are used in the reaction process are converted into fissionable materials. The long-range development of nuclear power production may be dependent upon whether or not breeder reactors can be made available soon. Since the types of *nuclear reactors* that are presently being used consume uranium-235, it is thought that, in the future, the supply of *this* fuel will become low, forcing its price to rise substantially. A price increase in this naturally available nuclear fuel would make nuclear power production less economically competitive with other systems.

Uranium fuel for nuclear fission reactors is produced from ore and is then purified and converted to the desired state through a series of processes. Most nuclear fuel elements are made into plates or rods, which are protected by a cladding of stainless steel, zirconium, or aluminum. The cladding must be capable of containing the nuclear fuel, so as not to allow the release of radioactive materials.

Used fuel is released from the fission reactors when it no longer produces heat effectively during the nuclear reaction. It is not depleted at *this* time; therefore, further processing can bring about the recovery of more fuel from the used fuel. The used *fuel*, which is released from a nuclear reactor, is usually stored underwater for a period of time to permit cooling and radioactive shielding. This type of storage reduces the radioactivity of the fuel. After the storage period has elapsed, the fuel may be reprocessed more safely and easily. The reprocessing of nuclear fuel is very expensive. A large factor contributing to *this* cost is the expense of constructing a reprocessing facility. These facilities must be extensively shielded for radiation protection, both internally and externally. The production and use of nuclear fuels in the United States is rigidly controlled. An agency of the federal government keeps a continuous account of all nuclear fuels produced, used, or reprocessed.

Nuclear Reactors

There is a variety of types of *nuclear reactors*. The major type used in the United States has been the water-moderated reactor. The fundamental difference between a nuclear power plant and a conventional power plant is the fuel that is employed. Most conventional power plants burn coal, oil, or gas to create heat, while the present nuclear plants "burn" uranium. Burning uranium has proved to be a very effective source of power production; however, there is much controversy over this source of power.

It is estimated that burning one ounce of uranium produces roughly the same energy output as burning 100 tons of coal. The "burning" that takes place in a nuclear reactor is referred to as *nuclear fission*. Nuclear fission is the method used in nuclear power generation, and it is quite different from ordinary combustion. The burning of coal results from the carbon combining with oxygen to form carbon dioxide, along with the release of heat. The fissioning or splitting of the uranium atom results in the uranium combining with a neutron and, subsequently, splitting into lighter elements. This process produces a massive quantity of heat.

The *reactors* used at nuclear power plants must be capable of controlling *fission* reactions. When nuclear fuels are bombarded by neutrons, they split and release energy, radiation, and other neutrons. This process is a sustained chain reaction, producing a great amount of heat energy, which is used for the production of steam, which is used to rotate a steam turbine-generator system. The nuclear fission power-generating system is about the same as a conventional fossil fuel steam plant, except that a nuclear reactor is used to produce the heat energy, rather than a burning fuel confined in a furnace.

Within the *nuclear reactor,* there is a mixture of fuel and a moderator material. There are three known nuclear fission fuels: *uranium-235, uranium-233,* and *plutonium-239*. Moderators are used to slow the speed of neutrons in fission reactions. Since the neutrons involved in the fission reaction have high energy levels, they are called fast neutrons. They are slowed by collisions with moderator materials such as water, deuterium oxide, beryllium, and other lightweight materials. Neutrons that have been slowed down possess an energy equilibrium and are referred to as thermal neutrons. These thermal neutrons aid in additional fission reactions. Thus, moderators playa significant role in sustaining nuclear fission reactions.

Nuclear reactors differ in several ways. Differences include the type of fuel and moderator, the thermal output capacity, and the type of coolant. Several classifications of nuclear reactors, according to types of coolant, are discussed in the following sections.

Moderating Nuclear Reactors

A uranium atom undergoes *fission* when it absorbs a neutron and, at the same time, produces two lighter elements and emits two or three neutrons. These neutrons, in turn, react with other uranium atoms, which will undergo fission and produce more neutrons. Heat is increased in the reactor as the number of neutrons is increased. If a *reactor* is left uncontrolled, it may destroy itself. *Moderating* a reactor, therefore, means controlling the multiplication of neutrons in the reactor core. There are several methods used for moderating nuclear reactors.

Boiling-Water Reactor (BWR)—Water is a popular coolant for reactors. In this type of reactor, shown in Figure 4-18, water is pumped into the reactor enclosure. The water is then converted into steam, which is delivered to a steam turbine. The water also serves as the moderator material of the reactor.

Pressurized-Water Reactor (PWR)—The pressurized-water reactor, shown in Figure 4-19 is similar to a boiling-water reactor, except that the coolant water is pumped through the reactor under high pressure. Steam is produced in an adjacent area from a separate stream of water, which is pumped through the steam-production system. Just as in the BWR, the water within the reactor serves as the moderator.

High-temperature Gas-cooled Reactor (HTGR)—The high-temperature gas-cooled reactor, shown in Figure 4-20 uses pressurized helium gas to transfer heat from the reactor to a steam-production system. The advan-

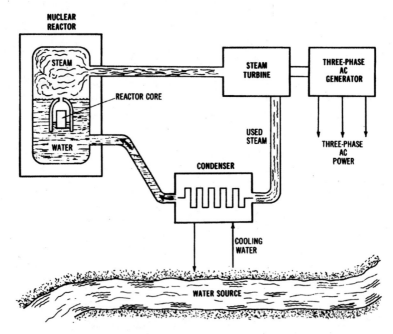

Figure 4-18. Simplified diagram of a boiling water reactor (BWR)

tage of helium gas over water is that the helium can operate at much higher temperatures.

Other types of reactors, such as the *liquid-metal fast-breeder reactor (LMFBR)* and the molten-salt-cooled reactor, have some potential in electrical power production systems. The LMFBR type is shown in Figure 4-21.

OPERATIONAL ASPECTS OF MODERN POWER SYSTEMS

There are several operational aspects of modern electrical power production systems that must be considered. These considerations include the location of power plants, electrical *load requirements,* and electrical *load demand* control. Each of these will be discussed in the following sections.

Location of Electrical Power Plants

A critical issue that now faces those involved in the production of electrical power is the location of power plants. Federal regulations associated with the National Environmental Policy Act (NEPA) have made the location of power plants more difficult. At present, there are a vast number of individual power plants throughout the country. However, the addition

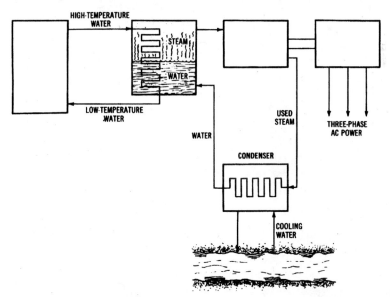

Figure 4-19. Simplified diagram of a pressurized water reactor (PWR)

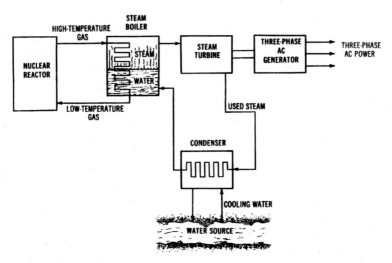

Figure 4-20. Simplified diagram of a high-temperature gas-cooled reactor (HTGR)

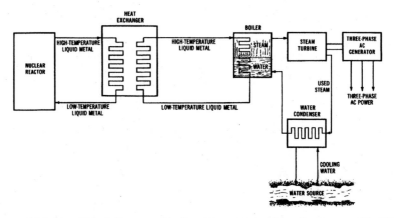

Figure 4-21. Simplified diagram of a liquid-metal fast-breeder reactor (LMFBR)

of new generating plants involves such current issues as air pollution, water pollution, materials handling (particularly with nuclear plants), fuel availability, and federal, state, and municipal regulations.

These issues have brought about some recent thought about the construction of "energy centers." Such systems would be larger and more standardized than the power plants of today. This concept would reduce the number of plants that are needed to produce a specific quantity of electrical power. Other advantages of this concept include better use of land resources, easier environmental control management, and more economical construction and management of facilities. These advantages may make centralized power production the best alternative, socially, economically, and technically, for meeting future electrical power requirements.

Electrical Load Requirements

The electrical power that must be produced by our power systems varies greatly according to several factors, such as the time of the year, the time of the week, and the time of the day. The level of electrical power *supply* and *demand* is much more difficult to predict than that of most quantities that are bought and sold. Electrical power must be readily available, in sufficient quantity, whenever it is required. The overall supply and demand situation is something most of us take for granted until our electrical power is interrupted. Electrical power systems in the United States must be *interconnected* on a regional basis, so that power stations can support one another in meeting the variable *load demands*.

The use of electrical power has been *forecasted* to increase every ten years at a rate that will cause a doubling of the kilowatt hours required. Some forecasts, however, show the rate of electrical power demand to have a "leveling-off" period in the near future. This effect may be due to a saturation of the possible uses of electrical power for home appliances, industrial processes, and commercial use. These factors, combined with greater conservation efforts and other social and economic factors, support the idea that the electrical *power demand* will increase at a slower rate in future years. The forecasting of the present demand by the electrical utility companies must be based on an analysis by regions. The demand varies according to the type of consumer supplied by the power stations that constitute the system. Different types of load are encountered when *residential, industrial,* and *commercial* systems are supplied by the electrical utility companies.

Industrial use of electrical power accounts for approximately 40 percent of the total kilowatt-hour (kWh) consumption, and the industrial use of electrical power is projected to increase at a rate similar to its present rate, in the near future. The shortage of natural gas should not significantly affect electrical power consumption by industry. Most of the conversions of gas systems will be to systems that use oil in place of gas.

The major increases in *residential power demand* have been due to an increased use by customers. A smaller increase is accounted for by an increase in the number of customers. Such variables as the type of heating used, the use or nonuse of air conditioning, and the use of major appliances (freezers, dryers, ranges) affect the residential electrical power demand. At present, residential use of electrical power accounts for approximately 30 percent of the total consumption. The rate of increase will probably taper off in the near future.

Commercial use of electrical power accounts for less than 25 percent of the total kWh usage. Commercial power consumption includes usage by office buildings, apartment complexes, school facilities, shopping establishments, and motel and hotel buildings. The prediction of the future electrical power demand by these facilities is somewhat similar to the prediction of future residential demand. The rate of increase in the commercial use of electrical power is also expected to decline in the future. These percentages are subject to change over time.

Electrical Load-demand Control

As the costs of producing power continue to rise, power companies must search for ways to limit the maximum rate of energy consumption. To cut down on *power usage*, industries have begun to initiate programs that will cut down on the load during *peak* operating periods. The use of certain machines may be limited while other large, power-consuming machines are operating. In larger industrial plants, and at power production plants, it would be impossible to manually control the complex regional switching systems, so computers are being used to control loads.

To prepare the *computers* for power-consumption control, power companies must determine the *peak demand patterns* of local industries, and the surrounding region, supplied by a specific power station. The load of an industrial plant may then be balanced, according to area demands, with the power station output. The computer may be programmed to act as a switch, allowing only those processes to operate that are within the load calculated for the plant for a specific time period. If the load drawn by an industry exceeds the limit, the computer may deactivate part of the system. When demand is decreased in one area, the computer can cause the power system to increase power output to another part of the system. Thus, the industrial load is constantly monitored by the power company to ensure a sufficient supply of power at all times.

Chapter 5

Alternative Power Systems

Several methods of producing electrical power are either in limited use or in the experimental stage at the present time. Some of these methods show promise as a possible electrical power production method for the future. *Alternative power systems* are discussed in Chapter 5.

IMPORTANT TERMS

Chapter 5 deals with alternative electrical power production systems. After studying this chapter, you should have an understanding of the following terms:

Solar cell
Solar energy system
Solar heating
Geothermal power system
Wind energy system
Magnetohydrodynamic (MHD) system
Nuclear fusion power system
Magnetic confinement fusion
Fuel cell system
Tidal power system
Coal gasification
Oil shale fuel production
Alternative nuclear power plants
Biomass

POTENTIAL POWER SOURCES

Solar power is one potential electrical power source. The largest energy source available today is the sun, which supplies practically limitless energy. The energy available from the sun far exceeds any foreseeable future need. *Solar cells* are now being used to convert light energy into small quantities of electrical energy. Possible solar-energy systems might include home heating or power production systems, orbiting space systems, and steam-driven electrical power systems. Each of these systems utilizes *solar collectors* that concentrate the light of the sun so that a large quantity of heat will be produced. Potentially, this heat could be used to drive a steam turbine in order to generate additional electrical energy.

Geothermal systems also have promise as future energy sources. These systems utilize the heat of molten masses of material in the interior of the earth. Thus, heat from the earth is a potential source of energy for power generation in many parts of the world. The principle of *geothermal* systems is similar to other steam turbine-driven systems. However, in this case, the source of steam is the heat obtained from within the earth through wells. These wells are drilled to a depth of up to two miles into the earth. *Geothermal* sources are used to produce electrical energy in certain regions of the western United States.

Wind systems have also been considered for producing electrical energy. However, winds are variable in most parts of our country. This fact causes wind systems to be confined to being used with *storage* systems, such as batteries. It is possible that wind machines may be used to rotate small generators which could, potentially, be located at a home. However, large amounts of power would be difficult to produce by this method.

Another energy source which has some potential for future use is *magnetohydrodynamics (MHD)*. The operation of an MHD system relies upon the flow of a conductive gas through a magnetic field, thus causing a direct current (DC) voltage to be generated. The electrical power developed depends upon the strength of the magnetic field that surrounds the conductive gas, and on the speed and conductivity of the gas. At the present time, only small quantities of electrical energy have been generated using the MHD principle; however, it does have some potential as a future source of electrical energy.

Still another possible energy source is *nuclear fusion*. This process has not been fully developed, due to the extremely high temperatures that are produced as fusion of atoms takes place. A *fusion reactor* could use *tritium* or *deuterium* (heavy hydrogen) as fuels. These fuels may be found in sea water in large quantities, thus reducing the scarcity of nuclear fuel. It is estimated that there is enough deuterium in the oceans to supply all the energy the world would ever need.

If *nuclear-fusion reactors* could be used in the production of electrical energy, the process would be similar to the nuclear-fission plants which are now in operation. The only difference would be in the nuclear reaction that takes place to change the circulating water into steam to drive the turbines. The major problem of the *nuclear-fusion* process is controlling the high temperatures generated, estimated to reach 100 million degrees Fahrenheit.

Another energy source which could be used in the future is the *fuel cell*. This type of cell converts the chemical energy of fuels into direct current electrical energy. A fuel cell contains two porous electrodes and an electrolyte. One type of fuel cell operates as hydrogen gas passes through one porous electrode and oxygen gas passes through the other electrode. The chemical reactions of the electrodes with the electrolyte either release electrons to an external circuit, or draw electrons from the external circuit, thus producing a current flow.

Still another possible alternative power production system utilizes *tidal energy*. Tidal systems would use the rise and fall of the water along a coastal area as a source of energy for producing electrical power. Coal gasification is yet another process that could be used for future power systems. This process is used to convert the poorer grades of coal into a gas. The use of oil shale to produce fuel is also being considered.

It should be pointed out that many of the future energy sources are *direct conversion* processes. For example, the fuel cell converts chemical energy directly to electrical energy, and the solar cell converts light energy directly to electrical energy. A more complex transformation of energy takes place in most power plants today. Heat energy is needed to produce mechanical energy, which produces electrical energy. This explains the inefficiency of our present systems of producing electrical energy. Perhaps advances in electrical power technology will bring about new and more efficient methods of producing electrical energy.

SOLAR ENERGY SYSTEMS

For many years, many have regarded the *sun* as a possible source of electrical power. However, few efforts to use this cheap source of energy have been made. We are now faced with the problem of finding alternative sources of energy, and solar energy is one possible alternative source. The sun delivers a constant stream of radiant energy. The amount of *solar energy* coming toward the earth through sunlight in one day equals the energy produced by burning many millions of tons of coal. It is estimated that enough solar energy is delivered to the United States in less than one hour to meet the power needs of the country for one year. This is why solar energy is a potential source for our ever-growing electrical power needs. However, there are still several problems to be solved prior to using solar energy. One major problem is in developing methods for controlling and utilizing the energy of the sun. There are two methods for collecting and concentrating solar energy presently in use. Both of these methods involve a mirror-like reflective surface.

The first method uses *parabolic mirrors* to capture the energy of the sun. These mirrors concentrate the energy from the sun by focusing the light onto an opaque receiving surface. If water could be made to circulate through tubes, the heat focused onto the tubes could turn the water into steam. The steam, then, could drive a turbine to produce mechanical energy. The mirrors could be rotated to keep them in proper position for the best light reflection.

The second method uses a *flat-plate solar collector.* Layers of glass are laid over a blackened metal plate, with an air space between each layer. The layers of glass act as a heat trap. They let the rays of the sun in, but keep most of the heat from escaping. The heated air could be used to warm a home.

The first widespread use of solar energy will probably be to heat homes and other buildings. Experiments in doing this are already underway in many areas. To heat a home, a flat-plate collector may be mounted on the part of the roof that slopes in a southward direction. It should be tilted at an angle to receive the greatest amount of sunlight possible. The sun would be used to heat a liquid (or air) that would be circulated through the collector. The heated liquid (air) would be stored in an insulated tank and, then, pumped into the house through pipes and radiators (air ducts). By adding a steam turbine, a generator, etc., to the solar collector just discussed, the heat could be used to drive the steam turbine to generate electrical energy. This *solar power system* is illustrated in Figure 5-1.

Another major problem in solar heating is in the *storage* of the heat produced by the sun. In areas that have several cloudy days each year, an auxiliary heating system is required. However, solar energy is being used

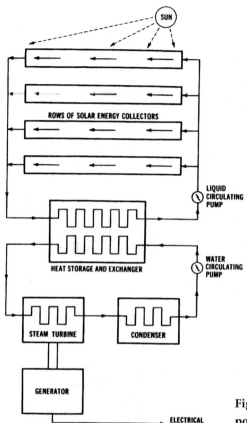

Figure 5-1. Block diagram of a solar power system

today on a limited basis. For instance, flashlights and radios can be powered by solar cells. The main advantage of this usage is that a solar battery can be used for an infinite period of time. Considering all of these aspects of solar energy and its potential, many feel that solar energy will be the next major form of energy to be utilized extensively in the United States. The development of large-scale power production systems which use solar energy, however, remains questionable.

GEOTHERMAL POWER SYSTEMS

About 20 miles below the crust of the earth is a molten mass of liquid and gaseous matter called *magma,* which is still cooling from the time that the earth was formed. When this magma comes close to the crust of the earth, possibly through a rupture, a volcano could be formed and erupt. Magma could also cause steam vents, like the ones at the *"Geysers"* area in California. These are naturally occurring vents that permit the escape of the steam formed by the water that comes in contact with the underground magma. A basic *geothermal power system* is shown in Figure 5-2, and the understructure for such a system is illustrated in Figure 5-3.

In the 1920s, an attempt was made to use the Geysers area as a power source, but the pipelines were not able to withstand the corrosive action of the steam and the impurities in it. Later, in the 1950s, stainless steel alloys were developed that could withstand the steam and its impurities, so the Pacific Gas and Electric Company started development of a power system to use the heat from within the earth as an energy source. The first generating unit at the *Geysers power plant* began operation in 1960. At present, more than 500 megawatts of electrical power are available from the generating units in the Geysers area.

In the *geothermal system,* steam enters a path (through a pipeline or a vent) to the surface of the earth. The pipelines that carry the steam are constructed with large expansion loops, causing small pieces of rock to be

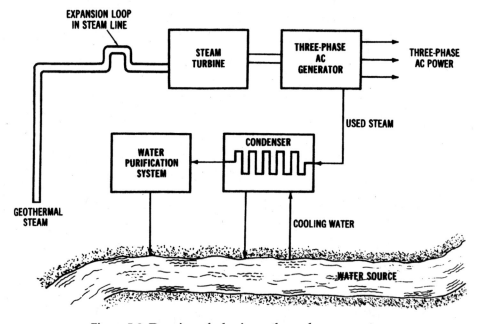

Figure 5-2. Drawing of a basic geothermal power system.

Alternative Power Systems 123

left in the loop. This system of loops avoids damage to the steam turbine blades. After the steam goes through the turbine, it goes to a condenser, where it is combined with cooler water. This water is pumped to cooling towers, where the water temperature is reduced. This part of the *geothermal* system is similar to conventional steam systems.

The *geothermal* power production method offers another alternative method of producing electrical power. This method makes it possible to control energy, in the form of steam or heated water, that is produced by natural geysers or underground channels. The high-pressure steam for power production is made without burning any of our fossil fuels. Thus, this method can be used to drive turbine-generator systems, such as the one at the Geysers system in California.

The setting up of a *geothermal power plant* requires the drilling of holes deep into the surface of the earth. One hole may be used to send cold water down into the tremendously hot material located under the surface of the earth. An adjacently drilled hole could be used to bring steam back to the surface. This method is capable of being used in any area of the world, but it requires drilling holes up to 10 miles deep. On the other hand, natu-

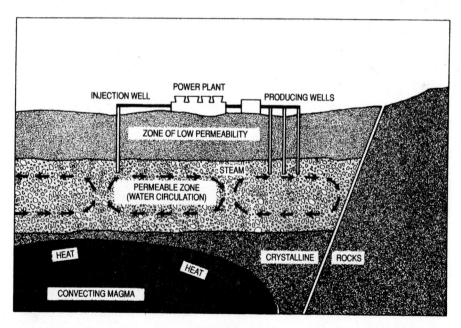

Figure 5-3. Drawing of a basic geothermal power system that illustrates the underground structure *(Courtesy Pacific Gas and Electric Co.)*

ral geothermal steam production is limited to active volcanic areas, such as the Geysers area of the United States. Since no fuel is burned, geothermal methods represent one way of saving our valuable and diminishing quantities of fossil fuels. The major problem associated with *geothermal power production* is the drilling of the deep holes into the ground.

WIND SYSTEMS

Nonpolluting, inexpensive power systems have been desired for many years. Now, technology has advanced to the point where there may be some inexpensive methods of producing electrical power that we have not used significantly before.

In the early 1900s, in many farm areas, particularly in the Midwest, electrical power had not been distributed to homes. It was conceived that the *wind* could be used to provide a mechanical energy, not only for water pumps, but also for generators with which to provide electrical power. The major problem was that the wind did not blow all the time and, then, when it did, it often blew so hard that it destroyed the windmill. Most of these units used fixed-pitch propellers as fans. They also used low-voltage DC electrical systems, and there was no way of storing power during periods of nonuse, or during times of low wind speeds. These early units used a rotating-armature system (see Chapter 6) and, therefore, along with other maintenance problems, had to have the brushes replaced very often.

The *wind-generating plants* used today have a system of storage batteries, rectifiers, and other components that provides a constant power output even when the wind is not blowing. They also have a 2-blade or 3-blade propeller system that can be "feathered" during periods of high winds so that the mill will not destroy itself. A simplified *wind-power system* is shown in Figure 5-4. Most of the systems that are in use today are individual units capable of producing 120-volt alternating current (AC) with constant power outputs in the low-kilowatt range. The generator output may be interconnected through a system of series-connected batteries with an automatic, solid-state voltage control that is designed to convert the DC voltage to 120-volt AC. The cost of these individual, electrical power-generating systems is fairly low, and these systems provide a complete, self-contained, nonpolluting, power source. Wind systems would be ideal for remote homes that have a low-power requirement. They should prove to be very dependable and have low maintenance costs. Larger power-generating plants could be possible when located in very windy places. These plants would probably be limited in their power output, due to the requirement for large storage batteries. Because of the varying wind speeds, it would be difficult to connect two or more units in parallel and keep them in a phase relationship. Compared with other inexpensive types of power production plants, such as solar or hydroelectric, *wind systems* are smaller in physical size per kilowatt output and have a lower initial cost. Also, a solar power plant can only produce power during the daylight hours, while a wind-generator will operate whenever there is a wind.

The primary disadvantage of a windmill is obvious. What do you do on days without wind? One answer is the use of *storage batteries.* Another disadvantage is the fact that windmills generally have a very low efficiency—about 50 percent. Although windmill power will probably never be a large contributor to the solving of any power crisis, some individuals have

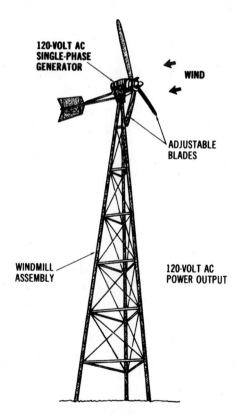

Figure 5-4. A simplified wind power system

calculated that several large-diameter windmills, with a fairly constant wind, could produce many kilowatts of electrical power. They could play an important role in reducing the use of our natural resources for electrical power production, particularly on an individual basis.

MAGNETO HYDRODYNAMIC (MHD) SYSTEMS

MHD stands for *magnetohydrodynamic,* a process of generating electricity by moving a conductor of small particles suspended in a superheated gas through a magnetic field. The process is illustrated in Figure 5-5. The metallic conductors are made of metals such as potassium or cesium, and can be recovered and used again. The gas is heated to a temperature much hotter than the temperature to which steam is heated in conventional power plants. This *superheated gas* is in what is called a plasma state at these high temperatures. This means that the electrons of many of the gas atoms have been stripped away, thus making the gas a good electrical conductor. The combination of metal and gas is forced through an electrode-lined channel which is under the influence of a superconducting magnet that has a tremendous field strength. The *magnet* must be of the superconducting type, since a regular electromagnet of that strength would require too much power. A superconducting magnet, therefore, is one of the key parts to this type of generation system.

With high operating temperatures and a high-speed gas flow, it becomes a difficult task to keep the conductive channel from becoming destroyed. *Cooling* is very important and is accomplished by circulating a suitable coolant throughout jackets built into the channel. Also, due to the high temperatures and the metal particles moving at high speed, erosion of the channel becomes a critical problem. This problem has been eased by using a coal slag injected into the hot gas and metal stream. The coal slag acts to replace the eroded material as it is lost.

Pollution problems are very few. The main one is the high levels of nitrogen oxides that are produced as a direct result of the high combustion temperatures inherent to the system. Sulfur oxides and ash are also a problem for any plants using coal or oil. An afterburner system has been proposed to eliminate the nitrogen oxides, while the sulfur oxides and the ash would be collected, separated chemically, and then recycled.

At this time, *MHD generators* are primarily experimental. There have been several large units made, but they have not operated for any signifi-

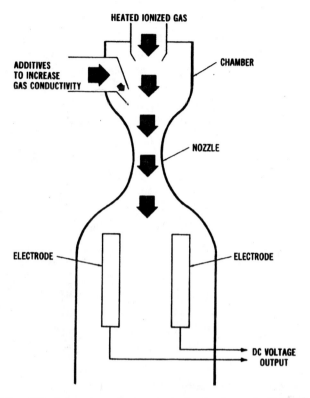

Figure 5-5. Simplified drawing of a magneto hydrodynamic (MHD) generator system

cant time periods. Efficiency is of importance for this type of generation system. While the conventional steam generation system (fired by coal) is at most 40 percent efficient (that is, it turns 40 percent of the coal energy into electricity), and most nuclear-powered plants are about 33 percent efficient, MHD plants could operate at 60 percent efficiency. Thus, fuel supplies could be enhanced by using this generation system. At present, the future of MHD systems for large-scale electrical power generation is questionable.

NUCLEAR-FUSION POWER SYSTEMS

Another *alternative power* production method which has been considered is *nuclear fusion*. *Deuterium,* the type of fuel used for this process, is very abundant. The supply of deuterium is considered to be unlimited, since it exists as heavy hydrogen in sea water. The use of such an abundant fuel could solve some of our problems that are related to the depletion of fossil fuels. Another outstanding advantage of this system is that its radioactive waste products would be minimal.

The *fusion process* results when two atomic nuclei collide under controlled conditions to cause a rearrangement of their inner structure and, thus, release a large amount of energy during the reaction. These nuclear reactions or fusings of atoms must take place under tremendously high temperatures. The energy released through nuclear fusion would be thousands of times greater per unit than the energy from typical chemical reactions, and considerably greater than that of a nuclear-fission reaction.

The *fusion reaction* involves the fusing together of two light elements to form a heavier element, with heat energy being released during the reaction. This reaction could occur when a deuterium ion and a tritium ion are fused together. A deuterium ion is a hydrogen atom with one additional neutron, and a tritium ion is a hydrogen atom with two additional neutrons. A temperature in the range of 100,000,000° C is needed for this reaction to produce a great enough velocity for the two ions to fuse together. Sufficient velocity is needed to overcome the forces associated with the ions. The deuterium-tritium fusion reaction produces a helium atom and a neutron. The neutron, with a high enough energy level, could cause another deuterium-tritium reaction of nearby ions, providing the time of the original reaction is long enough. A much higher amount of energy would be produced by a *nuclear-fusion reaction* than by a *fission reaction*. There are several different techniques being investigated for producing nuclear fusion. At this time, each is still in the theoretical development stage.

NUCLEAR-FUSION METHODS

Several methods are being considered today for using the heat from nuclear fusion to generate electrical power.

Magnetic-confinement Method

One method is called *magnetic confinement*. The proposed design of a magnetic-confinement power system is shown in Figure 5-6. At present, it is thought that fusion reactors could be economical if the reaction can be carried out in an intense magnetic field provided by superconducting magnets. The magnet and the magnetic-field designs are very complex,

Alternative Power Systems 129

however, and the forces associated with them are tremendous. Controlling the fusion reaction is still one of the major problems of trying to use this method; therefore, it remains to be studied further.

Laser-induced Fusion

Laser-induced fusion should also be mentioned. This method relies on inertia rather than an intense magnetic field for the confinement of the nuclear fuel. In this method, a small pellet of frozen fuel is injected into a combustion chamber. There, it is hit by several short, intense, laser beams coming into the chamber from several directions. This process happens very quickly and causes the pellet to collapse, due to the intensity of the beams. The fuel is rapidly heated. The fusion reaction takes place at an instant just before the material in the pellet can overcome its inertia and expand, due to the intense heating effect. This process uses a laser beam to produce a sufficient amount of heat to cause the nuclear-fusion reaction. A proposed design for a laser-induced fusion power system is shown in Figure 5-7. As with other fusion methods, this method has not been tested. It is being theoretically developed. However, since the magnetic-confinement method has not been developed beyond the conceptual stage, other methods, such as

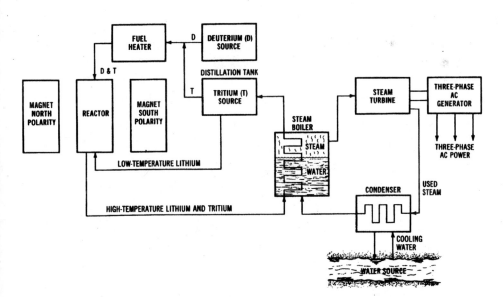

Figure 5-6. Proposed design of a magnetic-confinement nuclear fusion power system

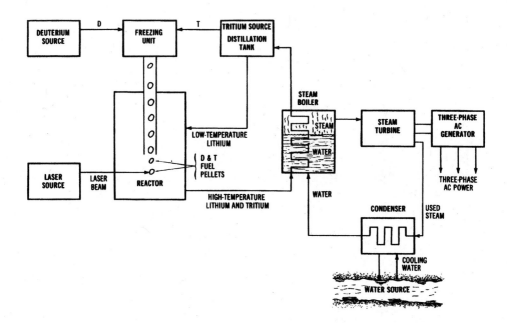

Figure 5-7. Proposed design of a laser-induced nuclear fusion power system

the laser-induced method, must be considered. It is thought that the pulsed-laser technique may be a viable alternative method for producing sustained nuclear-fusion reactions. This process is presently being developed with *deuterium* being considered as the proposed fuel.

FUTURE OF NUCLEAR FUSION

Since *nuclear fusion* is considered to be environmentally safe and would use a very abundant fuel, much research is being done to formulate ways of controlling the fusion-reaction process. Fusion reactors have not yet been developed beyond the theoretical stages. The problems involved in constructing a fusion reactor center around the vast amounts of heat produced. The fusion fuel must be heated to a high temperature and, then, the heated fuel must be confined for a long enough period of time that the energy released by the fusion process can become greater than the energy that was required to heat the fuel to its reaction temperature. This is necessary in order to sustain the fusion reaction and to produce continuous energy.

Alternative Power Systems 131

All present fusion power-plant designs have basic *problems*. Solutions to these problems, and development of economically attractive commercial systems, will take many years and much money. However, it seems possible that such systems may someday be developed.

FUEL-CELL SYSTEMS

Another alternative energy system which has been researched is the *fuel cell*. Fuel cells convert the chemical energy of fuels into electrical energy. An advantage of this method is that its efficiency is greater than steam turbine-driven production systems, since the conversion of energy is directly from chemical to electrical. Ordinarily, fuel cells use oxygen and hydrogen gas as fuels, as shown in Figure 5-8. Instead of consuming its electrodes, as ordinary batteries do, a *fuel cell* is supplied with chemical

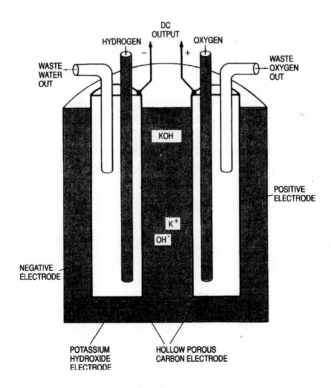

Figure 5-8. Simplified drawing of an oxygen-hydrogen fuel cell

reactants from an external source. Ordinarily, *hydrogen gas* is delivered to the anode of the cell and *oxygen gas* or air is delivered to the cathode. The reaction of these gases is similar to the reverse of electrolysis. (During the electrolysis process, an electric current decomposes water into hydrogen and oxygen.)

Fuel cells were developed long ago. It was known early that electrochemical reactions could be used to convert chemical energy directly to electrical energy. The first commercial fuel cells were used as auxiliary power sources for United States space vehicles. These were hydrogen-oxygen fuel cells. The development of fuel cells has brought about some cells with a power capacity of up to 500 watts. Some cells that are in the research stages have a power output of up to 100 kilowatts. These fuel cells produce low-voltage direct current. Several cells may be connected in series-parallel configurations to produce greater voltage and current levels. Several large companies are performing experiments with the design of *fuel cells.* Many new technological approaches to fuel cells are being developed. Large-scale systems are planned that may use phosphoric acid as an electrolyte. This system will operate with a variety of hydrocarbon fuels. Cells presently being developed use high-temperature carbon or alkaline electrolytes. Cells in the developmental stages might use synthetic fuels derived from coal.

Some problems have been encountered in the development of *fuel cells.* The chemical reaction brought about by the fuels requires a catalyst. The importance of the catalyst is greater at lower operating temperatures. At higher temperatures, cheaper and more abundant catalysts may be used. However, some designers speculate that the catalysts required for lower temperature operation may become unavailable or, if not, at least very expensive in the future. At present, the initial cost of a fuel-cell system is very great. A system, at this time, would not be competitive with other power systems in operation. Another developmental problem is that of water disposal. A vast amount of water is produced by the chemical reactions of the fuels.

In addition to space applications, other *applications* could potentially include mobile electrical power sources. It has been suggested that fuel cells could be used as power sources to drive electric cars. Also, fuel cells are being considered as power sources for trains, submarines, and military vehicles. However, the use of fuel cells for large-scale power systems would not be feasible.

Alternative Power Systems 133

TIDAL POWER SYSTEMS

The rise and fall of waters along coastal areas, which is caused by gravitational forces, is the basis for the tidal electrical power production method. There are presently some *tidal power systems* in operation. Tidal systems would be desirable, since they do not pollute the atmosphere, do not consume any natural resource, and do not drastically change the surrounding environment as some conventional *hydroelectric* systems do.

The depth of tidal water varies greatly at different times of the year. These depths are determined by changes in the sun and moon in relation to the earth. Tides are readily predictable, since the same patterns are established year after year. A *tidal power system* would have to be constructed where water in sufficient quantity could be stored with a minimum amount of dam construction. A tidal system could be made to operate during the rise of tides and the fall of tides. Also, the pumped-storage method could be used in conjunction with tidal systems to assure power output during peak load times. A potential tidal system site along the United States-Canada border has been studied; however, the economic feasibility of tidal systems at this time is not too promising. One tidal system that is in operation is at Normandy, France.

COAL-GASIFICATION FUEL SYSTEMS

The process of *coal gasification* has aroused interest in recent years. This process involves the conversion of coal or coke to a gaseous state through a reaction with air, oxygen, carbon dioxide, or steam. Many people feel that this process will be able to produce a natural gas substitute. Some of the methods of producing gas from coal include:

1. The *BI-CAS process*, in which a reaction of coal, steam, and hydrogen gas produces methane gas.
2. The *COCAS process*, in which a liquid fuel and a gaseous fuel are produced.
3. The CSG process (or *consolidated synthetic gas process*), which develops a very slow reaction to produce methane.
4. The *Hydrane process*, in which methane is produced by a direct reaction of hydrogen and coal, with no intermediate gas production.
5. The *Synthane process*, in which methane is produced by a method in-

volving several different steps.

Coal gasification may be one solution to the problem of our depleting natural gas supply. Since some electrical power systems use natural gas as the fuel, we should be concerned about coal gasification as an alternative method to aid in the production of electrical power.

OIL-SHALE FUEL-PRODUCTION SYSTEMS

Another method which should be mentioned as an alternative method that could aid in the production of electrical energy is fuel from *oil shale*. There is a potential fuel source located at the oil-shale deposits which exist primarily in Colorado, Wyoming, and Utah. This *oil shale* was formed by a process similar to that which created crude petroleum. However, there was never enough underground heat or pressure to convert the organic sediment to the same consistency as oil. Instead, a waxy hydrocarbon called *kerogen* was produced. This compound is mixed with fine rock and is referred to as oil shale.

Extracting crude petroleum from this substance is a very complex process that starts with either underground or strip mining. The mined rock is crushed and then heated to produce a raw oil, which, in turn, must be upgraded to a usable level. All of these operations could take place at the mining site. In addition, large amounts of *shale waste* must be removed. The resulting waste substance is about the consistency and color of fireplace soot. The waste occupies more volume than the original oil-shale formations. The potential impact of oil-shale development is questionable. Vast quantities of land might be disturbed. There are definitely many factors to consider in addition to our energy needs.

ALTERNATIVE NUCLEAR POWER PLANTS

A unique concept in electrical power production, considered in the 1970s, was called a *floating nuclear power plant*. Floating nuclear plants were proposed to be nuclear-fission plants mounted on huge floating platforms for operation on the water. These systems were planned for use on the Atlantic Ocean. Breakwaters would surround the power plant to protect it from waves and ship collisions. These units could be located on rivers, inlets, or in the ocean. The plants could be manufactured on land and then transported to the area where they would be used. The electrical power produced by the plant could be distributed by underwater cables to the shore. The power lines could then be connected to onshore overhead power-transmission lines. The *floating nuclear power plants* could be mass-produced, unlike conventional nuclear plants which are individually built.

Floating nuclear plants could have advantages over other power systems. There are obvious ecological benefits. A plant located on the ocean would have less thermal effect on the water due to heat dissipation over a large body of water. Also, these units would not require the use of land for locating power plants. They would be flexible, since they could be located on rivers, inlets, or oceans.

This discussion is included to *stimulate thought* about potential "alternative" systems for producing electrical power. Each of the systems discussed have potential problems; however, serious experimentation must be conducted to assure that electrical power can be produced economically. Our technology is dependent upon low-cost electrical power.

BIOMASS SYSTEMS

Another alternative system being considered as a potential method of producing electrical power is *biomass*. Biomass sources of energy for possible use as fuel sources for electrical power plants are wood, animal wastes, garbage, food processing wastes, grass, and kelp from the ocean. Many countries in the world use biomass sources as primary energy sources. In fact, the United States used some of the biomass sources of energy almost exclusively for many years. The potential amount of energy which could be produced in the United States by *biomass sources* is substantial at this time also. There are still several questions about using biomass sources of energy for producing electrical; however, as fossil fuels become less abundant, biomass sources may receive more active consideration.

Chapter 6

Alternating Current Power Systems

Some of the basics of *alternating current (AC) power systems* were discussed in Chapter 2. You should already have an understanding of the effects of resistance, inductance, and capacitance in AC circuits. Therefore, this chapter will deal primarily with the systems that are used to *produce* AC electrical power.

The vast majority of the electrical power produced in the United States is *alternating current*. Massive mechanical generators at power plants throughout our country provide the necessary electrical power to supply our homes and industries. Most generators produce *three-phase* alternating current; however, *single-phase* generators are also used for certain applications of a smaller nature. The operation of mechanical generators relies upon a fundamental principle of electricity called *electromagnetic induction*.

IMPORTANT TERMS

Chapter 6 deals with AC power systems. After studying this chapter, you should have an understanding of the following terms:

Electromagnetic Induction
Faraday's Law
Left-Hand Rule of Generation
Single-Phase AC Voltage
Generator
Stator
Rotor
Field Pole
Prime Mover

Slip Ring/Brush Assembly
AC Sine Wave
Single-Phase AC Generator
Rotating-Armature Method
Rotating-Field Method
Three-Phase AC Generator
Three-Phase Wye Connection
Three-Phase Delta Connection
High-Speed Generator
Low-Speed Generator
Frequency
Harmonics
Voltage Regulation
Efficiency

ELECTROMAGNETIC INDUCTION

The basic principle that allows electrical power to be produced by alternators was discovered in the early 1800s by Michael Faraday, an English scientist. *Faraday's Law* is the basis of electrical power production. The principle of *electromagnetic induction* was one of the most important discoveries in the development of modern technology. Without electrical power, our lives would certainly be different. Electromagnetic induction, as the name implies, involves electricity and magnetism. When electrical conductors, such as alternator windings, are moved within a magnetic field, an electrical current develops in the conductors. The electrical current produced in this way is caned an *induced current*. A simplified illustration showing how induced electrical current develops is shown in Figure 6-1

A *conductor* is placed within the magnetic field of a horseshoe magnet so that the left side of the magnet has a north polarity (N), and the right side has a south polarity (S). *Magnetic lines of force* travel from the north polarity of the magnet to the south polarity. The ends of the conductor are connected to a current meter to measure the induced current. The meter is the zero-centered type, so its needle can move either to the left or to the right. When the conductor is moved, current will flow through the conductor. *Electromagnetic induction* takes place whenever there is relative motion between the conductor and the magnetic field. Either the conduc-

Alternating Current Power Systems 139

tor can be moved through the magnetic field, or the conductor can be held stationary and the magnetic field can be moved past it. Thus, current will be induced as long as there is *relative motion* between the conductor and magnetic field.

If the conductor shown in Figure 6-1 is moved upward, the needle of the meter will move to the right. However, if the conductor is moved *downward*, the needle of the meter will deflect to the left. This shows that the direction of movement of the conductor within the magnetic field determines the direction of current flow. In one case, the current flows through the conductor from the front of the illustration to the back. In the other situation, the current travels from the back to the front. The direction of current flow is indicated by the direction of the meter deflection. The principle demonstrated here is the basis for electrical power generation.

In order for an *induced current* to be developed, the conductor must have a complete path or closed circuit. The meter in Figure 6-1 was connected to the conductor to make a complete current path. If there is no closed circuit, electromagnetic induction cannot take place. It is important to remember that an induced current causes an induced electromotive force (voltage) across the ends of the conductor.

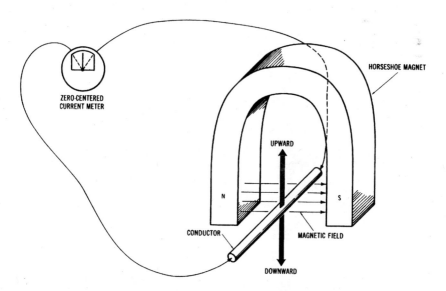

Figure 6-1. Electromagnetic induction

BASIC GENERATOR OPERATION

As stated previously, the operation of the electrical generators used today depends upon the principle of *electromagnetic induction*. When conductors move through a magnetic field, or when a magnetic field is moved past conductors, an induced current develops. The current that is induced into the conductors produces an induced electromotive force or voltage.

Left-hand Rule

We can determine the direction of current flow through a moving conductor within a magnetic field by using the *left-hand rule*. Refer to Figure 6-2 and use your left hand in the following manner:

1. Arrange your *thumb, forefinger,* and *middle finger* so that they are at approximately right angles to one another.
2. Point your *thumb* in the direction of conductor movement.

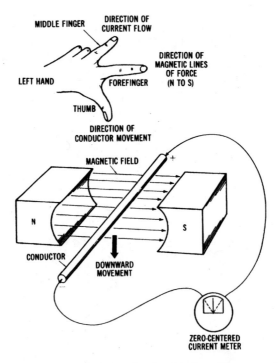

Figure 6-2. Left-hand rule of induced current

3. Point your *forefinger* in the direction of the magnetic lines of force (from north to south).
4. Your *middle finger* will now point in the direction of induced current flow (negative to positive).

Voltage Development in a Generator

It is known that when a conductor moves across a certain number of magnetic lines of force in one second, an *induced electromotive force (emf)* of one volt is developed across the conductor. Thus, the induced voltage value can be changed by modifying either the strength of the magnetic field, or the speed of conductor movement through the magnetic field. If the magnetic field is made stronger, more voltage will be induced. If the conductor is moved at a faster speed, more voltage will be induced. Likewise, if more conductors are concentrated within the magnetic field, a greater voltage will develop. These rules of electromagnetic induction are very important for the operation of mechanical generators that produce electrical power.

Sample Problem: Voltage Induced into A Conductor

In electrical generators, the coils move with respect to a magnetic field or flux. Electromagnetic induction occurs in accordance with *Faraday's Law*, which was formulated in 1831. This law states: 1) If a magnetic flux that links a conductor loop has relative motion, a voltage is induced, and 2) the value of the induced voltage is proportional to the rate of change of flux.

The voltage induced in a conductor of a generator is defined by Faraday's Law as follows:

$$V_i = B \times L \times v$$

where:
V_i = induced voltage in volts,
B = magnetic flux in teslas,
L = length of conductor within the magnetic flux in meters, and
v = relative speed of the conductor in meters per second.

Given: the conductors of the stator of a generator have a length of 0.5 M. The conductors move through a magnetic field of 0.8 teslas at a rate of 68 m/s.

Find: the amount of induced voltage in each conductor.
Solution:
$$V_i = B \times L \times v$$
$$= 0.8 \times 0.5 \times 60$$
$$V_i = 24 \text{ Volts}$$

SINGLE-PHASE AC POWER SYSTEMS

Electrical power can be produced by *single-phase generators*, commonly called *alternators*. The principle of operation of a single-phase alternator is shown in Figure 6-3. In order for a generator to convert mechanical energy into electrical energy, three conditions must exist:

1. There must be a *magnetic field* developed.
2. There must be a group of *conductors* adjacent to the magnetic field.
3. There must be *relative motion* between the magnetic field and the conductors.

These conditions are necessary in order for electromagnetic induction to take place.

Generator Construction

Generators used to produce electrical power require some form of *mechanical energy*. This mechanical energy is used to move electrical conductors through the magnetic field of the generator. Figure 6-3 shows the basic parts of a mechanical generator. A generator has a stationary part and a rotating part. The stationary part is called the *stator*, and the rotating part is called the *rotor*. The generator has magnetic field poles of north and south polarities. Also, the generator must have a method of producing a rotary motion, or a *prime mover*, connected to the generator shaft. There must also be a method of electrically connecting the rotating conductors to an external circuit. This is done by a *slip ring/brush assembly*. The stationary brushes are made of carbon and graphite. The slip rings used on AC generators are made of copper. They are permanently mounted on the shaft of the generator. The two slip rings connect to the ends of a *conductor loop*. When a *load* is connected, a closed external circuit is made. With all of these generator parts functioning together, electromagnetic induction can take place and electrical power can be produced.

Alternating Current Power Systems 143

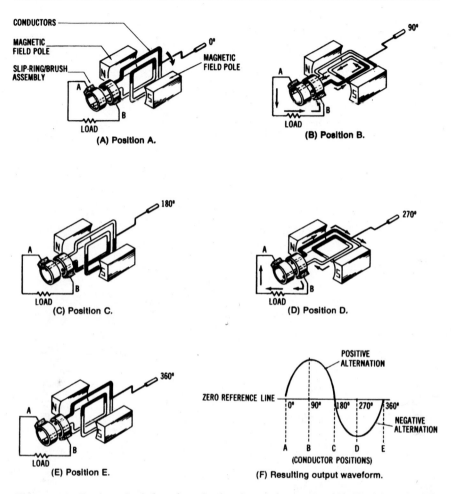

Figure 6-3. Basic principle of a single-phase alternator: (A) Position A, (B) Position B, (C) Position C, (D) Position D, (E) Position E, (F) Resulting output waveform

Generating AC Voltage

Figure 6-3 shows a magnetic field developed by a set of *permanent magnets*. *Conductors* that can be rotated are placed within the magnetic field, and they are connected to a *load* device by means of a *slip ring/brush assembly*. Figure 6-3 simulates a single-phase alternator.

In position A (Figure 6-3A), the conductors are positioned so that the minimum amount of *magnetic lines of force* is "cut" by the conductors as they rotate. No current is induced into the conductors at position A, and the resulting current flow through the load will be zero. If the conductors are rotated 90° in a clockwise direction to position B (Figure 6-3B), they will pass from the *minimum* lines of force to the most concentrated area of the magnetic field. At position B, the induced current will be maximum, as shown by the waveform diagram of Figure 6-3F. Note that the induced current rises gradually from the zero reference line to a *maximum* value at position B. As the conductors are rotated another 90° to position C (Figure 6-3C), the induced current becomes zero again. No current flows through the load at this position. Note, in the diagram in Figure 6-3F, how the induced current drops gradually from maximum to zero. This part of the induced AC (from 0° to 180°) is called the *positive alternation.* Each value of the induced current, as the conductors rotate from the 00 position to the 1800 position, is in a positive direction. This action could be observed visually if a meter were connected in place of the load.

When the conductors are rotated another 90° to position D (Figure 6-30), they once again pass through the most concentrated portion of the magnetic field. *Maximum* current is induced into the conductors at this position. However, the direction of the induced current is in the opposite direction from that of position B. At the 270 position, the induced current is maximum in a negative direction. As the conductors are rotated to position E (same as at position A), the induced current is minimum once again. Note, in the diagram of Figure 6-3E, how the induced current decreases from its maximum negative value back to zero again (at the 360° position). The part of the induced current from 180° to 360° is called the *negative alternation.* The complete output, which shows the induced current through the load, is called an AC waveform. As the conductors continue to rotate through the magnetic field, the *cycle* is repeated.

AC Sine Wave

The induced current produced by the method discussed above is in the form of a *sinusoidal* waveform or *sine wave.* This waveform is referred to as a sine wave because of its mathematical origin, based on the trigonometric sine function. The current induced into the conductors, shown in Figure 6-3, varies as the sine of the *angle of rotation* between the conductors and the magnetic field. This induced current produces a voltage. The instantaneous voltage induced into a single conductor can be expressed as:

$$V_i = V_{max} \times \sin \theta$$

Alternating Current Power Systems 145

where:
   V_i = the instantaneous induced voltage,
   V_{max} = the maximum induced voltage, and
   θ = the angle of conductor rotation from the zero reference.

For example, at the 30° position (Figure 6-4), if the maximum voltage is 100 volts, then

$$V_i = 100 \text{ V} \times \sin \theta$$

The sine of 30° = 0.5. Therefore:

$$V_i = 100 \text{ V} \times 0.5$$
$$= 50 \text{ volts}$$

SINGLE-PHASE AC GENERATORS

Although much *single-phase* electrical power is used, particularly in the home, very little electrical power is produced by single-phase *alternators*. The single-phase electrical power used in the home is usually devel-

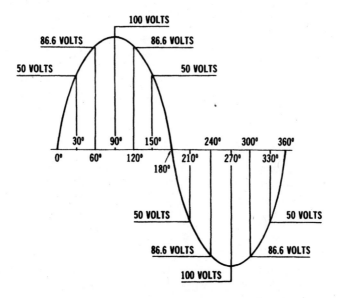

Fig 6-4. Mathematic origin of an AC sine wave.

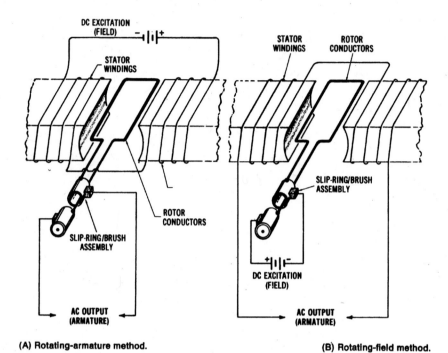

(A) Rotating-armature method.

(B) Rotating-field method.

Figure 6-5. The two basic methods of generating single-phase alternating current: (A) Rotating-armature method, (B) Rotating-field method

oped by three-phase alternators, and then converted to single-phase electricity by the power distribution system. There are two basic methods that can be used to produce single-phase AC. One method is called the *rotating-armature method*, and the other is the *rotating field method*. These methods are illustrated in Figure 6-5.

Rotating-armature Method

In the *rotating-armature method*, shown in Figure 6-4A, an AC voltage is induced into the conductors of the rotating part of the machine. The electromagnetic field is developed by a set of stationary pole pieces. Relative motion between the conductors and the magnetic field is provided by a prime mover, or mechanical energy source, connected to the shaft (which is a part of the rotor assembly). Prime movers may be steam turbines, gas turbines, or hydraulic turbines, gasoline engines, diesel engines, or possibly gas engines, or electric motors. Remember that all generators convert mechanical energy into electrical energy, as shown in Figure 6-6. Only

Alternating Current Power Systems

small power ratings can be used with the rotating-armature type of alternator. The major disadvantage of this method is that the AC voltage is extracted from a slip ring/brush assembly (see Figure 6-5A). A high voltage could produce tremendous sparking or arc-over between the brushes and the slip rings. The maintenance involved in replacing brushes and repairing the slip-ring commutator assembly would be very time-consuming and expensive. Therefore, this method is used only for alternators with low power ratings.

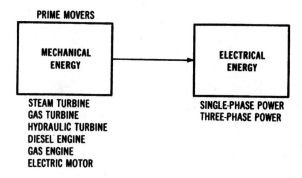

Figure 6-6. Energy conversion by generators

Rotating-field Method

The *rotating-field method,* shown in Figure 6-5B, is used for alternators capable of producing larger amounts of power. The direct current (DC) excitation voltage, which develops the magnetic field, is applied to the rotating portion of the machine. The AC voltage is induced into the stationary conductors of the machine. Since the DC-excitation voltage is a much lower value than the AC voltage that is produced, maintenance problems associated with the slip ring/brush assembly are minimized. In addition, the conductors of the stationary portion of the machine may be larger, so as to handle more current, since they do not rotate.

THREE-PHASE AC GENERATORS

The vast majority of electrical power produced in the United States is *three-phase* power. Because of their large power ratings, three-phase generators utilize the rotating field method. A typical three-phase generator in a power plant might have 250 volts DC excitation applied to the rotating field through the slip ring/brush assembly, while 13.8 kilovolts AC is induced into the stationary conductors.

Commercial power systems use many *three-phase alternators* connected in parallel to supply their regional load requirements. Normally, industrial loads represent the largest portion of the load on our power systems. The residential (home) load is somewhat less. Because of the vast load that has to be met by the power systems, three-phase generators have high power ratings. Nameplate data for a typical commercial three-phase alternator are shown in Figure 6-7. The nameplate of a generator specifies the manufacturer's rated values for the machine. It is usually a metal plate that is placed in a visible position on the generator frame. The following information would typically be listed on the nameplate:

TURBINE GENERATOR

STEAM TURBINE No. 128917

Rating 66000 kW	3600 RPM	21 Stages	
Steam: Pressure	1250 PSIG	*Temp* 950 F	*Exhaust Pressure* 1.5" HG. ABS.

GENERATOR

		Rating	Capability	Capability
No. 8287069 Hydrogen Cooled				
Type ATB 2 Poles 60 Cycles	Gas Pressure	30 PSIG	15 PSIG	0.5 PSIG
3 PH. Y Connected For 3800 Volts	KVA	88235	81176	70588
Excitation 250 Volts	Kilowatts	75000	69000	60000
Temp Rise Guaranteed Not To Exceed	Armature Amp	3691	3396	2953
45 C On Armature By Detector	Field Amp	721	683	626
74 C On Field By Resistance	Power Factor	0.85	0.85	0.85

Figure 6-7. Nameplate data for a commercial three-phase alternator: (A) Power rating (in kilowatts), (B) Voltage rating (in volts), (C) Current rating (in amperes), (D) Temperature rise (in degrees Centigrade)

Generation of Three-phase Voltage

The basic construction of a *three-phase AC generator* is shown in Figure 6-8, with its resulting output waveform given in Figure 6-10. Note that three-phase generators must have at least six stationary *poles,* or two poles for each phase. The three-phase generator shown in the drawing is a *rotating-field* type generator. The magnetic field is developed electromagnetically by a DC voltage. The DC voltage is applied from an external power

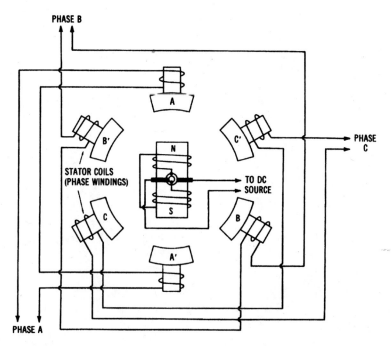

Figure 6-8. Simplified drawing showing the basic construction of a three-phase AC generator

source through a slip ring/brush assembly to the windings of the rotor. The magnetic polarities of the rotor, as shown, are north at the top and south at the bottom of the illustration. The magnetic lines of force develop around the outside of the electromagnetic rotor assembly.

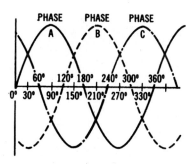

Figure 6-9. Output waveforms of a three-phase AC generator

Through *electromagnetic induction*, a current can be induced into each of the stationary (stator) coils of the generator. Since the beginning of phase A is physically located 120° from the beginning of phase B, the induced currents will be 120° apart. Likewise, the beginnings of phase B and phase C are located 120° *apart*. Thus, the voltages developed as a result of electromagnetic induction are 1200 apart, as shown in Figure 6-9.

Voltages are developed in each stator winding as the electromagnetic field rotates within the enclosure that houses the stator coils.

Three-Phase Connection Methods

In Figure 68, poles A', B', and C' represent the beginnings of each of the phase windings of the alternator. Poles A, B, and C represent the ends of each of the phase windings. There are two methods that may be used to connect these windings together. These methods are called *wye* and *delta connections*.

Three-phase Wye-connected Generators

The windings of a three-phase generator can be connected in a *wye configuration* by connecting either the beginnings or the ends of the windings together. The unconnected ends of the windings become the three-phase power lines from the generator. A three-phase wye-connected generator is illustrated in Figure 6-10. Notice that the beginnings of the windings (poles A', B', and C') are connected together. The other ends of the windings (poles A, B, and C) are the three-phase power lines that are connected to the load to which the generator will supply power.

Three-phase Delta-connected Generators

The windings of a three-phase generator may also be connected in a *delta arrangement,* as shown in Figure 6-11. In the *delta configuration,* the

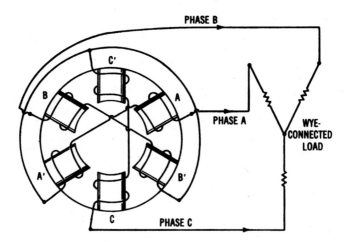

Figure 6-10. Simplified drawing of the stator of a three-phase generator that is connected in a *wye* configuration

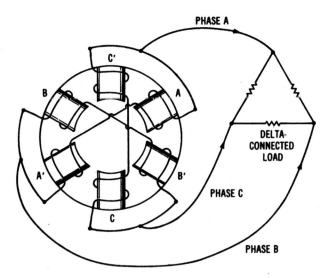

Figure 6-11. Simplified drawing of the stator of a three-phase generator that is connected in a delta configuration

beginning of one phase winding is connected to the end of the adjacent phase winding. Thus, the beginnings and ends of all adjacent phase windings are connected together. The voltage, current, and power characteristics of three-phase wye and delta connections were discussed earlier in Chapter 2.

Advantages of Three-phase Power

Three-phase power is used primarily for industrial and commercial applications. Many types of industrial equipment use three-phase AC power because the power produced by a three-phase voltage source, as compared to single-phase power, is less pulsating. You can see this effect by observing that a peak voltage occurs every 120° in the three-phase waveform in Figure 6-9. A single-phase voltage has a peak voltage only once every 360° (Figure 6-3). This comparison is somewhat similar to comparing the power developed by an eight-cylinder engine to the power developed by a four-cylinder engine. The eight-cylinder engine provides smoother, *less-pulsating* power. The effect of smoother power development on electric motors (with three-phase voltage applied) is that it produces a more *uniform torque* in the motor. This factor is very important for the large motors that are used in industry.

Three separate single-phase voltages can be derived from a three-phase transmission line, and three-phase power is more *economical* than single-phase power to distribute from plants to consumers that are located a considerable distance away. *Fewer conductors* are required to distribute the three-phase voltage. Also, the equipment that uses three-phase power is physically *smaller in size* than similar single-phase equipment.

HIGH-SPEED AND LOW-SPEED GENERATORS

Generators can also be classified as either *high-speed* or *low-speed* types (Figure 6-12). The type of generator used depends upon the prime mover used to rotate the generator. *High-speed generators* are usually driven by steam turbines. The high-speed generator is smaller in diameter and longer than a low-speed generator. The high-speed generator ordinarily has two stator poles per phase; thus, it will rotate at 3600 rpm to produce a 60-hertz frequency.

Low-speed generators are larger in diameter and not as long as high-speed machines. Typical low-speed generators are used at hydroelectric power plants. They have large-diameter revolving fields that use many poles. The number of stator poles used could, for example, be twelve for a 600-rpm machine, or eight for a 900-rpm generator. Notice that a much larger number of poles is required for low-speed generators.

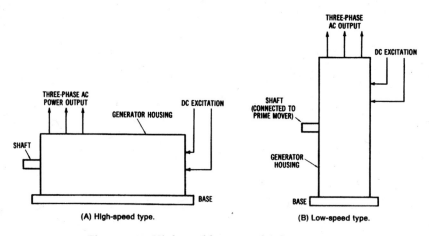

Figure 6-12. High- and low-speed AC generators

Alternating Current Power Systems

GENERATOR FREQUENCY

The *frequency* of the sinusoidal waveforms (sine waves) produced by an AC generator is usually *60 hertz*. One *cycle* of alternating current is generated when a conductor makes one complete revolution past a set of north and south field poles. A speed of 60 revolutions per second (3600 revolutions per minute) must be maintained to produce 60 hertz. The frequency of an AC generator (alternator) may be expressed as:

$$f = \frac{\text{speed of rotation (rpm)} \times \text{number of poles per phase}}{120}$$

where f is the frequency in hertz.

Sample Problem:

Given: a six-pole three-phase alternator rotates at a speed of 3600 rpm.

Find: the frequency of the alternator.

Solution:

$$f = \frac{N/3 \times \text{rpm}}{120}$$

$$\frac{6/3 \times 3600}{120}$$

$$= 60 \text{ Hz}$$

Note that if the number of *poles* is increased, the *speed of rotation* may be reduced while still maintaining a 60-Hz frequency.

HARMONICS

The voltage and current of power lines are often distorted because of the effect of *harmonics*. This distortion can be caused by magnetic effects in transformers, or by power control equipment. Harmonics are frequencies that are whole number multiples of the power line frequency.

Sample Problem:
Given: a power line frequency of 60 hertz is applied to an electric motor.
Find:
1. fundamental frequency
2. 3rd harmonic
3. 5th harmonic
4. 7th harmonic

Solution:
1. fundamental = 60 Hz
2. 3rd harmonic = 60 × 3 = 180 Hz
3. 5th harmonic = 60 × 5 = 300 Hz
4. 7th harmonic = 60 × 7 = 420 Hz

GENERATOR VOLTAGE REGULATION

As an increased electrical *load* is added to an alternator, it tends to slow down. The decreased speed causes the generated voltage to decrease. The amount of voltage change depends on the generator design and the type of load connected to its terminals. The amount of change in generated voltage from a no-load condition to a rated full-load operating condition is referred to as *voltage regulation*. Voltage regulation may be expressed as:

$$VR = \frac{V_{NL} - V_{FL}}{V_{FL}} \times 100$$

where:
VR = the voltage regulation in percent,
V_{NL} = the no-load terminal voltage, and
V_{FL} = the rated full-load terminal voltage.

Sample Problem:
Given: a single-phase alternator has a no-load output voltage of 122.5 volts and a rated full-load voltage of 120.0 volts.
Find: the voltage regulation of the alternator.
Solution:

Alternating Current Power Systems

$$VR = \frac{V_{NL} - V_{FL}}{V_{FL}} \times 100$$

$$= \frac{122.5 - 120 \times 100}{120} \times 100$$

$$VR = 0.02 = 2\%$$

GENERATOR EFFICIENCY

Generator *efficiency* is the ratio of the *power output* in watts to the *power input* in horsepower. The efficiency of a generator may be expressed as:

$$\text{Efficiency (\%)} = \frac{P_{out}}{P_{in}} \times 100$$

where:
P_{in} = the power input in horsepower, and
P_{out} = the power output in watts.

Sample Problem:
Given: a three-phase alternator has a power output of 22 MW and a power input of 35,000 horsepower.
Find: efficiency of the alternator.
Solution:

$$= (\%) \text{ Eff} = \frac{P_{out}}{P_{in}} \times 100$$

$$= \frac{22,000,000 \text{ W}}{35,000 \text{ hp} \times 746} \times 100$$

$$\% \text{Eff} = 84\%$$

To convert horsepower to watts, remember that 1 horsepower = 746 watts. The efficiency of a generator usually ranges from 70 percent to 85 percent.

Chapter 7

Direct Current Power Systems

Alternating Current (AC) is used in greater quantities than *direct current* (DC); however, many important operations are dependent upon DC power. Industries use DC power for many specialized processes. Electroplating and DC variable-speed motor drives are only two examples that show the need for DC power to sustain industrial operations. We use DC *power* to start our automobiles, and many types of portable equipment in the home use DC power. Most of the electrical power produced in the United States is *three-phase AC,* and this three-phase AC power may be easily converted to DC for industrial or commercial use. Direct current is also available in the form of primary and secondary *chemical cells.* These cells are used extensively. In addition, *DC generators* are also used to supply power for specialized applications.

IMPORTANT TERMS

Chapter 7 deals with DC power systems. After studying this chapter, you should have an understanding of the following terms:

Primary Cell
Secondary Cell
Electrolyte Electrode
Battery
Internal Resistance
Carbon-Zinc Cell
Mercury Cell
Alkaline Cell
Nuclear Cell
Lead-Acid Cell

Specific Gravity
Ampere-Hour Rating
Nickel-Iron (Edison) Cell
Nickel-Cadmium Cell
Silver Cell
DC Generator
Split-Ring Commutator
DC Permanent-Magnet Generator
DC Separately Excited Generator
DC Self-Excited Series-Wound Generator
DC Self-Excited Shunt-Wound Generator
DC Self-Excited Compound-Wound Generator
Armature Reaction
Interpoles
Rectification
Single-Phase, Half-Wave Rectifier
Single-Phase, Full-Wave Rectifier
Single-Phase, Bridge Rectifier
Three-Phase, Half-Wave Rectifier
Three-Phase, Full-Wave Rectifier
Silicon-Controlled Rectifier (SCR)
Rotary Converter
Filter Circuit
Capacitor Filter
RC Filter
Pi (π) Filter
Regulation
Zener Diode Regulator
Series Transistor Regulator
Shunt Transistor Regulator

DC PRODUCTION USING CHEMICAL CELLS

The conversion of chemical energy into electrical power can be accomplished by the use of *electrochemical cells*. A *cell* is composed of two dissimilar metals, which are immersed in a conductive liquid or paste called an electrolyte. Chemical cells are classified as either *primary cells* or *secondary cells. Primary cells* are ordinarily not usable after a certain period of time. After this period of time its chemicals can no longer produce electrical energy. *Secondary cells* can be renewed after they are used, by reactivating the chemical process that is used to produce electrical energy. This reactivation is known as charging. Both primary cells and secondary cells have many applications. When two or more cells are connected in series, they form a battery.

CHARACTERISTICS OF PRIMARY CELLS

The operational principle of a *primary cell* involves the placing of two unlike conductive materials called *electrodes* into a conductive solution called an electrolyte. When the chemicals that compose the cell are brought together, their molecular structures are altered. During this alteration, their atoms may either gain additional electrons or lose some of their electrons. These atoms then have either a positive or a negative electrical charge, and are referred to as ions. The *ionization* process thus develops a chemical solution capable of conducting an electrical current. A *carbon-zinc primary cell* is shown in Figure 7-1.

When an external *load* device, such as a lamp, is connected to a cell, a current will flow from one electrode to the other through the electrolyte material. Current leaves the cell through its *negative electrode*, flows through the load device, then reenters the cell through its *positive electrode*, as shown in Figure 7-2. Thus, a complete circuit is established between the cell (source) and the lamp (load).

The voltage developed by a *primary cell* is dependent upon the electrode materials and the type of electrolyte used. The familiar *carbon-zinc cell* of Figure 7-1 produces approximately 1.5 volts. The negative electrode of the cell is the zinc container, and the positive electrode is a carbon rod. A sal ammoniac paste, which acts as the electrolyte, is placed between the electrodes. This type of cell is usually called a *dry cell*.

Internal Resistance of Cells

An important characteristic of a chemical cell is its *internal resistance*. Since a cell conducts electrical current, its resistance depends on its cross-sectional area, the length of its current path, the type of materials used, and the operational temperature. The amount of *current* that a cell will deliver to a load is expressed as:

160   *Electrical Power Systems Technology*

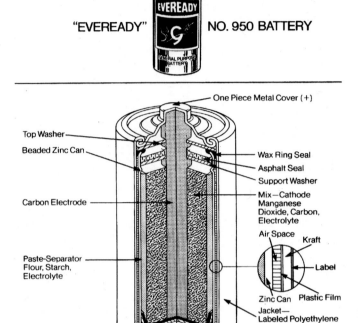

Figure 7-1. Batteries—Common sources of direct current (DC): Cutaway of a general purpose carbon-zinc cell *(Courtesy Eveready Corp.)*

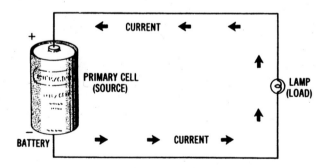

Figure 7-2. Current path for a primary cell

$$I = \frac{V}{R_i + R_L}$$

where:

    I = the current in amperes,
    V = the voltage or electromotive force (emf) of the cell in volts,
    R_i = the internal resistance of the cell in ohms, and
    R_L = the resistance of the load in ohms.

Sample Problem:

Given: a 1.5 volt battery has an internal resistance of 0.8 ohms.

Find: load current when 10.0 ohms of load resistance is connected to the battery.

Solution:

$$I = \frac{1.5 \text{ V}}{0.8 + 10.00 \Omega}$$

$$= 0.139 \text{ A}$$

This formula, and another circuit that illustrates it, are shown in Figure 7-3. The formula shows that when a cell delivers current to a load, some of the voltage developed must be used to overcome the internal resistance of the cell. As load current increases, the voltage drop (I × R) within the cell increases. This increased voltage drop causes the output voltage (V_o) of the cell to decrease. Thus,

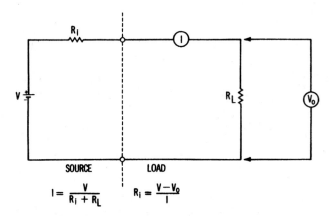

Figure 7-3. Illustrating the internal resistance of a cell

$$V_o = V - IR_i$$

where:
    V_o = the output voltage of the cell in volts,
    V = the no-load cell voltage in volts,
    I = the load current in amperes, and
    R_i = the internal resistance of the cell in ohms.

Sample Problem:
Given: the no-load voltage of a battery is 9.05 volts. Its rated load current is 200 mA, and its internal resistance is 0.1 ohms.
Find: the rated output voltage of the battery.
Solution:

$$\begin{aligned}V_o &= V - IR_i \\ &= 9.05\ V - (.2\ A \times 0.1\Omega) \\ Vo &= 9.03\ \text{Volts}\end{aligned}$$

Modifying this formula to solve directly for R_i, we obtain

$$R_i = \frac{V - V_o}{I}$$

Sample Problem:
Given: the rated output of a battery is 30.0 volts, and its no-load voltage is 30.15 volts. Its full-load current is 350 mA.
Find: the internal resistance of the battery.
Solution:

$$R_i = \frac{V - V_o}{I}$$

$$= \frac{30.15 - 30.0\ V}{0.35\ A}$$

$$R_i = 0.43\ \text{ohms}$$

The same method is used to determine the *internal resistance* of all voltage sources, including primary cells, secondary cells, and generators.

Applications of Primary Cells

There are many types of *primary cells* available today, with unlimited applications. The *carbon-zinc* (or Lechanche) cell, discussed previously, is the most often used type of dry cell. This cell has a low initial cost and is available in a variety of sizes. It is used primarily for portable equipment or instruments. For uses that require a higher voltage or current than one cell can deliver, manufacturers combine several cells in series, parallel, or series-parallel arrangements to form batteries suitable for specialized applications. *Carbon-zinc batteries* can be obtained in voltage ratings from 1.5 volts, at over I-ampere capability, up to about 3000 volts.

Mercury Cells—Another type of primary cell is the *mercury* or *zinc-mercuric oxide cell*, shown in Figure 7-4. This cell was developed as an improvement of the carbon-zinc cell. The mercury cell has a more constant voltage output, a longer active service time, and a smaller physical size. However, the mercury cell is more expensive and produces a voltage of 1.35 volts, which is lower than that of the carbon-zinc cell.

Alkaline Cells—An *alkaline or zinc-magnesium dioxide cell*, shown in Figure 7-5, is another type of primary cell. These cells have very low internal resistance and a voltage per cell of 1.5 volts. Because of their low internal resistance, alkaline cells will supply higher currents to the electrical loads connected to them. Alkaline cells are capable of a much longer service life than the equivalent carbon-zinc cells.

Nuclear Cells—A recent source of DC power that has been developed

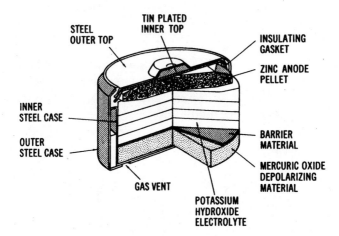

Figure 7-4. Cutaway drawing of a mercury primary cell

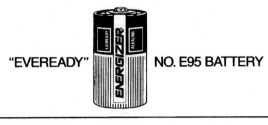

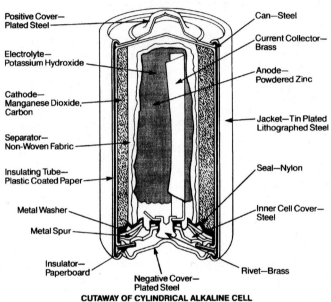

Figure 7-5. Alkaline cells: Cutaway of a cylindrical alkaline cell (Courtesy *Eveready* Corp.)

is the *nuclear cell*. Such cells convert the energy given off by atomic nuclei into electrical energy. At present, a high-voltage type (thousands of volts with a picoampere current capacity) and a low-voltage type (approximately 1 volt with a microampere current capability) are available.

The illustration in Figure 7-6 shows the design of one type of *nuclear cell*. The radioactive source is connected to one electrode of the cell. The source used here is tritium gas absorbed in zirconium metal, which emits beta particles. The carbon-collector electrode collects these beta particles as they are emitted from the *tritium* source, thus producing an electrical charge. The electrodes are separated by either a vacuum or some other type of dielectric, which electrically isolates the two electrodes. The nu-

Direct Current Power Systems

clear cell shown in Figure 7-6 has air removed through the center tube to form the vacuum inside it that isolates the positive and negative electrodes. The DC output produced by this cell is extracted through terminals connected to the tritium source (negative) and to the carbon connector (positive). The output of this cell is dependent upon the nuclear energy generated by the tritium source; therefore, the voltage output decreases as the cell ages.

CHARACTERISTICS OF SECONDARY CELLS

Regardless of the type of *primary cell* used, its usable time is limited. When its chemicals are expended, it becomes useless. This disadvantage is overcome by *secondary cells*. The chemicals of a secondary cell may be reactivated by a charging process. Secondary cells are also called *storage cells*. The most common types of secondary cells are the *lead-acid cell, nickel-iron (Edison) cell,* and *nickel-cadmium cell*

The Lead-acid Cell

The widely used *lead-acid cell* is one type of *secondary cell*. The electrodes of the lead-acid cell are made of lead and lead peroxide. The *positive plate* is lead peroxide (PbO_2), and the *negative plate* is *lead (Pb)*. The *electrolyte* is dilute sulfuric acid (H_2SO_4). When the lead-acid cell supplies cur-

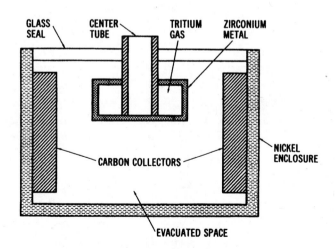

Figure 7-6. Cutaway drawing of a nuclear cell

rent to a load connected to it, the chemical process can be expressed as:

$$PbO_2 + Pb + 2H_2SO_4 \rightarrow 2PbSO_4 + 2H_2O.$$

The sulfuric acid ionizes to produce four positive hydrogen *ions* (H^+) and two negative sulfate (SO_{4-}) ions. A negative charge is developed on the lead plate when an SO_{4-} ion combines with the lead plate to form lead sulfate ($PbSO_4$). The positive hydrogen *ions* (H^+) combine with electrons of the lead-peroxide plate and become neutral hydrogen atoms. They next combine with the oxygen (O) of this plate to become water (H_2O). The lead-peroxide plate thus becomes *positively charged*. A fully charged lead-acid cell has an electrical potential developed between electrodes of approximately 2.0 volts.

After a cell discharges by supplying current to a load for a certain period of time, it is no longer able to develop an output voltage. The cell may then be charged by causing direct current to flow through the cell in the opposite direction.

The chemical process is thus reversed as follows:

$$2PbSO_4 + 2H_2O \rightarrow PbO_2 + Pb + 2H_2SO_4.$$

Thus, the original state of the chemicals is reestablished by the charging action.

Specific Gravity of Secondary Cells

The amount of charge may be measured by a specific-gravity test using a hydrometer to sample the electrolyte. The *specific gravity* of a liquid provides an index of how much heavier than water the liquid is. Pure sulfuric acid has a specific gravity of 1.840, while the dilute sulfuric acid of a *fully charged* lead-acid cell varies from 1.275 to 1.300. During *discharge* of a cell, water is formed, reducing the specific gravity of the electrolyte. A specific gravity of between 1.120 and 1.150 indicates a *fully discharged* cell. Recharging is accomplished using battery chargers.

Ampere-hour Rating of Secondary Cells

The capacity of a battery composed of lead-acid cells is given by an *ampere-hour rating*. A 60-ampere hour battery could theoretically deliver 60 amperes for 1 hour, 30 amperes for 2 hours, or 15 amperes for 4 hours. However, this is an approximate rating dependent upon the rate of discharge and the operating temperature of the battery. The normal operating temperature is considered to be 800 Fahrenheit.

Nickel-iron Cells

The *nickel-iron* (or *"Edison"*) *cell*, shown in Figure 7-7 is a secondary cell constructed with an unbreakable case of welded nickel-plated sheet steel. The positive plate is made of nickel tubes about V. inch in diameter and 4-1/4 inches long. Nickel oxide (NiO_2) is contained in the tubes, which are mounted into steel grids and forced into position under high pressure. The *positive plates* are assembled into groups. The *negative plates* of the Edison cell are made of flat nickel-plated steel pockets, which are composed of iron oxide. They are built into steel grids in groups, in the same way as the positive plates. The *electrolyte* is a solution of potassium hydroxide (KOH) with lithium hydroxide added.

Chemical action of the *nickel-oxide cell* is very complex as compared to that of the lead-acid cell; however, they are similar in many respects. The potassium hydroxide electrolyte (KOH) breaks up into negative and positive *ions*. The *negative ions* move toward the iron plate, oxidize the iron, and give up excess electrons to the plate. This plate becomes negatively charged. The positive ions move to the nickel-oxide plate and take electrons from the plate. The deficiency of electrons causes the nickel-oxide plate to become positively charged. The voltage produced by the chemical action of a nickel-iron cell is about 1.4 volts.

The *internal resistance* of the nickel-iron (Edison) cell is higher than that of the lead-acid cell. This resistance increases quickly as the cell discharges. Unlike the lead-acid cell, the nickel-iron cell cannot be tested using a hydrometer to show its discharge level. A voltmeter is used to indi-

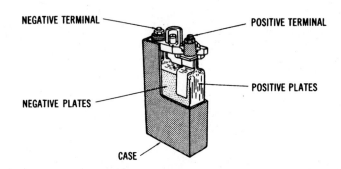

Figure 7-7. Cutaway drawing of a nickel-iron (Edison) secondary cell

cate the state of charge of the nickel-iron cell. This cell is also charged by applying a DC voltage to the *positive* and *negative* terminals and causing current within the cells to flow in reverse. The chemicals are thus reactivated so that the cells may be used again.

Other Types of Secondary Cells

Another type of secondary cell that has gained recent popularity is the *nickel-cadmium cell*, sometimes called a *"Ni-cad" battery*. These cells are available in large or small sizes with capacities up to 2000 ampere-hours. Figure 7-8 shows the small *rechargeable* type of nickel-cadmium cell, which is used extensively in portable equipment. The *positive plate* of this cell is nickel hydroxide, the *negative plate* is cadmium hydroxide, and the *electrolyte* is potassium hydroxide. These cells are extremely reliable, operate over a wide range of temperatures, and have a long life expectancy. A fully charged nickel-cadmium cell has a voltage of approximately 1.35 volts per cell.

There are a few other types of *secondary cells* in use today. Most of these types are for specialized applications. Other cells include the *silver-oxide-zinc cell* and the *silver-cadmium cell*. These cells have the highest output per physical size and the longest life, but they are more expensive than other cells.

Applications of Secondary Cells

Secondary cells, in the form of storage batteries, are used in industry and commercial buildings to provide emergency power in the event of a power failure. Such standby systems are necessary to sustain lighting and some critical operations when power is not available. Industrial trucks and loaders use storage batteries for their everyday operation. Many types of instruments and portable equipment rely upon batteries for power. Several of these instruments get their power from rechargeable secondary cells rather than primary cells. Railway cars use batteries for lighting when they are not in motion. Of course, automotive systems of all kinds use secondary batteries to supply DC power for starting and lighting.

DC GENERATING SYSTEMS

Mechanical generators are used in many situations to produce direct current. These generators convert mechanical energy into DC electrical energy.

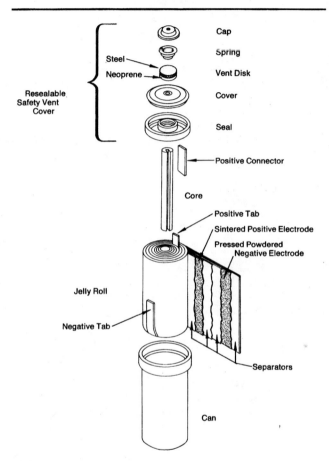

Figure 7-8. Sealed nickel-cadmium rechargeable cylindrical type cell-exploded view (Courtesy Eveready Corp.)

Construction of DC Generators

The parts of a simple *DC generator* are shown in Figure 7-9. The principle of operation of a DC generator is similar to that of the AC generator, which was discussed previously. A rotating *armature* coil passes through a *magnetic field* that develops between the north and south polarities of permanent magnets or electromagnets. As the coil rotates, *electromagnetic induction* causes a current to be induced into the coil. The current produced is an *alternating current*. However, it is possible to convert the alternating current that is induced into the armature into a form of *direct current*. This conversion of AC into DC is accomplished through the use of a *split-ring commutator*. The conductors of the armature of a DC generator are connected to split-ring commutator segments. The split-ring commutator shown in Figure 7-12 has two segments, which are insulated from one another and from the shaft of the machine on which it rotates. An end of each armature conductor is connected to each commutator segment. The purpose of the *split-ring commutator* is to reverse the armature-coil connection to the external load circuit at the same time that the current induced in the armature coil reverses. This causes DC of the correct polarity to be applied to the load at all times.

Voltage Output of DC Generators

The voltage developed by the single-coil generator shown in Figure 7-9 would appear as illustrated in Figure 7-10. This pulsating DC is not suitable for most applications. However, using many turns of wire around the armature and several *split-ring commutator* segments causes the voltage developed to be a smooth, direct current like that produced by a battery. This type of output is shown in Figure 7-10b. The voltage developed by a DC generator depends upon the strength of the *magnetic field*, the number of *coils* in the armature, and the *speed* of rotation of the armature. By increasing any of these factors, the voltage output can be increased.

Sample Problem: Voltage Output of a DC Generator

Voltage output of a DC generator can be expressed as:

$$V_o = \frac{Z \times n \times \Phi}{60}$$

where:
- V_o = voltage developed across the generator brushes in volts,
- Z = total number of armature conductors,
- n = speed of rotation in r/min, and
- Φ = magnetic flux per pole in webers.

Given: a four-pole DC generator rotates at 1200 r/min. The armature has 36 slots, and each coil has four turns of wire. The magnetic flux per pole is 0.05 webers.

Find: the voltage output of the generator.

Direct Current Power Systems

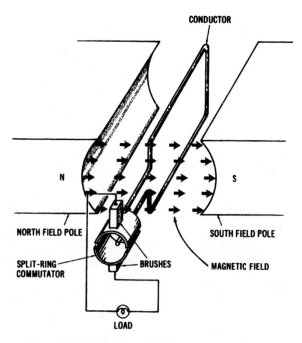

Figure 7-9. Simplified drawing of the basic parts of a DC generator

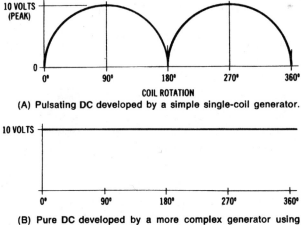

Figure 7-10. Output waveforms of a DC generator: (A) Pulsating DC developed by a simple single-coil generator, (6) Pure DC developed by a more complex generator using many turns of wire and many commutator segments

Solution: since each turn has 2 conductors, and 36 slots are used in the armature core, Z = 36 coils × 2 coils per turn × 4 turns of wire per coil = 288 conductors.

$$V_o = \frac{288 \times 1200 \times 0.05}{60}$$

$V_o = 288$ volts

Types of DC Generators

DC generators are classified according to the way in which a magnetic field is developed in the stator of the machine. One method is to use a permanent-magnet field. It is also possible to use electromagnets to develop a magnetic field by applying a separate source of DC to the electromagnetic coils. However, the most common method of developing a magnetic field is for part of the generator output to be used to supply DC power to the field of the machine. Thus, there are three basic classifications of DC generators: (1) *permanent-magnet field*, (2) *separately excited field*, and (3) *self-excited field*. The self-excited types are further subdivided according to the method used to connect the armature windings to the field circuit. This can be accomplished by the following connection methods: (1) *series*, (2) *parallel (shunt)*, or (3) *compound*.

Permanent-magnet DC Generator—A simplified diagram of a *permanent-magnet DC generator* is shown in Figure 7-11. The *conductors* shown in this diagram are connected to the *split-ring commutator and brush assembly* of the machine. The *magnetic field* is established by using permanent magnets made of Alnico (an alloy of aluminum, nickel, cobalt, and iron), or some other naturally magnetic material. It is possible to group several permanent magnets together to create a stronger magnetic field.

The *armature* of the permanent-magnet DC generator consists of many turns of insulated conductors. Therefore, when the armature rotates within the permanent-magnetic field, an induced voltage develops that can be applied to a load circuit. Applications for this type of DC generator are usually confined to those that require low amounts of power. A *magneto* is an example of a permanent-magnet DC generator.

Separately Excited DC Generator—Where large amounts of DC electrical energy are needed, generators with electromagnetic fields are used. Stronger fields can be produced by electromagnets. It is possible to control the strength of the electromagnetic field by varying the current through

Direct Current Power Systems

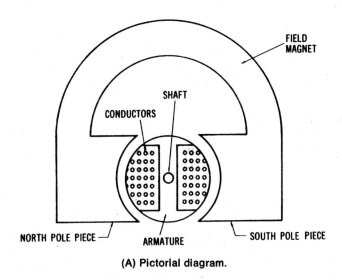

(A) Pictorial diagram.

Figure 7-11. Simplified drawing of a permanent-magnet DC generator: (A) Pictorial diagram, (B) Schematic diagram

the field coils. The output voltage of generators that use *electromagnetic fields* can be controlled with ease.

The direct current used to establish the electromagnetic field is referred to as the *excitation* current. When DC excitation current is obtained from a source separate from the generator, the generator is called a *separately excited DC generator*. This type of generator is shown in Figure 7-12. Storage batteries are often used to supply the DC excitation current to this type of generator. The field current is independent of the armature current. Therefore, the separately excited generator maintains a very stable output voltage. Changes in load of the external circuit affect the armature current, but do not vary the strength of the field. The voltage output of a separately excited DC generator can be varied by adjusting the current through the field. A high-wattage *rheostat* (variable resistance) connected in series with the field coils will accomplish field control of a separately excited DC generator.

Separately excited DC generators are used only in certain applications where precise voltage control is essential. Automatic control processes in industry often require such precision. However, the cost of a separately excited DC generator is often prohibitive, and, therefore, other means of obtaining DC electrical power are usually used.

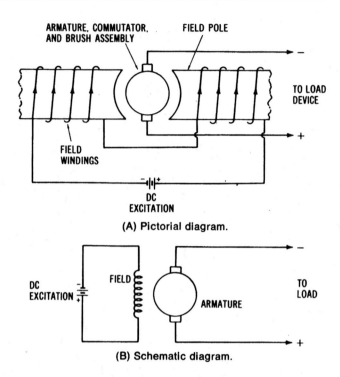

Figure 7-12. Simplified illustration of a separately excited DC generator: (A) Pictorial diagram, (B) Schematic diagram

Self-excited Series-wound DC Generators—Since DC generators produce direct current, it is possible to extract part of the output of a generator to obtain *excitation* current for the field coils. Generators that use part of their own output to supply the excitation current for the electromagnetic field are called *self-excited DC generators*. The method used to connect the armature windings to the field windings determines the characteristics of the generator. It is possible to connect armature and field windings in *series, parallel (shunt),* or *series-parallel (compound)*.

The *self-excited series-wound DC generator* has its armature windings connected in *series* with the field coils and the load circuit, as shown in Figure 7-13. In this generator, the total current flowing through the load also flows through the field coils. The field coils are, therefore, wound using only a *few turns* of *low-resistance* large-diameter wire. A sufficient electromagnetic field can then be produced by the large current flowing through the coils.

Direct Current Power Systems

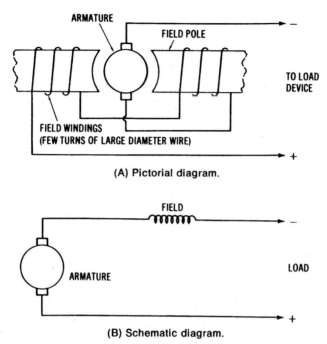

Figure 7-13. Simplified illustration of a self-excited, series-wound DC generator: (A) Pictorial diagram, (8) Schematic diagram

If the load circuit is disconnected, no current flows through the generator. However, the field coils retain a small amount of magnetism known as *residual magnetism*. Because of residual magnetism, a current will begin to flow through the generator and the load circuit when a load circuit is connected. The current will continue to increase, thus causing an increase in the magnetic strength of the field. The output voltage rises proportionally with increases in current flow through the load and generator circuits. The output curve of a *series-wound DC generator* is shown in Figure 7-14. The peak of the curve is reached when magnetic saturation of the field occurs. *Magnetic saturation* prohibits any increase in output voltage. After saturation is reached, any increase in load current causes a rapid decline in the output voltage due to circuit losses.

The output voltage of a *self-excited series-wound DC generator* varies appreciably with changes in load current. However, beyond the peak of its output curve, the load current remains fairly stable, even with large variations in voltage. There are specific applications, such as arc welding,

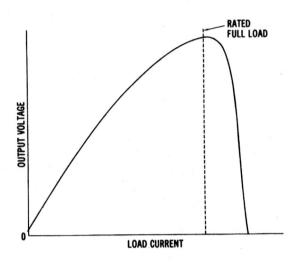

Figure 7-14. The output curve of a series-wound DC generator

that require stable load current as the voltage changes. However, the self-excited series-wound generator has very few applications.

Self-excited Shunt-wound DC Generator—connecting the field coils, the armature circuit, and the load circuit in parallel produces a shunt-wound DC generator configuration. Figure 7-15 shows this type of generator. The *output current* developed by the *generator* (I_A) has one path through the *load circuit* (I_L) and another through the *field coils* (I_F). The generator is usually designed so that the field current is not more than 5 percent of the total *armature current* (I_A).

In order to establish a *strong electromagnetic field* and to limit the amount of field current, the field coils are wound with many turns of small diameter wire. These *high-resistance* coils develop a strong field that is due to the number of turns, and, therefore, they rely very little on the amount of field current to develop a strong magnetic field.

With no load circuit connected to the shunt-wound DC generator, an induced voltage is produced as the armature rotates through the electromagnetic field. Again, the presence of *residual magnetism* in the field coils is critical to the operation of the machine. A current flows in the armature and field circuit as long as there is residual magnetism. As the generated current increases, the output voltage also increases, up to a peak level.

When a *load circuit* is connected to the shunt generator, the *armature current* (I_A) increases, because of the additional parallel path. This, then,

Direct Current Power Systems

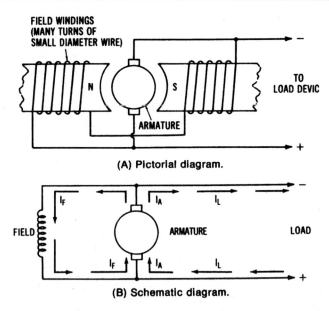

Figure 7-15. Simplified illustration of a self-excited, shunt-wound DC generator: (A) Pictorial diagram, (B) Schematic diagram

will increase the I × R drop in the armature, resulting in a smaller output voltage. Further increases in load current cause corresponding decreases in output voltage, as shown in the output curve of Figure 7-16. With small load currents, the output voltage remains nearly constant as load current

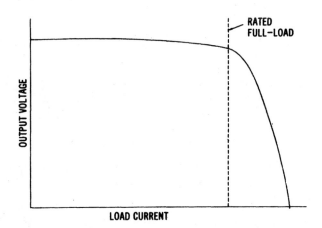

Figure 7-16. Output curve of a shunt-wound DC generator

varies. A large load causes the armature current to decrease. This decrease is desirable because it provides the generator with a built-in protective feature, in case of a short circuit.

The *self-excited shunt-wound DC generator* is used when a constant output voltage is needed. It may be used to supply excitation current to a large AC generator or to charge storage batteries. However, in applications where initial expense is not critical, the compound-wound generator described next may be more desirable.

Self-excited Compound-wound DC Generator—The *compound-wound DC generator* has two sets of field windings. One set is made of *low-resistance windings* and is connected in series with the armature circuit. The other set is made of *high-resistance* wire and is connected in parallel with the armature circuit. A *compound-wound DC generator* is illustrated in Figure 7-17.

As discussed previously, the *output voltage* of a series-wound DC generator increases with an increase in load current, while the output voltage of a shunt-wound DC generator decreases with an increase in load current. It is possible to produce a DC compound-wound generator, utilizing both series and shunt windings, that has an almost constant voltage output under changing loads. A constant voltage output can be obtained with varying loads, if the series-field windings have the proper characteristics to set up a sufficient magnetic field to counterbalance the voltage reduction caused by the I × R drop in the armature circuit.

A constant output voltage is produced by a *flat-compounded DC generator*. The no-load voltage is equal to the rated full-load voltage in this type of machine, as shown by the output curves in Figure 7-18. If the series windings produce a stronger field, the generator will possess a series characteristic. The voltage output will increase with an increase in load. A compound-wound generator whose full-load voltage is greater than its no-load voltage is called an *overcompounded DC generator*. Likewise, if the shunt windings produce a stronger field, the output will be more characteristic of a shunt generator. Such a generator, whose full-load voltage is less than its no-load voltage, is called an *undercompounded DC generator*.

Compound-wound DC generators can be constructed so that the series and shunt fields either aid or oppose one another. If the magnetic polarities of adjacent fields are the same, the magnetic fields aid each other and are said to be *cumulatively wound*. Opposing polarities of adjacent coils produce a *differentially wound* machine. For almost all applications of compound-wound machines, the cumulative-wound method is used. A generator wound in this way maintains a fairly constant voltage output with

Direct Current Power Systems

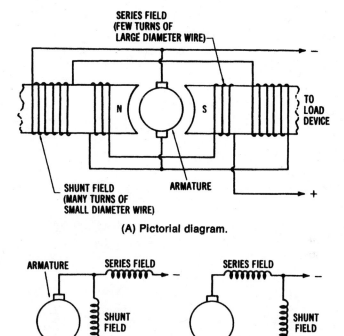

Figure 7-17. Simplified illustration of a compound-wound DC generator,: (A) Pictorial diagram, (B) Schematic diagram

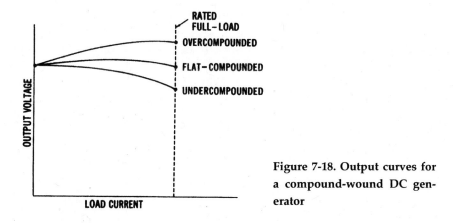

Figure 7-18. Output curves for a compound-wound DC generator

variations in load current. The compound-wound DC generator is used more extensively than other DC generators, because of its constant voltage output and its design flexibility, which allows it to obtain various output characteristics.

Common DC Generator Characteristics

DC generators are used primarily for operation with mobile equipment. In industrial plants and commercial buildings, they are used for standby power, for battery charging, and for specialized DC operations such as electroplating. In many situations, rectification systems that convert AC to DC have replaced DC generators, since they are cheaper to operate and maintain.

DC generators supply power to a load circuit by converting the mechanical energy of some prime mover, such as a gasoline or diesel engine, into electrical energy. The *prime mover* must rotate at a definite speed in order to produce the desired voltage.

When DC generators operate, a characteristic known as *armature reaction* takes place. The current through the armature windings produces a magnetic flux that reacts with the main field flux, as shown in Figure 7-19. The result is that a force is created that tends to rotate the armature in the opposite direction. As load current increases, the increase in armature current causes a greater amount of armature reaction to take place. This condition can cause a considerable amount of sparking between the brushes and the commutator. However, armature reaction may be reduced by placing windings called *interpoles* between the main field windings of the stator. These windings are connected in series with the armature windings. Thus, an increase in armature current creates a stronger magnetic field around the interpoles, which counteracts the main field distortion created by the armature conductors.

Rating of DC Generator—The *output* of a DC generator is usually rated in kilowatts, which is the electrical power capacity of a machine. Other ratings, which are specified by the manufacturer on the *nameplate* of the machine, are current capacity, output voltage, speed, and temperature. DC generators are made in a wide range of physical sizes, and with various electrical characteristics.

DC Generator Applications—DC generator use has declined rapidly since the development of the low-cost silicon rectifier. However, there are still certain applications where DC generators are used. These applications include railroad power systems, synchronous motor-generator units,

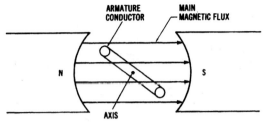

(A) Main magnetic flux with no current in the armature windings.

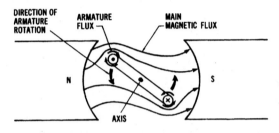

NOTE: ⊙ Indicates electron flow toward the observer, and a clockwise magnetic flux.

⊗ Indicates electron flow away from the observer, and a counterclockwise magnetic flux.

(B) Distortion of main magnetic flux with current in the armature windings.

Figure 7-19. Illustration of armature reaction: (A) Main magnetic flux with no current flow in the armature windings, (B) Distortion of main magnetic flux with current flow in the armature windings

power systems for large earth-moving equipment, and DC motor-drive units for precise equipment control.

DC CONVERSION SYSTEMS

Most of the electrical power produced is 60-Hz *three-phase alternating current*. However, the use of direct current is necessary for many applications. For instance, DC motors have more desirable speed control and torque characteristics than AC motors. We have already discussed two methods of supplying direct current-batteries and DC generators. The fact that batteries and DC generators have been used as DC power sources for many years has been discussed. In many cases today, where large amounts of DC power are required, it is more economical to use a system for converting AC to DC. DC load devices may be powered by systems, called either *rectifiers* or *converters*, that change AC into a suitable form of DC.

Single-phase Rectification Systems

The simplest system for converting AC to DC is *single-phase rectification*. A single-phase rectifier changes AC to *pulsating* DC. The most common and economical method is the use of low-cost, silicon, semiconductor rectifiers.

Single-phase Half-wave Rectification

A *single-phase half-wave rectifier* circuit, such as shown in Figure 7-20, converts an AC source voltage into a pulsating direct current. Let us assume that during the positive alternation of the AC cycle, the anode of the diode is positive (Figure 7-20). The diode will then conduct, since it is *forward biased* and the pn junction is low-resistant. The positive half cycle of the alternating current will then appear across the load device (represented by a resistor). When the negative portion of the AC alternation is input to the circuit, the anode of the diode becomes negative. The diode is now *reverse biased*, and no significant current will flow through the load device (Figure 7-20B). Therefore, there will be no voltage across the load. The input and resulting *output* waveforms of the half-wave rectifier circuit are shown in Figure 7-20C. The *pulsating* direct current of the output has an average DC level. The average value of the pulsating DC produced by single-phase half-wave rectification is expressed as:

$$V_d = 0.318 \times V_{max}$$

where:
 V_{dc} = the average value of rectified voltage, and
 V_{max} = the peak (maximum) value of applied AC voltage.

Sample Problem:

Given: a single-phase, half-wave rectifier has 15 volts (rms) applied to its input.
Find: DC output voltage of the rectifier.
Solution:
$$V_{dc} = 0.318 \times V_{max}$$
$$= 0.318 \times (15 \text{ V} \times 1.41)$$
$$V_{dc} = 6.72 \text{ volts}$$

Certain diode ratings should be considered for half-wave rectifier circuits. The maximum *forward current* (I_{max}) is the largest current that can flow through the diode while it is forward biased, without damaging the

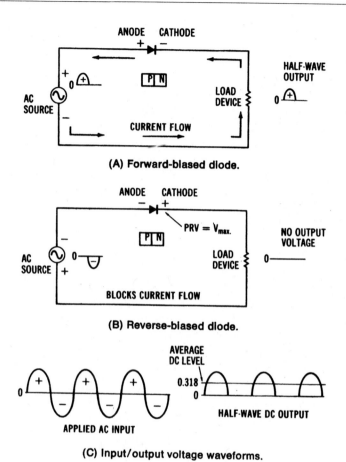

Figure 7-20. Single-phase half-wave rectification: (A) Forward-biased diode, (8) Reverse-biased diode, (C) Input/output voltage waveforms

device. The *peak inverse voltage (piv)* is the maximum voltage across the diode while it is reverse biased. For the half-wave rectifier, the maximum voltage developed across the diode is V_{max} of the applied AC. As shown in the preceding formula, the piv of a diode used in a half-wave circuit must be much larger than the DC voltage developed.

Single-phase Full-wave Rectification

In order to obtain a more pure form of DC energy, it is possible to improve upon half-wave rectification systems. Figure 7-21 shows a *single-phase full-wave rectifier* that uses two diodes and produces a DC output voltage during each alternation of the AC input. The rectified *output* of the full-wave rectifier has twice the DC voltage level of the half-wave rectifier.

The *full-wave rectifier* utilizes a *center-tapped transformer* to transfer AC source voltage to the diode rectifier circuit. During the *positive half cycle* of AC source voltage, the instantaneous charges on the transformer secondary are as shown in Figure 7-21. The peak voltage (V_{max}) is developed across each half of the transformer secondary. At this time, diode D1 is *forward biased*, and diode D2 is reverse biased. Therefore, conduction occurs from the center-tap, through the load device, through D1, and back to the outer terminal of the transformer secondary. The positive half cycle is developed across the load, as shown.

During the *negative half cycle* of the AC source voltage, diode D1 is *reverse biased*, and diode D2 is forward biased by the instantaneous charges shown in Figure 7-21. The current path is from the center-tap, through the load device, through O_2, and back to the outer terminal of the transformer secondary. The negative half cycle is also produced across the load, developing a full-wave output, as illustrated in Figure 7-21.

Each diode in a full-wave rectifier circuit must have a *piv rating* of twice the value of the peak voltage developed at the output, since twice the peak voltage ($2V_{max}$) is present across the diode when it is reverse biased. The *average voltage* for a full-wave rectifier circuit is:

$$V_{dc} = 2\,(0.318 \times V_{max})$$
$$= 0.636 \times V_{max}$$

Sample Problem:

Given: a single-phase, half-wave rectifier has 120 volts (rms) applied to its input.

Find: DC output voltage of the rectifier.

Solution:
$$Vdc = 0.636 \times V_{max}$$
$$= 0.636 \times (120\ V \times 1.41)$$
$$Vdc = 107.61\text{ volts}$$

This type of rectifier circuit produces twice the DC voltage output of a half-wave rectifier circuit. However, it requires a bulky center-tapped transformer, as well as diodes that have a piv rating of twice the peak value of applied AC voltage.

Direct Current Power Systems

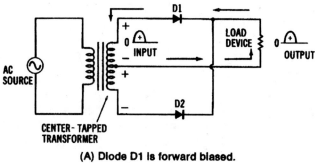

(A) Diode D1 is forward biased.

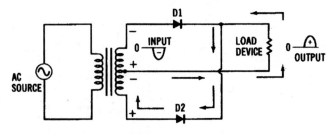

(B) Diode D2 is forward biased.

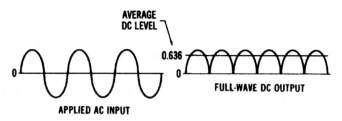

(C) Input/output voltage waveforms.

Figure 7-21. Single-phase full-wave rectification: (A) Diode D1 is forward biased, (B) Diode D2 is forward biased, (C) Input/output voltage waveforms

Single-phase Bridge Rectification

One disadvantage of the *full-wave rectifier* discussed previously is the requirement for a large center-tapped transformer. To overcome this disadvantage, four diodes may be used to form *a full-wave bridge rectifier*, as shown in Figure 7-22. Additionally, the diode piv rating is only required to be the peak output voltage value (V_{max}).

During the operation of a *bridge rectifier,* two diodes are forward biased during each alternation of the AC input. When the positive half cycle occurs, as shown in Figure 7-22A, diodes D1 and D3 are *forward biased*, while diodes D2 and D4 are *reverse biased*. This biasing condition is due to the instantaneous charges occurring during the positive alternation. The conduction path is from the instantaneous negative side of the AC source, through diode D3, through the load device, through diode D1, and back to the instantaneous positive side of the AC source.

During the *negative alternation* of the AC input, diodes D2 and D4 are *forward biased*, while diodes 01 and 03 are *reverse biased*. Conduction occurs, as shown in Figure 7-22B, from the instantaneous negative side of the source, through diode D2, through the load device, through diode 04, and back to the instantaneous positive side of the AC source. Since a voltage is developed across the load device during both half-cycles of the AC input, a full-wave output is produced, as shown in Figure 7-22, that is similar to that of the full-wave rectifier discussed previously. For high values of DC output voltage, the use of a bridge rectifier is desirable, since the diode piv rating is one-half that of other single-phase rectification methods. A typical design of a bridge rectifier unit is shown in Figure 7-22.

Three-phase Half-wave Rectification

Most industries are supplied with three-phase AC. It is therefore beneficial, because of the inherent advantages of three-phase power, to use *three-phase rectifiers* to supply DC voltage for industrial use. Single-phase rectifiers are ordinarily used where low amounts of DC power are required. To supply larger amounts of DC power for industrial requirements, a three-phase rectifier circuit, such as the one shown in Figure 7-23, could be employed. Three-phase rectifier circuits produce a purer DC voltage output than single-phase rectifier circuits, thus wasting less AC power.

Figure 7-23 shows a *three-phase half-wave rectifier* circuit that does not use a transformer. Phases A, B, and C of the wye-connected three-phase source supply voltage to the anodes of diodes D1, D2, and D3. The load device is connected between the cathodes of the diodes and the neutral point of the *wye-connected* source. Maximum conduction occurs through diode D1, since it is forward biased, when phase A is at its peak positive value. No conduction occurs through diode D1 during the negative alternation of phase A. The other diodes operate in a similar manner, conducting during the positive AC input alternation and not conducting during the associated negative AC alternation. In a sense, this circuit combines three single-phase, half-wave rectifiers to produce a half-wave DC out-

Direct Current Power Systems

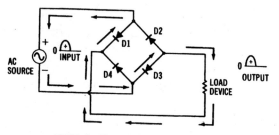

(A) Diodes D1 and D3 are forward biased.

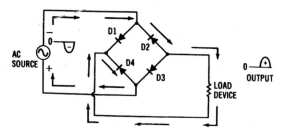

(B) Diodes D2 and D4 are forward biased.

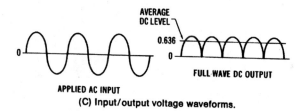

(C) Input/output voltage waveforms.

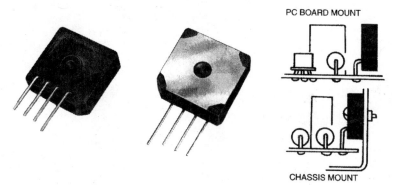

Figure 7-22. Single-phase full-wave bridge rectification: (A) Diodes D1 an D3 and forward biased, (B) Diodes D2 and D4 ore forward biased, (C) Input/output voltage waveforms, (D) Printed circuit board rectifier unit (Courtesy *Electronic Devices, Inc.*)

put, as shown in Figure 7-23B. Of course, the voltages appearing across the diodes are 120° out of phase. There is a period of time during each AC cycle when the positive alternations overlap one another, as shown in the shaded areas of the diagram (Figure 7-23B). During the overlap time period t_1, the phase A voltage is more positive than the phase B voltage, while during the t_2, interval, the phase B voltage is more positive. Diode D1 will conduct until the time period t_1 ends, then diode D_2 will conduct, beginning at the end of t_1 until the next area of overlapping is reached.

Note that *voltage* across the load device rises to a peak value twice during each phase alternation of the AC input voltage. These peaks are

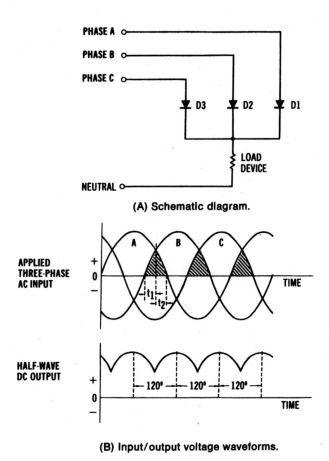

Figure 7-23. Three-phase half-wave rectification: (A) Schematic diagram, (B) Input/output voltage waveforms

120° *apart*. Since the DC output voltage never falls to zero, less AC ripple is present, which results in a purer form of DC than single-phase rectifiers produce. The *average DC output voltage* (V_{dc}) is expressed as:

$$V_{dc} = 0.831 \times V_{max}$$

which compares very favorably with single-phase full-wave rectifier circuits.

Sample Problem:
Given: a three-phase, half-wave rectifier circuit has 240 volts (rms) applied to its input.
Find: the DC output voltage of the rectifier.
Solution:
$$V_{dc} = 0.831 \times V_{max}$$
$$= 0.831 \times (240 \text{ V} \times 1.41)$$
$$V_{dc} = 281.2 \text{ volts}$$

A disadvantage of this type of three-phase rectifier is that the AC lines are *not isolated*. Isolation is a direct connection to the AC lines; the lack of isolation could be a safety hazard. To overcome this disadvantage, a transformer may be used, as shown in Figure 7-24, to form a similar three-phase, half-wave rectifier. The secondary voltage may be either increased or decreased by the proper selection of the transformer, thus providing a variable DC voltage capability. The circuit illustrated in Figure 7-24 has a delta-to-wye-connected transformer. The operation of this circuit is identical to that of the three-phase, half-wave rectifier previously discussed; however, line *isolation* has been accomplished.

Three-phase Full-wave Rectification
The *full-wave* counterpart of the three-phase, half-wave rectifier circuit is shown in Figure 7-24A. This type of rectifier circuit is popular for many industrial applications. Six rectifiers are required for operation of the circuit. The anodes of D4, D5, and D6 are connected together at point A, while the cathodes of D1, D2, and D3 are connected together at point B. The load device is connected across these two points. The three-phase AC lines are connected to the anode-cathode junctions of D1 and D4, D2 and D5, and D3 and D6. This circuit does not require the neutral line of the three-phase source; therefore, a delta-connected source could be used.

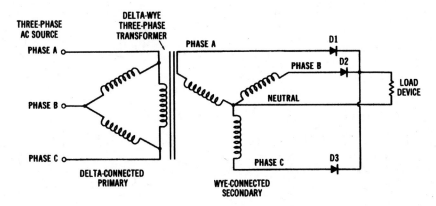

Figure 7-24. Three-phase half-wave rectification using a transformer

The resulting DC *output voltage* of the *three-phase, full-wave rectifier* circuit is shown in Figure 7-25B. The operation of the circuit is similar to that of a single-phase bridge rectifier in many respects. At any single instant of time during the three-phase AC input cycle, the anode voltage of one of the diodes is more positive than that of all the others, while the cathode voltage of another diode is more negative than that of all the others. These two diodes will then form the conduction path for that time period. This conducting action is similar to a bridge rectifier, since two diodes conduct during a time interval. Each rectifier in this circuit conducts during one-third of an AC cycle (120°). Peak positive DC output voltage occurs during every 60° of the three-phase AC input.

Rotary Converters

Another method that has been used to convert AC to DC is the use of a *rotary converter*. Rotating AC-to-DC converters are seldom used today. However, a motor-driven generator unit, such as shown in Figure 7-26 can be used to convert DC to AC. This system is called an *invertor*. When operated as a converter to produce DC, the machine is run off an AC line. The AC is transferred to the machine windings through slip rings and converted to DC by a split-ring commutator located on the same shaft. The amount of DC voltage output is determined by the magnitude of the AC voltage applied to the machine. Converters may be designed as two units, with motor and generator shafts coupled together, or as one unit housing both the motor and generator.

Direct Current Power Systems

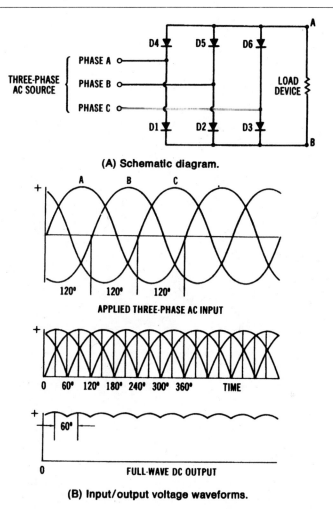

(B) Input/output voltage waveforms.

Figure 7-25. Three-phase full-wave rectification: (A) Schematic diagram, (B) Input/output voltage waveforms

DC FILTERING METHODS

The *pulsating direct current* produced by both single-phase and three-phase rectifier circuits is not *pure* DC. A certain amount of AC ripple is evident in each type of rectifier. For many applications, a smooth DC output voltage, with the AC ripple removed, is required. Circuits used to remove AC variations of rectified DC are called *filter circuits*.

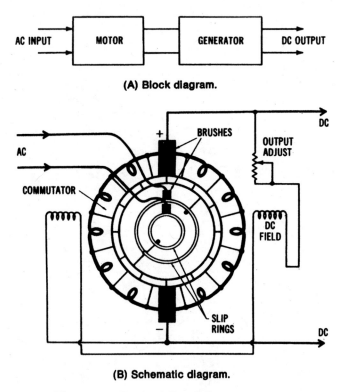

Figure 7-26. Rotary AC-to-DC converter

The output of a rectifier has a DC value and an AC ripple value, as shown in Figure 7-27. To gain a relative index of the amount of AC variation, the *ripple factor* of a rectifier output waveform may be determined. Ripple factor is expressed as:

$$r = \frac{V_{r(rms)}}{V_{dc}}$$

where:
- r = the ripple factor,
- $V_{r(rms)}$ = the rms value of the AC component, and
- V_{dc} = the average value of the rectified DC voltage.

Another index used to express the amount of AC ripple is the percent of ripple of a rectified DC voltage. *Ripple percentage* is expressed as:

$$\% \text{ ripple} = \frac{V_{r(rms)}}{V_{dc}} = 100$$

Sample Problem:
Given: a power supply has an AC input of 24 volts (rms) and a p-p ripple of 1.5 volts at its output, as measured with an oscilloscope.
Find: DC voltage output of the power supply.
Solution:

$$V_{DC} = V_{max} - \frac{V_{r(p-p)}}{2}$$

$$= (24 \text{ V} \times 1.41) - \frac{1.5 \text{ V}}{2}$$

$$V_{DC} = 33.09 \text{ volts}$$

A full-wave rectified voltage has a lower percentage of ripple than a half-wave rectified voltage. When a DC supply must have a low amount of ripple, a full-wave rectifier circuit should be used.

Capacitor Filter

A simple *capacitor filter* may be used to smooth the AC ripple of a rectifier output. Figure 7-28 shows the result of adding a capacitor across the output of a single-phase, full-wave bridge rectifier. The output waveform after the capacitor has been added is shown in Figure 7-28C.

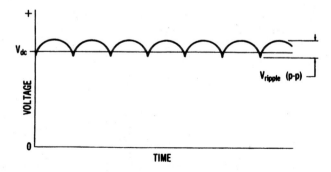

Figure 7-27. Rectifier output voltage

The *ideal* filtered DC voltage would be one with *no AC ripple* and a value equal to the peak voltage (V_{max}) from the rectifier output. Note that in Figure 7-28, the value of V_{dc} is approaching that of V_{max}, and compare this to the full-wave rectified voltage of Figure 7-28. There are two time intervals shown in Figure 7-28. Time period t_1 represents diode conduction that charges the filter capacitor (C) to the peak rectified voltage (V_{max}). Time period t_2 is the time required for the capacitor to discharge through the load (R_L).

If a different value of filter capacitor were put into the circuit, a change in the rate of discharge would result. If capacitor C discharged a very small amount, the value of V_{dc} would be closer to the value of V_{max}. With *light loads (high resistance)*, the capacitor filter will supply a high DC voltage with little ripple. However, with a *heavy (low resistance) load* connected, the DC voltage would drop, because of a greater ripple. The increased ripple is caused by the lower resistance discharge path for the filter capacitor. The effect of an increased load on the filter capacitor is shown in Figure 7-28.

By utilizing the value indicated on the waveforms of Figure 7-28, it is possible to express V_{dc} and $V_{r(rms)}$ as:

$$V_{dc} = V_{max} - \frac{V_{r(p-p)}}{2}$$

and

$$V_{r(rms)} = \frac{V_{r(p-p)}}{2 \times \sqrt{3}}$$

Sample Problem:

Given: the measured AC ripple voltage at the output of a filtered power supply is 0.8V (rms), and the DC output voltage is 15.0 volts.

Find: the percent ripple of the power supply.

Solution:

$$\%r = \frac{0.8V \times 100}{15.0 \text{ V}}$$

$$= 5.33\%$$

Sample Problem:

Given: a power supply has a p-p ripple of 1.2 volts at its output.
Find: the rms ripple voltage of the power supply.

Direct Current Power Systems

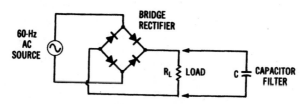

(A) Bridge rectifier circuit with filter capacitor.

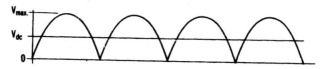

(B) Full-wave output waveform with no filter capacitor.

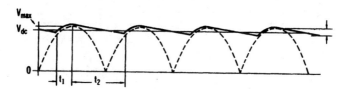

(C) Full-wave output waveform with filter capacitor across load.

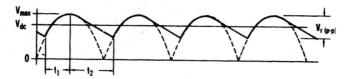

Figure 7-28. Filter capacitor operation: (A) Bridge rectifier circuit with filter capacitor, (8) Full-wave output waveform with no filter capacitor, (C) Full-wove output waveform with filter capacitor across load, (D) Full-wave output waveform with filter capacitor across heavier (lower resistance) load

Solution:

$$V_{r(rms)} = \frac{V_{r(p-p)}}{2 \times \sqrt{3}}$$

$$= \frac{1.2\ V}{3.46}$$

$$V_{r(rms)} = 0.34\ V$$

We can also express the amount of *ripple* of a 60-Hz full-wave filter-capacitor circuit with a light load as:

$$V_{r(rms)} = \frac{2.4 I_{dc}}{C}$$

Sample Problem:
Given: a power supply has a DC load current of 800 mA, and a capacitor filter of 300 uF is used.
Find: the value of ripple voltage.
Solution:

$$V_{r(rms)} = \frac{2.4 I_{dc}}{C}$$

$$= \frac{2.4 \times 800 \text{ mA}}{300 \text{ uF}}$$

$$V_{r(rms)} = \frac{2.4}{R_L C}$$

$$r = \frac{2.4}{R_L C}$$

where:
I_{dc} = the load current in milliamperes,
C = the filter capacitor value in microfarads, and
R_L = the load resistance in kilohms.

Sample Problem:
Given: a power supply has a DC load resistance of 500 ohms connected to its output, and a capacitor filter 100 uF is used.
Find: the value of ripple voltage.
Solution:

$$r = \frac{2.4}{R_L C}$$

$$= \frac{2.4}{0.5 \text{ K} \times 10.0 \text{ uF}}$$

$$r = 0.00048 = 0.048\%$$

Direct Current Power Systems

The *average DC voltage* output can be expressed as:

$$V_{dc} = V_{max} - \frac{4.16 I_{dc}}{C}$$

Sample Problem:
Given: the peak AC voltage applied to a power supply is 28.2 volts. A 500 uF filter capacitor is used.
Find: the DC output voltage when 300 mA of load current flows.
Solution:

$$V_{dc} = V_{max} - \frac{4.16 \times I_{dc}}{C}$$

$$= 28.2 - \frac{4.16 \times 300 \text{ mA}}{500 \text{ uF}}$$

$$V_{dc} = 25.7 \text{ volts}$$

Again, the value of V_{dc} is for a full-wave filter-capacitor circuit with a light load.

With a *heavier load (lower resistance)* connected to the filter circuit, more *current* (I_{dc}) is drawn. As I_{dc} increases, V_{dc} decreases. However, if the value of filter capacitor C is made larger, the value of V_{dc} becomes closer to that of V_{max}. The value of C for a full-wave rectifier can be determined by using the equation:

$$C = \frac{2.4 \times I_{dc}}{V_{r(rms)}}$$

Sample Problem:
Given: a power supply is rated at 950 mA of load current and maximum ripple of 0.5 volts (rms).
Find: the minimum value of capacitor filter needed for a filter using a full-wave rectifier.
Solution:

$$C = \frac{2.4 \times I_{dc}}{V_{r(rms)}}$$

$$= \frac{2.4 \times 950 \text{ mA}}{0.5 \text{ V}}$$

$$C = 4550 \text{ uF}$$

It should be pointed out that as the value of C increases, the value of the peak current through the diodes will also increase. There is, therefore, a practical limit for determining the value of C that is reflected in the foregoing equation.

The *capacitor filter* produces a high DC voltage with low AC ripple for light loads. However, its major disadvantages are higher ripple and lower Vdc at heavier loads, poor voltage regulation, and high peak current through the diodes.

RC Filter

It is possible to improve upon the previous filter circuit by using an *RC filter*. Figure 7-29 shows an RC filter stage. This filter has lower ripple than a capacitor filter, but has a lower average DC voltage, because the voltage is dropped across R1.

The purpose of R1 and C2 is to add another filter network, in addition to C1, to further reduce the ripple. This circuit also operates best with light loads connected. It is possible to have many stages of RC filters to further reduce AC ripple.

Pi- or π-type Filter

The use of the resistor (R1) in the RC filter is not desirable in many cases, since it reduces the average DC output of the circuit. To compensate for the DC voltage reduction, a *pi-type (or π-type) filter* may be used. Figure 7-30 shows this type of filter. The advantage of the choke coil (L1) over the resistor (R1) of the RC filter is that it offers only a small DC series re-

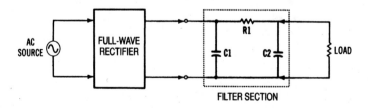

Figure 7-29. An RC filter circuit

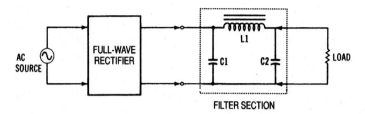

Figure 7-30. A pi-type filter circuit

sistance, but its AC impedance is much larger. It therefore passes DC and blocks the AC component of the rectified voltage.

DC REGULATION METHODS

The concept of *voltage regulation* can best be understood by referring to the formula:

$$VR = \frac{V_{NL} - V_{FL} \times 100}{V_{FL}}$$

where:
  VR = the voltage regulation in percent,
  V_{NL} = the no-load (open-circuit) voltage, and
  V_{FL} the rated, full-load voltage.

Sample Problem:
Given: a power supply has a no-load voltage of 24.85 volts and a rated full-load voltage of 24.0 volts.
  Find: the voltage regulation of the power supply.
  Solution:

$$\%VR = \frac{V_{NL} - V_{FL} \times 100}{V_{FL}}$$

$$= \frac{24.85 - 24.0 \times 100}{24.0}$$

$$= \ \%\ VR = 3.5\%$$

In an ideally regulated circuit, V_{NL} would equal V_{FL}.

In all types of power supplies that convert AC to DC, the DC output levels are affected by variations in the load. The lower the percentage of *regulation* (approaching 0 percent), the better regulated the circuit is. For instance, power supplies are capable of having a voltage regulation of less than 0.01 percent, which means that the value of the load has little effect on the DC output voltage produced. A well-regulated DC *power supply* is necessary for many industrial applications.

Voltage Regulators

A simple *voltage regulation* circuit that uses a zener diode is shown in Figure 7-31. This circuit consists of a series resistor (Rs) and a zener diode (D1) connected to the output of a rectifier circuit. However, we should review zener diode operation before discussing the circuit operation.

Zener diodes are similar to conventional diodes when they are *forward biased*. When they are *reverse biased*, no conduction takes place until a specific value of *reverse breakdown voltage (or "zener" voltage)* is reached. The zener is designed so that it will operate in the reverse breakdown region of its characteristic curve (see Figure 7-31B). The reverse breakdown voltage is predetermined by the manufacturer. When used as a voltage regulator, the zener diode is reverse biased so that it will operate in the breakdown region. In this region, changes in current through the diode have little effect on the voltage across it. The constant-voltage characteristic of a zener diode makes it desirable for use as a regulating device.

The circuit of Figure 7-31A is a *zener diode shunt regulator*. The zener establishes a constant voltage across the load resistance within a range of rectified DC voltages and output load currents. Over this range, the voltage drop across the zener remains constant. The current through the zener (I_Z) will vary to compensate for changes in load resistance, since $I_Z = I_T - I_L$. Thus, the output voltage will remain constant.

Transistor Voltage Regulators

An improvement over the zener voltage regulator is a *transistorized regulator*, as shown in the circuit in Figure 7-32. This regulator has transistor Q1 placed in series with the load device (R_L). Transistor Q1, then, acts to produce variable resistance to compensate for changes in the input voltage. The collector-emitter resistance of Q1 varies automatically with changes in the circuit conditions. The zener diode establishes the DC bias placed on the base of transistor Q1. When this circuit is operating proper-

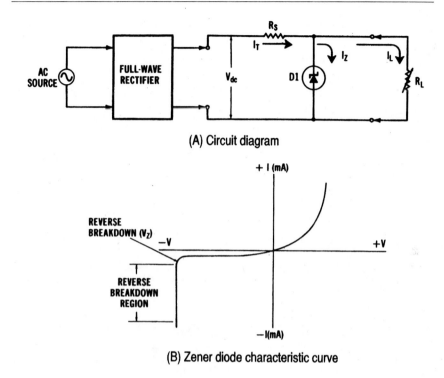

Figure 7-31. Zener diode voltage regulator: **(A)** Circuit diagram, **(B)** Zener diode characteristic curve

ly, if the voltage across the load increases, the rise in emitter voltage makes the base less positive. The current through Q1 will then be reduced, which results in an increase in the collector-emitter resistance of Q1. The increase in resistance will cause a larger voltage drop across transistor Q1, which will now compensate for the change in voltage across the load. Opposite conditions would occur if the load voltage were to decrease. Many variations of this circuit are used in regulated power supplies today.

Shunt regulators are also used in DC power supplies. The circuit of Figure 7-33 is a shunt voltage regulator. Again, the zener diode (D1) is used to establish a constant DC bias level. Therefore, voltage variations across the DC output will be sensed only by resistor R2. If the DC output voltage rises, an increased positive voltage will be present at the base of transistor Q2. The increased forward bias on transistor Q2 will cause it to conduct more. This makes the base of transistor Q1 more positive, and Q1 will then conduct more heavily. Increased current flow through both transistors causes an increase in the voltage drop across resistor R1. This increased voltage drop across R1 will counterbalance the rise in output voltage. Thus, the DC output voltage will remain stabilized. Decreases in DC output voltage would cause the circuit action to reverse. In addition, *integrated circuit voltage regulators* are now used extensively in power supplies that convert AC to DC.

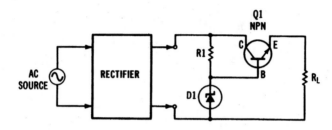

Figure 7-32. Transistor series voltage regulator circuit

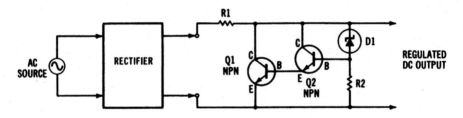

Figure 7-33. Transistor shunt voltage regulator circuit

UNIT III
Electrical Power Distribution Systems

Once electrical power has been produced, it must be distributed to the location where it is used. Unit III deals with *electrical power distribution systems*. The fundamentals of distribution systems are discussed in Chapter 8. This chapter provides an overview of distribution systems, *transformer* operation, and *conductor* characteristics. Chapter 9 examines specialized power distribution *equipment* such as protective devices, high-voltage equipment, and switch gear. Then, Chapter 10, *Single-Phase and Three-Phase Distribution Systems*, discusses the types of distribution systems used today. *Ground-fault interrupters* are also discussed extensively in this chapter, since their use is becoming widespread in electrical power distribution systems.

Figure III shows the *electrical power systems model* used in this book and the major topics of Unit III, Electrical Power Distribution Systems.

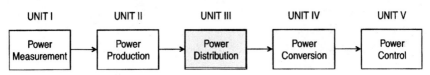

Figure III. Electrical power systems model

Power Distribution Fundamentals (Chapter 8)
Power Distribution Equipment (Chapter 9)
Single-phase and Three-phase Distribution Equipment (Chapter 10)

UNIT OBJECTIVES

Upon completion of this unit, you should be able to:
1. List several purposes of transformers.
2. Describe the construction of a transformer.
3. Explain transformer action.

4. Calculate turns ratio, voltage ratio, current ratio, and the power and efficiency of transformers.
5. Explain the purpose of isolation transformers, autotransformers, and current transformers.
6. Explain factors that cause losses in transformer efficiency.
7. Investigate the characteristics of transformers.
8. Describe the transmission and distribution of electrical power.
9. Compare underground, overhead, and HVDC power transmission systems.
10. Explain parallel operation of electrical power systems.
11. Describe ring, radial, and network power distribution systems.
12. Explain circular mil sizing of copper and aluminum electrical conductors.
13. Calculate cross-sectional area, resistance, or ampacity of round, square, or rectangular copper or aluminum conductors.
14. Describe the types of insulation used for conductors in electrical distribution systems.
15. Describe the following equipment used at substations for electrical power distribution:
High-Voltage Fuses
High-Voltage Circuit Breakers
High-Voltage Disconnect Switches
Lightning Arrestors
High-Voltage Insulators
High-Voltage Conductors
Voltage Regulators
16. Describe the following types of protective equipment used in electrical distribution systems:
Plug Fuses
Cartridge Fuses
Time-Lag Fuses
Low-Voltage Circuit Breakers
Protective Relays
Motor Fault Current Protection
Overheating Protection
Undervoltage Protection
17. Describe the following in relation to electrical power distribution systems:
Feeder Lines

Unit Objectives

205

    Branch Circuits
    Safety Switches
    Distribution Panelboards
    Low-Voltage Switchgear
    Service Entrance

18. Explain the operation of single-phase/two-wire and single-phase/three-wire power distribution systems.
19. Explain the operation of three-phase/three-wire, three-phase/four-wire, and three-phase/three-wire with neutral power distribution systems.
20. Describe and sketch diagrams of delta-delta, delta-wye, wye-wye, wye-delta, and open delta three-phase transformer connection methods.
21. Describe the following in relation to grounding of electrical power systems:
    System Ground
    Equipment Ground
    Ground Fault Circuit Interrupters
22. Explain the use of the National Electric Code (NEC) as a standard for electrical power distribution design in the United States.
23. Calculate voltage drop in an electrical circuit using a conductor table or circular mil formula.
24. Calculate voltage drop in a branch circuit or feeder circuit for a single-phase or three-phase distribution system.
25. Determine grounding conductor size by using the proper tables.
26. Describe the following in relation to electrical power systems:
    Nonmetallic Sheathed Cable
    "Hot" Conductor
    Neutral Conductor
    Safety Ground
    AWG and MCM Conductor Sizing
    Metal-Clad Cable
    Rigid Conduit
    Electrical Metallic Tubing (EMT)
    Uninterruptible Power Supply (UPS)
    Power Line Filter
    Power Conditioner
    Floor-Mounted Raceway
    Conduit Connectors

Wire Connectors
Plastic Conduit and Enclosures
Power Outlet Design
International Power Sources

Chapter 8

Power Distribution Fundamentals

Power distribution systems are a very important part of electrical power systems. In order to transfer electrical power from an alternating current (AC) or a direct current (DC) source to the place where it will be used, some type of distribution network must be utilized. The method used to distribute power from where it is produced to where it is used can be quite simple. For example, a battery can be connected directly to a motor, with only a set of wires.

More complex *power distribution systems* are used, however, to transfer electrical power from the power plant to industries, homes, and commercial buildings. Distribution systems usually employ such equipment as transformers, circuit breakers, and protective devices.

IMPORTANT TERMS

Chapter 8 deals with electrical power distribution fundamentals. After studying this chapter, you should have an understanding of the following terms:

Power Transmission System
Underground Distribution System
High-Voltage Direct Current (HVDC) Distribution System
Cryogenic Cable
Parallel Operation
Radial Distribution System
Ring Distribution System
Network Distribution System
Transformer
Primary Winding

Secondary Winding
Transformer Core
Closed-Core Transformer Construction
Shell-Core Transformer Construction
Efficiency
Power Losses
Step-Up Transformer
Step-Down Transformer
Transformer Voltage Ratio
Transformer Current Ratio
Multiple Secondary Transformer
Autotransformer
Transformer Polarity
Transformer Volt-Ampere Rating
Transformer Malfunction
Conductor
Circular Mil (cmil)
American Wire Gauge (AWC)
Resistivity
Ampacity
Insulation

OVERVIEW OF ELECTRICAL POWER DISTRIBUTION

The *distribution* of electrical power in the United States is normally in the form of *three-phase, 60-Hz alternating current*. This power, of course, can be manipulated or changed in many ways by the use of electrical circuitry. For instance, a rectification system is capable of converting the 60-Hz AC into a form of DC, as was discussed in Chapter 7. Also, *single-phase* power is generally suitable for lighting and small appliances, such as those used in the home or residential environment. However, where a large amount of electrical power is required, three-phase power is more economical.

The *distribution* of electrical power involves a very complex system of *interconnected* power transmission lines. These *transmission lines* originate at the electrical power-generating stations located throughout the United States. The ultimate purpose of these power transmission and distribution systems is to supply the electrical power necessary for *industrial, residential,* and *commercial* use. From the point of view of the systems, we may say that the overall electrical power system delivers power from the source to the load that is connected to it. A typical *electrical power distribution system* is shown in Figure 8-1.

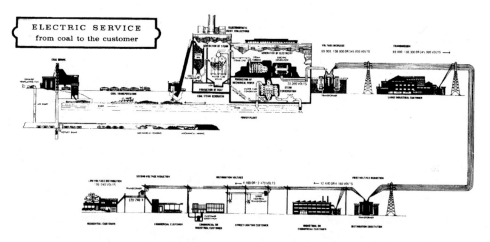

Figure 8-1. A typical electrical power distribution system (*Courtesy Kentucky Utilities Company*)

Industries use almost 50 percent of all the electrical power produced, so three-phase power is distributed directly to most industries. Electrical *substations* use massive *transformers* and associated equipment, such as oil-filled *circuit breakers*, high-voltage *conductors*, and huge strings of *insulators*, in distributing power to industry. From these substations, power is distributed to the industrial sites to energize the *industrial* machinery, to *residential* homes, and to *commercial* users of electrical power.

POWER TRANSMISSION AND DISTRIBUTION

Power *transmission* and *distribution* systems are used to interconnect electrical power production systems and to provide a means of delivering electrical power from the generating station to its point of utilization. Most electrical power systems east of the Rocky Mountains are interconnected with one another in a parallel circuit arrangement. These interconnections of power production systems are monitored and controlled, in most cases, by a computerized *control center*. Such control centers provide a means of data collection and recording, system monitoring, frequency control, and signaling. Computers have become an important means of assuring the efficient operation of electrical power systems.

The *transmission* of electrical power requires many long, interconnected power lines, to carry the electrical current from where it is produced to where it is used. However, overhead power transmission lines require much planning to ensure the best use of our land. The location of *overhead transmission lines* is limited by zoning laws and by populated areas, highways, railroads, and waterways, as well as other topographical and environmental factors. Today, an increased importance is being placed upon environmental and aesthetic factors. *Power transmission lines* ordinarily operate at voltage levels from 12 kV to 500 kV of AC. Common transmission line voltages are in the range of 50 to 150 kV of AC. *High-voltage direct current* overhead transmission lines may become economical, although they are not being used extensively at the present time.

Another option is ultra high-voltage *transmission lines*, which use higher AC transmission voltages. Also, underground transmission methods for urban and suburban areas must be considered, since the right-of-way for overhead transmission lines is limited. AC overhead transmission voltages have increased to levels in the range of 765 kV, with research now dealing with voltages of over 1000 kV. One advantage of overhead cables is their ability to dissipate heat. The use of *cryogenic cable* may bring about a solution to heat dissipation problems in conductors.

Underground Distribution

The use of *underground cable* is ordinarily confined to the short lengths required in congested urban areas. The cost of underground cable is much more than that of aerial cable. To improve underground cable power-handling capability, research is being done in forced-cooling techniques, such as circulating-oil and compressed-gas insulation. Another possible method is the use of *cryogenic cables* or *superconductors*, which operate at extremely low temperatures and have a large power-handling capability.

High-voltage Direct Current Transmission

An alternative to transmitting AC voltages for long distances is *high-voltage direct current (HVDC)* power transmission. HVDC is suitable for long-distance overhead power lines, or for underground power lines. DC power lines are capable of delivering more power per conductor than equivalent AC power lines. Because of its fewer power losses,

HVDC is even more desirable for underground distribution. The primary disadvantage of HVDC is the cost of the necessary AC-to-DC conversion equipment. There are, however, some HVDC systems in operation in the United States. At present, HVDC systems have been designed for transmitting voltages in the range of 600 kV. The key to the future development of HVDC systems may be the production of solid state power conversion systems with higher voltage and current rating. With a continued developmental effort, HVDC should eventually playa more significant role in future electrical power transmission systems.

Cryogenic Cable

There are some *problems* involved in installing an overhead electrical power distribution system, particularly in urban areas. One of these problems is obtaining a right-of-way for the overhead cable through heavily populated areas. The difficulty is caused primarily by the unattractive appearance of the lines and the potential danger of the high voltage. The problems associated with overhead transmission lines have led to the development of *cryogenic cable* for underground power distribution. *Cryogenic cables* are not considered to be superconductive, but they do have greater electrical conductivity at very low temperatures. These cables, which are still in the developmental stage, will use a metallic conductor cooled to the temperature of liquid nitrogen. One advantage of cryogenic cable over conventional cable is that its greater conductive characteristics will give it a lower line loss (I × R loss).

One design of a *cryogenic cable* is shown in Figure 8-2. This design involves the use of three separate cables, each having a hollow center for cooling purposes. The conductive portion of the cables is stranded aluminum. The aluminum conductors are wrapped with an insulating material that contains liquid nitrogen. Cryogenic cable has considerable potential in any future development of electrical power systems.

Parallel Operation of Power Systems

Electrical power *distribution systems* are operated in a *parallel* circuit arrangement. When more power sources (generators) in parallel are added, a greater load demand or current requirement can be met. On a smaller scale, this is like connecting two or more batteries in parallel to provide greater current capacity. Two *parallel-connected three-phase alternators* are depicted in Figure 8-3. Most power plants have more than one alternator connected to any single set of power lines inside the plant. These power

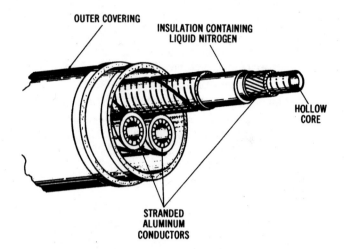

Figure 8-2. Construction of a cryogenic cable

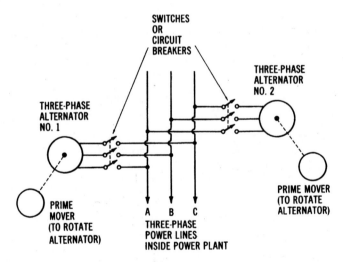

Figure 8-3. Parallel connection of two three-phase alternators

lines, or *"bus" lines,* are usually large, copper bar conductors that can carry very high amounts of current. At low-load demand times, only one alternator would be connected to the bus lines.

Figure 8-4 expands the concept of *parallel-connected* systems. An illustration of two power plants joined together through a distribution substation is shown. The two power plants might be located 100 miles apart,

Electrical Power Distribution Systems

yet they are connected in parallel to supply power to a specified region. If, for some reason (such as repairs on one alternator), the output of one power plant is reduced, the other power plant is still available to supply power to the requesting localities. It is also possible for power plant number one to supply part of the load requirement ordinarily supplied by power plant number two, or vice versa. These regional *distribution systems* of parallel-connected power sources provide automatic compensation for any increased load demand in any area.

The major *problem* of parallel-connected distribution systems occurs when excessive *load demands* are encountered by several power systems in a single region. If all of the power plants in one area are operating near their peak power-output capacity, there is no back-up capability. The *equipment-protection* system for each power plant, and also for each alternator in the power plant, is designed to disconnect it from the system when its maximum power limits are reached. When the power demand on one part of the distribution system becomes excessive, the protective equipment will disconnect that part of the system. This places an even greater load on the remaining parts of the system. The excessive load now could cause other parts of the system to disconnect. This cycle could continue until the entire system is inoperative. This is what occurs when *blackouts* of power systems take place. No electrical power can be supplied to any part of the

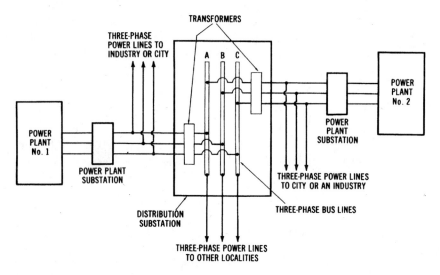

Figure 8-4. Joining two power plants in parallel as part of a regional power system

system until most of the power plants are put back in operation. The process of putting the output of a power plant back *on-line,* when the system is down during power outages, can be a long and difficult procedure.

RADIAL, RING, AND NETWORK DISTRIBUTION SYSTEMS

There are three general classifications of electrical power distribution systems. These are the *radial, ring,* and *network* systems shown in Figures 8-5 through 8-7. Radial systems are the simplest type, since the power comes from one power source. A generating system supplies power from the substation through radial lines that are extended to the various areas of a community (Figure 8-5). *Radial systems* are the least reliable in terms of continuous service, since there is no back-up distribution system connected to the single power source. If any power line opens, one or more loads are interrupted. There is more likelihood of power outages. However, the radial system is the least expensive. This system is used in remote areas where other distribution systems are not economically feasible.

Ring distribution systems (Figure 8-6) are used in heavily populated areas. The distribution lines encircle the service area. Power is delivered from one or more power sources into substations near the service area. The power is then distributed from the substations through the radial power lines. When a power line is opened, no interruption to other loads occurs. The *ring system* provides a more continuous service than the radial system. Additional power lines and a greater circuit complexity make the ring system more expensive.

Network distribution systems (Figure 8-7) are a combination of the radial and ring systems. They usually result when one of the other systems is expanded. Most of the distribution systems in the United States are network systems. This system is more complex, but it provides very reliable service to consumers. With a network system, each load is fed by two or more circuits.

USE OF TRANSFORMERS FOR POWER DISTRIBUTION

The heart of a *power distribution system* is an electrical device known as a *transformer.* This device is capable of controlling massive amounts of power for efficient distribution. Transformers are also used for many other

Electrical Power Distribution Systems

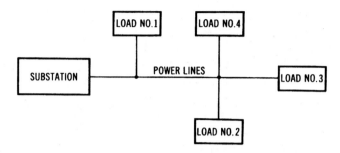

Figure 8-5. A radial power distribution system

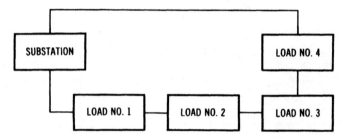

Figure 8-6. A ring power distribution system

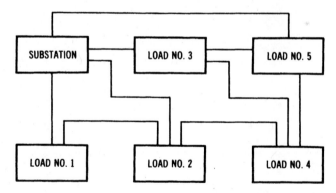

Figure 8-7. A network power distribution system

applications. A knowledge of transformer operation is essential for understanding power distribution systems. The distribution of AC power is dependent upon the use of transformers at many points along the line of the power distribution system.

It is economically feasible to transmit electrical power over long distances at high voltages since less current is required at high voltages, and therefore line loss (I^2R) is reduced significantly. A typical high-voltage transmission line may extend a distance of 50 to 100 miles from the generating station to the first substation. These high-voltage power transmission lines typically operate at 100,000 to 500,000 volts by using *step-up transformers* to increase the voltage produced by the AC generators at the power station. Various *substations* are encountered along the power distribution system, where transformers are used to reduce the high transmission voltages to a voltage level such as 480 volts, which is suitable for industrial motor loads, or to 120/240 volts for residential use.

Transformers provide a means of converting an AC voltage from one value to another. The basic construction of a *transformer* is illustrated in Figure 8-8. Notice that the transformer shown consists of two sets of windings that are not physically connected. The only connection between the primary and secondary windings is the magnetic coupling effect known as *mutual induction,* which takes place when the circuit is energized by an AC voltage. The laminated iron core plays an important role in transferring *magnetic flux* from the *primary winding* to the *secondary winding*.

Transformer Operation

If an AC current, which is constantly changing in value, flows in the

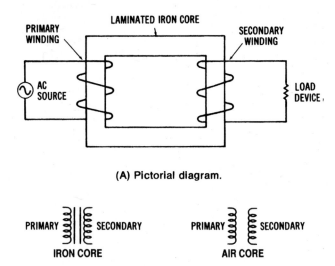

Figure 8-8. Basic transformer construction: (A) Pictorial design, (B) Schematic diagram

primary winding, the magnetic field produced around the primary winding will be transferred to the secondary winding. Thus, an *induced voltage* is developed across the *secondary winding.* In this way, electrical energy can be transferred from the source (primary-winding circuit) to a load (secondary-winding circuit).

The efficient transfer of energy from the primary to the secondary windings of a transformer depends on the *coupling* of the magnetic field between these two windings. Ideally, all magnetic lines of force developed around the primary winding would be transferred by magnetic coupling to the secondary winding. However, a certain amount of *magnetic loss* takes place as some lines of force escape to the surrounding air.

Transformer Core Construction

In order to decrease the amount of *magnetic loss,* transformer windings are wound around *iron cores.* Iron cores concentrate the magnetic lines of force, so that better coupling between the primary and secondary windings is accomplished. Two types of transformer cores are illustrated in Figure 8-9. These cores are made of laminated iron to reduce undesirable *eddy currents,* which are induced into the core material. These eddy currents cause power losses. The diagram of Figure 8-9A shows a *closed-core transformer* construction. The transformer windings of the closed-core type are placed along the outside of the metal core. Figure 8-9B shows the *shell-core* type of construction. The shell-core construction method produces better magnetic coupling, since the transformer windings are surrounded by metal on both sides. Note that the primary and secondary windings of both types are placed adjacent to one another for better magnetic coupling.

Transformer Efficiency and Losses

Transformers are very efficient electrical devices. A typical efficiency rating for a transformer is around 98 percent. *Efficiency* of electrical equipment is expressed as:

$$\text{Efficiency (\%)} = \frac{P_{out}}{P_{in}} \times 100$$

where:
P_{out} is the power output in watts, and
P_{in} is the power input in watts.

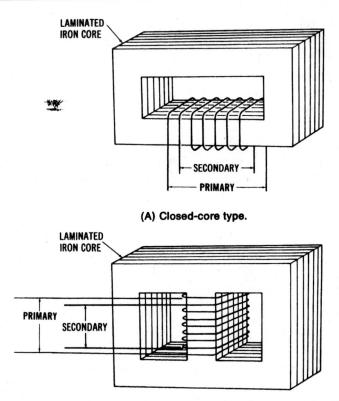

Figure 8-9. Types of iron-core transformers: (A) Closed-core type, (B) Shellcore type

Sample Problem:

Given: a transformer circuit has a power output of 2.5 kW and a power input of 2,550 W.

Find: the efficiency of the transformer.

Solution:

$$\% \text{ Eff} = \frac{2,500 \text{ W}}{2,550 \text{ W}} \times 100$$

$$\% \text{ Eff} = 98\%$$

The losses that reduce efficiency, in addition to *flux leakage*, are copper and iron losses. *Copper loss* is the I^2R loss of the windings, while *iron losses* are those caused by the metallic core material. The insulated laminations of the iron core help to reduce iron losses.

Step-Up and Step-Down Transformers

Transformers are functionally classified as *step-up* or *step-down* types. These types are illustrated in Figure 8-10. The *step-up transformer* in Figure 8-10 has fewer turns of wire on the primary than on the secondary. If the primary winding has 50 turns of wire and the secondary has 500 turns, a turns ratio of 1:10 is developed. Therefore, if 12-volts AC is applied to the primary from the source, 10 times that voltage, or 120 volts ac, will be transferred to the secondary load (assuming no losses).

The example in Figure 8-10B is a *step-down transformer*. The step-down transformer has more turns of wire on the primary than on the secondary. The primary winding of the example has 200 turns, while the secondary winding has 100 turns—or a 2:1 ratio. If 120-volts AC are applied to the primary from the source, then one-half that amount, or 60-volts AC, will be transferred to the secondary load.

Transformer Voltage and Current Relationships

In the preceding examples, a *direct* relationship is shown between the primary and secondary turns and the voltages across each winding. This relationship may be expressed as:

$$\frac{V_P}{V_S} = \frac{N_P}{N_S}$$

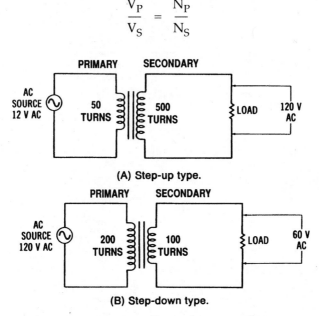

(A) Step-up type.

(B) Step-down type.

Figure 8-10. Step-up and step-down transformers

where:
- V_P = the voltage across the primary winding,
- V_S = the voltage across the secondary winding,
- N_P = the number of turns in the primary winding, and
- N_S = the number of turns in the secondary winding.

Sample Problem:
Given: a transformer circuit has the following values:
V_P = 240 volts
N_P = 1000 turns
V_S = 120 volts

Find: the number of secondary turns of wire required to accomplish this step-down of voltage.

Solution:

$$\frac{V_P}{V_S} = \frac{N_P}{N_S}$$

$$\frac{240 \text{ V}}{120 \text{ V}} = \frac{1000}{N_S}$$

$$N_S = 500 \text{ turns}$$

The transformer is a *power-control device;* therefore, the following relationship can be expressed:

$$P_P = P_S + \text{losses}$$

where:
- P_P = the primary power, and
- P_S = the secondary power.

Sample Problem:
Given: the power output (secondary power) of a transformer is 15 kW, and its losses are as follows:
  iron loss -200 W
  copper loss -350 W

Find: the power input (primary power) required. Solution:

$$\begin{aligned} P_P &= P_S + \text{losses} \\ &= 15{,}000 \text{ W} + (200 \text{ W} + 350 \text{ W}) \end{aligned}$$

$$P_P = 15{,}550 \text{ watts}$$

The *losses* are those that ordinarily occur in a transformer. In transformer theory, an *ideal* device is usually assumed, and losses are not considered. Thus, since $P_P = P_S$, and $P = V \times I$, then

$$V_P \times I_P = V_S \times I_S$$

where:
    I_P = the primary current, and
    I_S = the secondary current.

Therefore, if the voltage across the secondary is stepped up to twice the voltage across the primary, then the secondary current will be stepped down to one-half the primary current. The *current relationship* of a transformer is thus expressed as:

$$\frac{I_P}{I_S} = \frac{N_S}{N_P}$$

Note that whereas the voltage-turns ratio is a direct relationship, the current ratio is an *inverse* relationship.

Sample Problem:

Given: an ideal transformer circuit has the following values:
V_P = 600V
V_S = 2400V
I_S = 80A

Find: the primary current drawn by the step-up transformer
Solution:

$$\frac{I_P}{I_S} = \frac{N_S}{N_P}$$

$$\frac{I_P}{80 \text{ A}} = \frac{2400 \text{ V}}{600 \text{ V}}$$

$$I_P = 3{,}200 \text{ A}$$

Multiple Secondary Transformers

It is also possible to construct a transformer that has *multiple secondary windings,* as shown in Figure 8-11. This transformer is connected to a

120-volt AC source, which produces the primary magnetic flux. The secondary has two step-down windings and one step-up winding. Between points 1 and 2, a voltage of 5-volts AC could be supplied. Between point 5 and point 6, 30-volts AC may be obtained, and between points 3 and 4, a voltage of 360-volts AC can be supplied to a load. This type of transformer is used for the power supply of various types of electronic equipment and instruments.

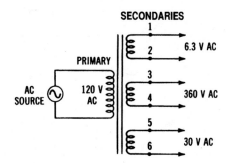

Figure 8-11. Transformer with multiple secondary windings

Autotransformers

Another specialized type of transformer is the *autotransformer*, shown in Figure 8-12. The autotransformer has only *one winding*, with a common connection between the primary and secondary. The principle of operation of the autotransformer is similar to that of other transformers. Both the *step-up* and *step-down* types are shown in Figure 8-12. Another type of control device is a *variable autotransformer*, in which the winding tap may be adjusted along the entire length of the winding to provide a variable AC voltage to a load.

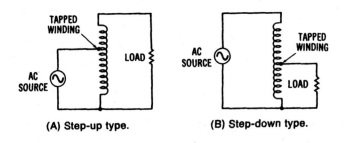

Figure 8-12. Autotransformers: (A) Step-up type, (B) Step-down type

Current Transformers

Current transformers are often used to reduce a large value of line current to a smaller value, for measurement or control purposes. These transformers are used to measure the current magnitude of high-current systems. Since most *metering* systems respond linearly to current changes, the current transformer principle can also be used to measure quantities other than current in high-power systems.

Transformer Polarity and Ratings

Power distribution transformers usually have *polarity markings*, so that their windings may be connected in parallel to increase their current capacity. The standard markings are H_1, H_2, H_3, et cetera, for the high-voltage windings, and X_1, X_2, X_3, et cetera, for the low-voltage windings. Many power transformers have two similar primary windings and two similar secondary windings to make them adaptable to different voltage requirements simply by changing from a series to a parallel connection. The voltage combinations available from this type of transformer are shown in Figure 8-13.

The *ratings* of power transformers are very important. Usually, transformers are rated in *kilovolt-amperes (kVA)*. A kilowatt rating is not used, since it would be misleading, because of the various power-factor ratings of industrial loads. Other power transformer ratings usually include *frequency*, rated *voltage* of each winding, and an *impedance* rating.

Power transformers located along a power distribution system operate at very *high temperatures*. *Cooling equipment* is necessary for large power transformers. The purpose of the cooling equipment is to conduct heat away from the transformer windings. Several power transformers are of the *liquid-immersed* type. The windings and core of the transformer are immersed in an insulating liquid, which is contained in the transformer enclosure. The liquid insulates the windings, and conducts heat away from them as well. One insulating liquid that is used extensively is called *Askarel*. Some transformers, called *dry types*, use forced air or inert gas as coolants. Some locations, particularly indoors, are considered hazardous for the use of liquid-immersed transformers. However, most transformers, rated at over 500 kVA, are liquid filled.

Transformer Malfunctions

Transformer *malfunctions* result when a circuit problem causes the insulation to break down. *Insulation breakdown* permits electrical arcs to

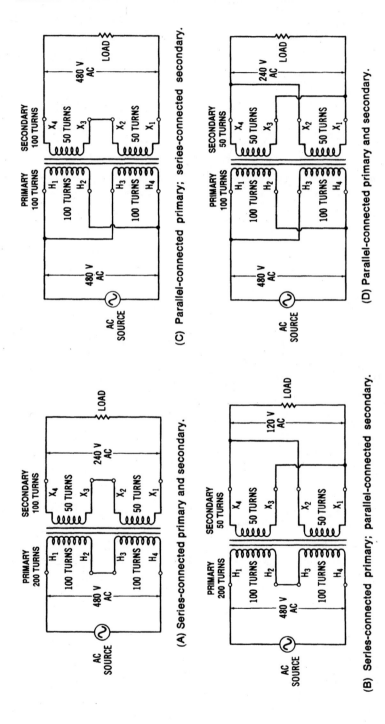

Figure 8-13. Some transformer connection methods for various voltage combinations: (A) Series-connected primary and secondary, (B) Series-connected primary, parallel-connected secondary, (C) Parallel, connected primary, series-connected secondary, (D) Parallel-connected primary and secondary

flow from one winding to an adjacent winding. These *arcs*, which may be developed throughout the transformer, cause a decomposition of the *paper* or *oil* insulation used in the transformer. This can be a particularly hazardous problem for larger power transformers, since the reaction of the electric arc and the insulating material may produce a gas. For this reason, it is very important for circuit protection to be provided for transformers. They should have power removed promptly whenever some type of fault develops. *Current-limiting fuses* may also be used to respond rapidly to any circuit malfunction.

CONDUCTORS IN POWER DISTRIBUTION SYSTEMS

The portions of the electrical distribution system that carry current are known as *conductors*. Conductors may be in the form of *solid* or *stranded* wires, cable assemblies, or large metallic bus-bar systems. A conductor may have insulation, or, in some cases, it may be bare metal.

Conductor Characteristics
Round conductors are measured by using an *American wire gage (AWG)* (see Figure 8-14). The sizes range from No. 36 (smallest) to No. 0000 (largest), with 40 sizes within this range. The *cross-sectional area* of a conductor doubles with each increase of three sizes and the diameter doubles with every six sizes. The area of conductors is measured in *circular mils (cmil)*.

Almost all conductors are made of either *copper* or *aluminum*. Both of these metals possess the necessary flexibility, current-carrying ability, and economical cost to act as efficient and practical conductors. *Copper* is a better conductor; however, *aluminum* is 30 percent lighter in weight. Therefore, aluminum conductors are used when weight is a factor in conductor selection. One specialized overhead power line conductor is the *aluminum-conductor, steel-reinforced (ACSR)* type used for long-distance power transmission. This type of conductor has stranded aluminum wires.

Conductor Types
Copper is still the most widely used conductor material, both for *solid* and for *stranded* electrical wire. The availability of a variety of thermosetting and thermoplastic insulating materials offers great flexibility in meeting the requirements for most conductor applications. The *operating temperature* ranges for various types of insulation are given in Table 8-1.

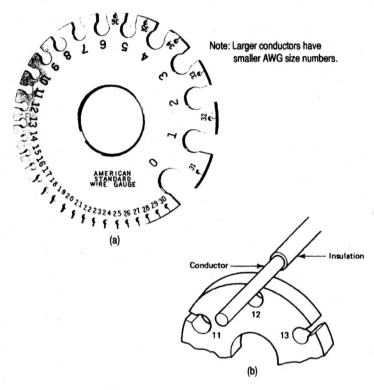

Figure 8-14. Using the AWG: (A) The American wire gage (AWG) (B) Using the AWG to measure conductor size

Table 8-1. Operating Temperature
Ranges of Various Types of Insulation

Type of Insulation	Temperature Range (ac)
Neoprene®	–30° to 90°
Teflon®	–70° to 200°
Polyethylene	–60° to 80°
Rubber	–40° to 75°
Vinyl	–20° to 80°
Polypropylene	–20° to 105°

Copper has a combination of various properties, such as malleability, strength, and high electrical and thermal conductivity. It is also capable of being alloyed or coated with other metals. *Copper* may be plated with silver to produce a conductor with better solderability and, also, a conductor that has better high-frequency characteristics. This is due to the high conductivity of the silver and the *"skin effect"* present at higher frequencies.

Where little vibration and no flexing are required of a wire or cable, single-strand conductors may be used. The advantage of a *single-strand conductor* is its lower cost compared to that of equivalent types of stranded wire. Wire and cable with solid conductors may be used as interconnection wires for electrical instruments and similar equipment. *Stranded conductors* are used to provide more flexibility. They also have a longer usage life than do solid conductors. If a solid conductor were cut by wire strippers during its installation, it would probably break after being bent a few times. However, stranded wire would not break in this situation. Wires having from 26 to 41 strands may be used where much flexibility is needed, while wires with from 65 to 105 strands may be used for special purposes.

Flat or *round braided conductors* are occasionally used for certain applications where they are better suited than solid or stranded cables. These conductors are seldom insulated, since this would hinder their flexibility.

CONDUCTOR AREA

The unit of measurement for conductors is the *circular mil (cmil)*, since most conductors are round. One mil is equal to 0.001 inch (0.0254 mm); thus, one cmil is equal to a circle whose diameter is 0.001 inch. The *cross-sectional area* of a conductor (in cmils) is equal to its diameter (D), in mils squared, or cmil = D^2. For example, if a conductor is 1/4 inch (6.35 mm) in diameter, its circular mil area can be found as follows. The decimal equivalent of 1/4 inch is 0.250 inch, which equals 250 mils. Inserting this value into the formula for the cross-sectional area of a conductor gives you:

$$\begin{aligned} \text{Area} &= D^2 \text{ (in mils)} \\ &= (250)^2 \\ &= 62{,}500 \text{ cmils} \end{aligned}$$

If the conductor is not round, its area may still be found by applying the following formula:

$$\text{Area (in cmils)} = \frac{\text{Area (in square mils)}}{0.7854}$$

This formula allows us to convert the dimensions of a conductor to *square mils*, and then to an equivalent value in cmils. For example, if a conductor is 1/2 inch × 3/4 inch (12.7 mm × 19.05 mm), the cmil area may be found using the following method. Again, convert the fractional inches into decimal values, and then into equivalent mil values. Thus, 1/2 inch equals 0.5 inch, which equals 500 mils, and 3/4 inch is 0.75 inch, which equals 750 mils. Using the formula for *square mils* and substituting, you have:

$$\text{Area} = \frac{\text{Area (in square mils)}}{0.7854}$$

$$= \frac{500 \text{ mils} \times 750 \text{ mils}}{0.7854}$$

$$= \frac{375{,}000 \text{ mils}^2}{0.7854}$$

$$= 477{,}463.7 \text{ cmils}$$

We can also use the *cmil area* of a conductor, which may be found in any standard conductor table, to find the diameter of a conductor. If a table shows that the cmil area of a conductor is equal to 16,510 cmils (a No. 8 AWG conductor), its *diameter* is found by the following method. Since

$$D^2 = \text{cmil}$$

then:

$$D = \sqrt{\text{cmil}}$$

$$= \sqrt{16{,}510 \text{ cmils}}$$

$$= 128.5 \text{ mils}$$

$$= 0.1285 \text{ inch}$$

RESISTANCE OF CONDUCTORS

The *resistance* of a conductor expresses the amount of opposition it will offer to the flow of electrical current. The unit of measurement for resistance is the ohm (Ω). The *resistivity (p)* of a conductor is the resistance for a specified cross-sectional area and length. This measurement is given in *circular mil-feet (cmil-ft)*. The resistivity of a conductor changes with the temperature, so resistivity is usually specified at a temperature of 20° Celsius. The *resistivity* for some common types of conductors is listed in Table 8-2.

Table 8-2. Resistivity of Common Conductors

Conductor	Resistivity in ohms per cmil-ft
Silver	9.8
Copper	10.4
Aluminum	17.0
Tungsten	33.0
Nickel	50.0
Iron	60.0

We can use Table 8-2 to calculate the *resistance* of any size conductor. We know that resistance increases as the length increases and decreases as the cross-sectional area increases. The following method can be used to find th0e resistance of 500 feet (152.4 meters) of aluminum conductor that is V. inch (6.35 mm) in diameter. According to Table 8-2, aluminum has a *resistivity* of 17 ohms. The *diameter* (D) equals 1/4 inch, which equals 0.250 inch, which is the equivalent of 250 mils. Using the formula and substituting, we have:

$$\text{Resistance} = \frac{\text{Resistivity} \times \text{Length (in feet)}}{\text{Diameter}^2 \text{ (in mils)}}$$

$$= \frac{17 \times 500}{(250)^2}$$

$$= \frac{8{,}500}{62{,}500}$$

$$= 0.136 \text{ ohms}$$

CONDUCTOR SIZES AND TYPES

Table 8-3 lists the sizes of *copper* and *aluminum* electrical conductors. The *American wire gage (AWG)* is the standard used to measure the diameter of conductors. The sizes range from No. 40 A WG, which is the smallest, to No. 0000 A We. Sizes larger than No. 0000 AWG are expressed in thousand circular mil (MCM) units.

Note, in Table 8-3, that as the *A WG size* number becomes smaller, the conductor becomes larger. Sizes up to No.8 A WG are solid conductors, while larger wires have from 7 to 61 strands. Table 8-3 also lists the DC *resistance* (in ohms per 1000 feet) of the copper and aluminum conductors. These values are used to determine conductor *voltage drop* in power distribution systems.

AMPACITY OF CONDUCTORS

A measure of the ability of a conductor to carry electrical current is called *ampacity*. All metal materials will conduct electrical current to some extent; however, copper and aluminum are the two most desirable types used. *Copper* is used more often than aluminum, since it is the better conductor of the two and is physically stronger. However, aluminum is usually used where weight is a factor, such as for long-distance overhead power lines. The weight of copper is almost three times that of a similar volume of *aluminum;* however, the resistance of aluminum is over 150 percent that of copper. The ampacity of an aluminum conductor is, therefore, less than that of a similar size copper conductor. A wiring design will ordinarily use aluminum conductors that are one size larger than the copper conductors necessary to carry a specific amount of current, to allow for this difference.

The *ampacity* of conductors depends upon several factors, such as the type of material, the cross-sectional area, and the type of area in which the conductors are installed. Conductors in the open, or in *"free air,"* dissipate heat much more rapidly than those that are enclosed in a metal *raceway*, or plastic cable. When several conductors are contained within the same enclosure, heat dissipation is a greater problem.

The National Electric Code (NEC)

The *National Electrical Code (NEC)* is a very important document to understand. All industrial equipment and wiring must conform to the

Table 8-3. Sizes of Copper and Aluminum Conductors

	Size (AWG or MCM)	Area (cmil)	Number of Wires	Diameter of Each Wire (in.)	DC Resistance ($\Omega/1000$ ft) 25°C	
					Copper	Aluminum
AWG sizes	18	1,620	1	0.0403	6.51	10.7
	16	2,580	1	0.0508	4.10	6.72
	14	4,110	1	0.0641	2.57	4.22
	12	6,530	1	0.0808	1.62	2.66
	10	10,380	1	0.1019	1.018	1.67
	8	16,510	1	0.1285	0.6404	1.05
	6	26,240	7	0.0612	0.410	0.674
	4	41,740	7	0.0772	0.259	0.424
	3	52,620	7	0.0867	0.205	0.336
	2	66,360	7	0.0974	0.162	0.266
	1	83,690	19	0.0664	0.129	0.211
	0	105,600	19	0.0745	0.102	0.168
	00	133,100	19	0.0837	0.0811	0.133
	000	167,800	19	0.0940	0.0642	0.105
	0000	211,600	19	0.1055	0.0509	0.0836
MCM sizes	250	250.000	37	0.0822	0.0431	0.0708
	300	300,000	37	0.0900	0.0360	0.0590
	350	350,000	37	0.0973	0.0308	0.0505
	400	400,000	37	0.1040	0.0270	0.0442
	500	500,000	37	0.1162	0.0216	0.0354
	600	600,000	61	0.0992	0.0180	0.0295
	700	700,000	61	0.1071	0.0154	0.0253
	750	750,000	61	0.1109	0.0144	0.0236
	800	800,000	61	0.1145	0.0135	0.0221
	900	900,000	61	0.1215	0.0120	0.0197
	1000	1,000,000	61	0.1280	0.0108	0.0177

NEC standards. The NEC is not difficult to use. The user should become familiar with the comprehensive *index* contained in the NEC, and with the organization of the various *sections*. For instance, if you wish to review the standards related to "system grounding," you should look in the index and locate this term. The index will refer you to the sections in the NEC that discuss "system grounding." A *table of contents* for an NEC is shown below. This listing provides an overview of the organization of the NEC. It is important for electrical technicians to learn to use the NEC.

National Electrical Code (NEC)
Table of Contents

Chapter 1—*General*
100 Definitions
110 Requirements for Electrical Installations

Chapter 2—*Wiring Design and Protection*
200 Use and Identification of Grounded Conductors
210 Branch Circuits
215 Feeders
220 Branch Circuit and Feeder Calculations
225 Outside Branch Circuits and Feeders
230 Services
240 Overcurrent Protection
250 Grounding
280 Lightning Arresters

Chapter 3—*Wiring Methods and Materials*
300 Wiring Methods
305 Temporary Wiring
310 Conductors for General Wiring
318 Cable Trays
320 Open Wiring on Insulators
324 Concealed Knob-and-Tube Wiring
326 Medium Voltage Cable
330 Mineral-Insulated Metal-Sheathed Cable
333 Armored Cable
334 Metal-Clad Cable
336 Nonmetallic-Sheathed Cable
337 Shielded Nonmetallic-Sheathed Cable
338 Service Entrance Cable
339 Underground Feeder and Branch Circuit Cable
340 Power and Control Tray Cable
342 Nonmetallic Extensions
344 Underplaster Extensions
345 Intermediate Metal Conduit
346 Rigid Metal Conduit
347 Rigid Nonmetallic Conduit
348 Electrical Metallic Tubing
349 Flexible Metallic Tubing
350 Flexible Metal Conduit
351 Liquidtight Flexible Metal Conduit
352 Surface Raceways
353 Multioutlet Assembly

Electrical Power Distribution Systems 233

354 Underfloor Raceways
356 Cellular Metal Floor Raceways
358 Cellular Concrete Floor Raceways
362 Wireways
363 Flat Cable Assemblies
364 Busways
365 Cablebus
366 Electrical Floor Assemblies
370 Outlet, Switch and Junction Boxes, and Fittings
373 Cabinets and Cutout Boxes
374 Auxiliary Gutters
380 Switches
384 Switchboards and Panelboards

Chapter 4—*Equipment for General Use*
400 Flexible Cords and Cables
402 Fixture Wires
410 Lighting Fixtures, Lampholders, Lamps, Receptacles, and Rosettes
422 Appliances
424 Fixed Electric Space Heating Equipment
426 Fixed Outdoor Electric De-Icing and Snow-Melting Equipment
427 Fixed Electric Heating Equipment for Pipelines and Vessels
430 Motors, Motor Circuits, and Controllers
440 Air-Conditioning and Refrigerating Equipment
445 Generators
450 Transformers and Transformer Vaults
460 Capacitors
470 Resistors and Reactors
480 Storage Batteries

Chapter 5—*Special Occupancies*
500 Hazardous (Classified) Locations
501 Hazardous Locations—Class I Installations
502 Hazardous Locations—Class II Installations
503 Hazardous Locations—Class III Installations
510 Hazardous (Classified) Locations-Specific
511 Commercial Garages, Repair, and Storage
513 Aircraft Hangars
514 Gasoline Dispensing and Service Stations
515 Bulk-Storage Plants
516 Finishing Processes
517 Health Care Facilities
518 Places of Assembly
520 Theaters and Similar Locations
530 Motion Picture and Television Studios and Similar Locations
540 Motion Picture Projectors
545 Manufactured Buildings
547 Agricultural Buildings
550 Mobile Homes and Mobile Home Parks
551 Recreational Vehicles and Recreational Vehicle Parks
555 Marinas and Boatyards

Chapter 6—*Special Equipment*
600 Electric Signs and Outline Lighting
610 Cranes and Hoists
620 Elevators, Dumbwaiters, Escalators, and Moving Walks
630 Electric Welders
640 Sound-Recording and Similar Equipment
645 Data Processing Systems
650 Organs
660 X-ray Equipment
665 Induction and Dielectric Heating Equipment
668 Electrolytic Cells
670 Metal-Working Machine Tools
675 Electrically Driven or Controlled Irrigation Machines
680 Swimming Pools, Fountains, and Similar Installations

Chapter 7—*Special Conditions*
700 Emergency Systems
710 Over 600 Volts, Nominal—General
720 Circuits and Equipment Operating at Less than 50 Volts
725 Class I, Class II, and Class III Remote—Control, Signaling, and Power Limited Circuits
750 Stand-By Power Generation Systems
760 Fire Protective Signaling Systems

Chapter 8—*Communications Systems*
800 Communication Circuits
810 Radio and Television Equipment
820 Community Antenna Television and Radio Distribution Systems

Chapter 9—*Tables and Examples*
Tables
Examples

Index

AMPACITY TABLES

Tables 8-4 through 8-6 are used for *conductor ampacity* calculations for electrical wiring design. These tables are simplified versions of those given in the NEC. Table 8-4 is used to determine conductor ampacity when a single conductor is mounted in *free air*. Table 8-5 is used to find the ampacity of conductors when not more than three are mounted in a *raceway* or cable. These two tables are based on *ambient temperatures* of 30° Celsius (86° Fahrenheit). Table 8-6 lists the *correction factors* that are used for temperatures over 30°C. As an example, we will find the *ampacity* of three No. 10 copper conductors with RHW insulation that are mounted in a raceway. They will be located in a foundry area where temperatures reach 50°C. The ampacity for No. 10 RHW copper wire is 30 A (from Table 8-5). The *correction factor* for an RHW-insulated conductor at 50°C ambient temperature is 0.75 (Table 8-6). Therefore,

$$30 \text{ amperes} \times 0.75 = 22.5 \text{ amperes}$$

Table 8-4. Allowable Ampacities of Single Conductance in Free Air

	Wire Size	Copper		Aluminum	
		With R, T, TW Insulation	With RH, RHW TH, THW Insulation	With R, T, TW Insulation	With RH, RHW, TH THW Insulation
AWG	14	20	20		
	12	25	25	20	20
	10	40	40	30	30
	8	55	65	45	55
	6	80	95	60	75
	4	105	125	80	100
	3	120	145	95	115
	2	140	170	110	135
	1	165	195	130	155
	0	195	230	150	180
	00	225	265	175	210
	000	260	310	200	240
	0000	300	360	230	280
MCM	250	340	405	265	315
	300	375	445	290	350
	350	420	505	330	395
	400	455	545	355	425
	500	515	620	405	485
	600	575	690	455	545
	700	630	755	500	595
	750	655	785	515	620
	800	680	815	535	645
	900	730	870	580	700
	1000	780	935	625	750

USE OF INSULATION IN POWER DISTRIBUTION SYSTEMS

Synthetic *insulation* for wire and cable is classified into two broad categories—*thermosetting* and *thermoplastic*. The mixtures of materials within each of these categories are so varied as to make the available number of

Table 8-5. Ampacities of Conductors in a Raceway of Cable (3 or less)

		Copper		Aluminum	
	Wire Size	With R, T, TW Insulation	With RH, RHW TH, THW Insulation	With R, T, TW Insulation	With RH, RHW, TH THW Insulation
AWG	14	15	15		
	12	20	20	15	15
	10	30	30	25	25
	8	40	45	30	40
	6	55	65	40	50
	4	70	85	55	65
	3	80	100	65	75
	2	95	115	75	90
	1	110	130	85	100
	0	125	150	100	120
	00	145	175	115	135
	000	165	200	130	155
	0000	195	230	155	180
MCM	250	215	255	170	205
	300	240	285	190	230
	350	260	310	210	250
	400	280	335	225	270
	500	320	380	260	310
	600	355	420	285	340
	700	385	460	310	375
	750	400	475	320	385
	800	410	490	330	395
	900	435	520	355	425
	1000	455	545	375	445

Table 8-6. Correction Factors for Temperatures about 300°C

Ambient Temperature		Conductor Correction Factor	
C°	F°	TW	R, T, RH, RHW, TH, THW
40	104	0.82	0.88
45	113	0.71	0.82
50	122	0.58	0.75
55	131	0.41	0.67
60	140	—	0.58
70	158	—	0.35

insulations almost unlimited. Most insulation is composed of compounds made of synthetic rubber polymers (thermosetting) and from synthetic materials (thermoplastics). These synthetic materials are combined to provide specific physical and electrical properties.

Thermosetting materials are characterized by their ability to be stretched, compressed, or deformed within reasonable limits under mechanical strain, and then to return to their original shape when the stress is removed.

Thermoplastic insulation materials are best known for their excellent electrical characteristics and relatively low cost. These materials are popular, since they allow much thinner insulation thicknesses to be used to obtain good electrical properties, particularly at the higher voltages.

There are many types of *insulation* used today for electrical conductors. Some new materials have been developed that will last for exceptionally long periods of time and will withstand very high operating temperatures. The operating conditions where the conductors are used mainly determine the type of insulation required. For instance, system voltage, heat, and moisture affect the type of insulation required. Insulation must be used that will withstand both the heat of the surrounding atmosphere and the heat developed by the current flowing through the conductor. Exceptionally large currents will cause excessive heat to be developed in a conductor. Such heat could cause insulation to melt or burn. This is why overcurrent protection is required as a safety factor to prevent fires. The *ampacity* or current-carrying capacity of a conductor depends upon the type of insulation used. The NEC has developed a system of *abbreviations* for identifying various types of insulation. Some of the abbreviations are shown in Table 8-7.

Table 8-7. Common Abbreviations for Types of Electrical Insulation

Abbreviation	Type of Insulation
R	Rubber—140° F
RH	Heat-Resistant Rubber—167° F
RHH	Heat-Resistant Rubber—194° F
RHW	Moisture and Heat-Resistant Rubber—167° F
T	Thermoplastic-140° F
THW	Moisture and Heat-Resistant Thermoplastic—167° F
THWN	Moisture and Heat-Resistant Thermoplastic With Nylon—194° F

Chapter 9

Power Distribution Equipment

In order to distribute electrical power, it is necessary to use many types of specialized *equipment*. The electrical power system consists of such specialized equipment as power *transformers*, high-voltage *fuses* and *circuit breakers*, *lightning arresters*, power-factor-correcting *capacitors*, and *power-metering* systems. Some types of specialized power distribution equipment will be discussed in this chapter.

IMPORTANT TERMS

In Chapter 9, power distribution equipment is discussed. After studying this chapter, you should have an understanding of the following terms:

Substation
High-Voltage Transmission Line
High-Voltage Fuse
High-Voltage Circuit Breaker
High-Voltage Disconnect Switch
Lightning Arrester
High-Voltage Insulator
High-Voltage Conductor
ACSR Conductor
Voltage Regulator
Plug Fuse
Cartridge Fuse
Time Delay Fuse
Renewable-Element Fuse
Low-Voltage Circuit Breaker

Protective Relay
Fault Current
Bimetallic Overload Relay
Melting Alloy Overload Relay
Undervoltage Protection
Raceway
Feeder Circuit
Branch Circuit
Underground Distribution
Safety Switch
Panelboard
Low-Voltage Switchgear
Load Center
Service Entrances
Grounding Electrode
Ground
Service Entrance Conductors
Uninterruptible Power Supply
Power Line Filter
Power Conditioner
Floor-Mounted Raceway
Conduit Connectors
Wire Connectors
Plastic Conduit and Enclosures
Power Outlet Design
International Power Source

EQUIPMENT USED AT SUBSTATIONS

Substations are very important parts of the electrical power systems. The link between the high-voltage transmission lines and the low-voltage power distribution systems is the substation. The function of a *distribution substation,* such as the one shown in Figure 8-2, is to receive electrical power from a high-voltage system and convert it to voltage levels suitable for industrial, commercial, or residential use. The major functional component of a substation is the transformer, whose basic characteristics were discussed previously. However, there are many other types of specialized equipment required for the operation of a substation.

High-voltage Fuses

Since power lines are frequently short circuited, various protective equipment is used to prevent damage to both the power lines and the equipment. This *protective equipment* must be designed to handle high voltages and currents. Either *fuses* or *circuit breakers* may be used to protect high-voltage power lines. High-voltage fuses (those used for over 600 volts) are made in several ways. An *expulsion-type fuse* has an element that will melt and vaporize when it is overloaded, causing the power line it is connected in series with to open. *Liquid fuses* have a liquid-filled metal enclosure that contains the fuse element. The liquid acts as an arc-suppressing medium. When the fuse element melts from an excessive current in a power line, the element is immersed in the liquid to extinguish the arc. This type of fuse reduces the problem of high-voltage arcing. A *solid-material fuse* is similar to a liquid fuse, except that the arc is extinguished in a chamber filled with solid material.

Ordinarily, *high-voltage fuses* at substations are mounted adjacent to air-break disconnect switches. These switches provide a means of switching power lines and disconnecting them for repair. The fuse and switch enclosure is usually mounted near the overhead power lines at a substation.

High-voltage Circuit Breakers

Circuit breakers that control high voltages are also located at electrical substations. In this type of circuit breaker, the contacts are immersed in an insulating oil contained in a metal enclosure. Another type of high-voltage circuit breaker is the *magnetic air breaker* in which the contacts separate, in the air, when the power line is overloaded. Magnetic blowout coils are used to develop a magnetic field that causes the arc (which is produced when the contacts break) to be concentrated into arc chutes where it is extinguished. A modification of this type of circuit breaker is the compressed-air circuit breaker. In this case, a stream of compressed air is concentrated on the contacts when the power line is opened. The compressed air aids in extinguishing the arc that is developed when the contacts open. It should be pointed out that large arcs are present whenever a high-voltage circuit is interrupted. This problem is not encountered to any great extent in low-voltage protective equipment. There are two major types of *high-voltage circuit breakers—oil filled* and *oilless*.

These circuit breakers are designed to operate on voltages of 1000 volts to over 500,000 volts. *Oil-filled* circuit breakers are used primarily for outdoor substations, except for very high voltages in the range of 500,000 volts and higher. *Oilless* circuit breakers are ordinarily used for indoor operation.

High-voltage Disconnect Switches

High-voltage disconnect switches are used to disconnect electrical equipment from the power lines that supply the equipment. Ordinarily, disconnect switches are not operated when current is flowing through them. A high-voltage arcing problem would occur if disconnect switches were opened while current was flowing through them. They are opened mainly to isolate equipment from power lines for safety purposes. Most disconnect switches are the *"air-break"* type, which is similar in construction to knife switches. These switches are available for indoor or outdoor use in both manual and motor-operated designs.

Lightning Arresters

The purpose of using *lightning arresters* on power lines is to cause the conduction to ground of excessively high voltages that are caused by lightning strikes or other system problems. Without lightning arresters, power lines and associated equipment could become inoperable when struck by lightning. Arresters are designed to operate rapidly and repeatedly, if necessary. Their response time must be more rapid than that of the other protective equipment used on power lines.

Lightning arresters must have a rigid connection to ground on one side. The other side of the arrester is connected to a power line. Sometimes, they are connected to transformers or the insides of switchgear. Lightning is a major cause of power-system failures and equipment damage, so lightning arresters have a very important function. *Lightning arresters* are also used at outdoor substations. The lightning arrester is used to provide a path to ground for lightning strikes or hits. This path eliminates the *flashover* between power lines, which causes short circuits. *Valve-type* lightning arresters are used frequently. They are two-terminal devices in which one terminal is connected to the power line, and the other is connected to ground. The path from line to ground is of such high resistance that it is normally open. However, when lightning, which is a very high voltage, strikes a power line, it causes conduction from line to ground. Thus, voltage surges are conducted to ground before flashover between the lines occurs. After the lightning surge has been conducted to ground, the valve assembly then causes the lightning arrester to become nonconductive once more.

High-voltage Insulators

All power transmission lines must be isolated so as not to become safety hazards. Large strings of *insulators* are used at substations, and at other points along the power distribution system, to isolate the current-carrying conductors from their steel supports or any other ground-mounted equipment. Insulators may be made of *porcelain, rubber,* or a *thermoplastic* material.

Power transmission lines require many *insulators* in order to electrically isolate the power lines from the steel towers and wooden poles that support the lines. Insulators must have enough mechanical strength to support power lines under all weather conditions. They must also have sufficient insulating properties to prevent any arcing between the power lines and their support structures. High-voltage insulators are usually made of *porcelain*. Insulators are constructed in *"strings,"* which are suspended from steel or wooden towers. The design of these insulators is very important, since design affects their capacitance and their ability to withstand weather conditions.

High-voltage Conductors

The *conductors* used for power distribution are, ordinarily, uninsulated *aluminum* wires or *aluminum-conductor steel-reinforced (ACSR)* wires for long-distance transmission, and insulated copper wires for shorter distances.

Voltage Regulators

Voltage regulators are an important part of the power distribution system. They are used to maintain the voltage levels at the proper value, as a constant voltage must be maintained in order for the electrical equipment to function properly. For instance, motors do not operate properly when a reduced or an excessive voltage is applied to them. Transformer *tap-changers*, illustrated in Figure 9-1 may be used as voltage regulators. The secondary tap can be changed, either manually or automatically, to change the voltage output, in order to compensate for changes in the load voltage. As load current increases, line loss (I × R) also increases. Increased line loss causes the secondary voltage (Vs) to decrease. If the secondary tap is initially connected to tap No.4, the secondary voltage can be boosted by reconnecting to either tap No. 3, No. 2, or No. 1. This can be done automatically with a motor-controlled tap changer. There are various other types of automatic voltage regulators that can be used with electrical power distribution systems.

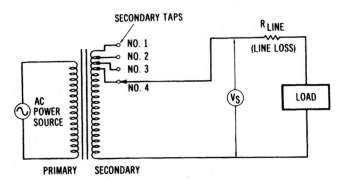

Figure 9-1. Transformer tap-changer voltage regulator

POWER SYSTEM PROTECTIVE EQUIPMENT

There are many devices that are used to *protect* electrical power systems from damage due to abnormal conditions. For instance, *switches, fuses, circuit breakers, lightning arresters,* and *protective relays* are all used for this purpose. Some of these devices automatically disconnect the equipment from power lines before any dam age can occur. Other devices sense variations from the normal operation of the system and make the changes necessary to compensate for abnormal circuit conditions. The most common electrical problem that requires protection is *short circuits*. Other problems include *overvoltage, undervoltage,* and *changes ill frequency*. Generally, more than one method of protection is used to protect electrical circuits from faulty conditions. The purpose of any type of protective device is to cause a current-carrying conductor to become inoperative when an excessive amount of current flows through it.

Types of Fuses

The simplest type of protective device is a *fuse*. Fuses are low-cost items and have a fast operating speed. However, in three-phase systems, since each hot line must be fused, two lines are still operative if only one fuse burns out. Three-phase motors will continue to run with one phase removed. This condition is undesirable, in most instances, since motor torque is greatly reduced, and overheating may result. Another obvious disadvantage of fuses is that replacements are required. All protective devices, including fuses, have an *operating-characteristic time curve,* such as the one shown in Figure 9-2, prevent any possible damage to equipment,

Power Distribution Equipment   245

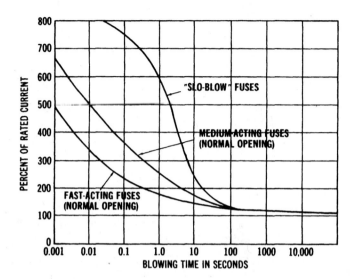

Figure 9-2. Typical operating-characteristic time curves for three different types of fuses *(Courtesy Littelfuse, Inc.)*

circuit protection should be planned utilizing these curves. They show the response time required for a protective device to interrupt a circuit, when an overload occurs.

Plug Fuses—Fuses are used in safety switches and power distribution panels. The *plug fuse* is a common type of fuse. Standard sizes for this fuse are 10, 15, 20, 25, and 30 amperes at voltages of 125 volts or below. These fuses have a zinc or metallic-alloy-fusible element enclosed in a case made of an insulating material. Their most common use is in safety switches and fuse panelboards.

Cartridge Fuses—*Cartridge fuses* are commonly used in power distribution systems for voltages up to 600 volts. They have a zinc- or alloy-fusible element, which is housed in a round fiber enclosure. One type has a *nonrenewable element,* while another type has a *renewable element.* Cartridge fuses may be used to protect high-current circuits, since they come in sizes of 60, 100, 200, 400, 600, and 1000 amperes.

Time Delay Fuses—A modification of the plug or cartridge fuse is called a *time delay fuse.* This type of fuse is used to delay the circuit-interrupting action. It is useful where momentary high currents exist periodically, such as motor-starting currents. The fuse element melts only when an excessive current is sustained over the time-lag period; thus, sufficient circuit protection is still provided.

Time delay fuse are used to limit current on systems including electric motors, which draw higher currents during their start cycle than during normal operation. These devices allow the system to start up at a higher than normal current; they then protect the system during normal operation without disrupting the distribution system.

Fuse Metals—The type of metal used in fuses is ordinarily an *alloy* material or, possibly, *aluminum*. All metals have resistance, so when current flows through metal, heat energy is produced. As the current increases, more heat is produced, causing the temperature of the metal to increase. When the *melting point* of the fuse metal is reached, the fuse will open, causing the circuit to which it is connected to open. Metals that decompose rapidly are used, rather than ones that produce small metallic globules when they melt. This reduces the likelihood of any *arc-over* occurring after the fuse metal has melted. The current rating of fuses depends upon the melting temperature of the fuse metal, as well as its shape, size, and the type of enclosure used.

Low-voltage Circuit Breakers

Circuit breakers are somewhat more sophisticated overload devices than are fuses. Although their function is the same as that of fuses, circuit breakers are much more versatile. In three-phase systems, circuit breakers can open all three hot lines when an *overload* occurs. They may also be activated by remote-control relays. *Relay systems* may cause circuit breakers to open in response to changes in frequency, voltage, current, or other circuit variables. Circuit breakers are used in industrial plants, and are usually of the low-voltage variety (less than 600 volts). They are not nearly as complex as their high-voltage counterparts (which were discussed previously). Most low-voltage circuit breakers are housed in molded-plastic cases that mount in metal power distribution panels. *Circuit breakers* are designed so that they will automatically open when a current occurs that exceeds the rating of the breaker. Ordinarily, the circuit breakers must be reset manually. Most circuit breakers employ either a *thermal tripping* element or a *magnetic trip* element. Ratings of circuit breakers extend into current ranges that are as high as 800 to 2000 amperes.

Protective Relays

Protective relays provide an accurate and sensitive method of protecting electrical equipment from short circuits and other abnormal conditions. Overcurrent relays are used to cause the rapid opening of electrical power lines when the current exceeds a predetermined value. The *response time* of the relays is very important in protecting the equipment from damage. Some common types of faults that relays protect against are *line-to-ground short circuits, line-to-line short circuits, double line-to-ground short circuits, and three-phase line short circuits.* Each of these conditions is caused by faulty circuit conditions that draw abnormally high current from the power lines.

Motor Fault Current Protection

Motor fault currents are excessive currents that occur in motors as the result of some unnatural malfunction. Since motor fault currents cannot be withstood for any duration of time, some type of protection must be provided to disconnect the motor from the power distribution system when a fault condition occurs. Such protection may be provided by *motor starters, circuit breakers,* or *fuses.* The type of protection used is dependent upon several characteristics of the power distribution system and the motor.

Motor Protective Devices

The distribution of electrical power to *motors* is a very important function of industrial and commercial power distribution. Distribution of energy to industrial electric motors is particularly important. Basic *functions* that motors are expected to perform are *starting, stopping, reversing,* and *speed variation.* These functions may be manually or automatically controlled. Various types of *protective devices* are used to provide for the efficient distribution of power to electric motors.

Overload protection is the most important motor protection function. Such protection should serve the motor, its branch circuit, and the associated control equipment. The major cause of motor overload is an excessive mechanical load on the motor, which causes it to draw too much current from the power source. A block diagram of a motor-protection system is shown in Figure 9-3.

Thermal overload relays are often used as protective devices. Thermal relays may be reset either manually or automatically. One type of thermal overload relay uses a *bimetallic heater* element. The bimetallic element bends as it is heated by the current flowing through it. When the current reaches the rating of the element, the relay opens the branch circuit. Another type of element is the *melting alloy* type. This device has contacts

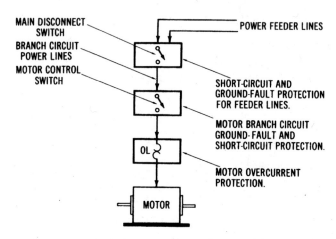

Figure 9-3. Block diagram of a motor-protection circuit

held closed by a ratchet wheel. At the current capacity of the device, the fusible alloy melts, causing the ratchet wheel to turn. A spring then causes the device to open the circuit.

Overheating Protection for Motors

Motors must be protected from excessive *overheating*. This protection is provided by *magnetic* or *thermal* protective devices, which are ordinarily within the motor-starter enclosure. Protective relays or circuit breakers can also perform this function. When an operational problem causes the motor to overheat, the protective device is automatically used to disconnect the motor from its power supply.

Undervoltage Protection

Motors do not operate efficiently when less than their rated voltage is applied, and some types of motors can be destroyed if they are operated continuously at reduced voltages. Magnetic contactors (see Chapter 15) may be used effectively to protect against *undervoltages*. A specific level of voltage is required to cause magnetic contactors to operate. If the voltage is reduced below a specified level, the magnetic contactor will open, thus disconnecting the circuit between the power source and the motor, and stopping the motor before any damage can be done.

POWER DISTRIBUTION INSIDE INDUSTRIAL AND COMMERCIAL BUILDINGS

Electrical power is delivered to the location where it is to be used, and then distributed *within a building* by the power distribution system. Various types of circuit breakers and switchgear are employed for power distribution. Another factor involved in power distribution is the distribution of electrical energy to the many types of loads that are connected to the system. This part of the distribution system is concerned with the *conductors, feeder* systems, *branch circuits, grounding* methods, and *protective* and *control* equipment that is used.

Raceways

Most electrical distribution to industrial and commercial loads is through wires and cables contained in *raceways*. These raceways carry the conductors, which carry the power to the various equipment throughout a building. Copper *conductors* are ordinarily used for indoor power distribution. The physical size of each conductor is dependent upon the current rating of the branch circuit. Raceways may be large metal *ducts* or rigid metal *conduits*. These raceways provide a compact and efficient method of routing cables, wires, et cetera, throughout an industrial complex. A *cable tray raceway* design for industrial applications is shown in Figure 9-4.

Feeder Lines and Branch Circuits

The conductors that carry current to the electrical load devices in a building are called *feeders* and *branch circuits*. Feeder lines supply power to branches, which are connected to them. *Primary feeder lines* may be either overhead or underground. Usually, overhead lines are preferred because they permit flexibility for future expansion. *Underground* systems cost more, but they help to maintain a more attractive environment. Secondary feeders are connected to the primary feeder lines, to supply power to individual sections within the building. Either aluminum or copper feeder lines may be used, depending on the specific power requirements. The distribution is from the feeder lines, through individual protective equipment, to *branch circuits*, which supply the various loads. Each branch circuit has various protective devices according to the needs of that particular branch. The overall feeder-branch system may be a very complex network of switching equipment, transformers, conductors, and protective equipment.

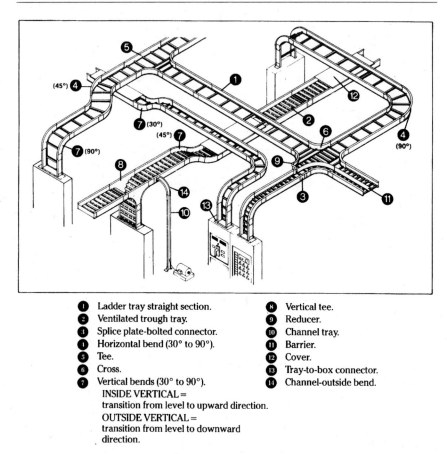

- ① Ladder tray straight section.
- ② Ventilated trough tray.
- ③ Splice plate-bolted connector.
- ④ Horizontal bend (30° to 90°).
- ⑤ Tee.
- ⑥ Cross.
- ⑦ Vertical bends (30° to 90°).
  INSIDE VERTICAL =
  transition from level to upward direction.
  OUTSIDE VERTICAL =
  transition from level to downward direction.
- ⑧ Vertical tee.
- ⑨ Reducer.
- ⑩ Channel tray.
- ⑪ Barrier.
- ⑫ Cover.
- ⑬ Tray-to-box connector.
- ⑭ Channel-outside bend.

Figure 9-4. Industrial cable tray raceway design (*Courtesy Chalfant Manufacturing Co.*)

Switching Equipment

In addition to circuit protection, power distribution systems must have equipment that can be used to connect or disconnect the entire system or parts of the system. Various types of *switching devices* are used to perform this function. A simple type of switch is the *safety switch*. This type of switch is mounted in a metal enclosure and operated by means of an external handle. Safety switches are used only to turn a circuit off or on; however, fuses are often mounted in the same enclosure with the safety switch.

Distribution Panelboards

Another type of switch is the kind used in conjunction with a circuit breaker *panelboard*. Panelboards are metal cabinets that enclose the main disconnect switch and the branch circuit protective equipment. Distribution panelboards are usually located between the power feed lines within a building and the branch circuits that are connected to it.

Low-voltage Switchgear

Metal-enclosed *low-voltage switchgear* is used in many industrial and commercial buildings as a distribution control center to house the circuit breakers, bus bars, and terminal connections that are part of the power distribution system. Ordinarily, a combination of switchgear and distribution transformers is placed in adjacent metal *enclosures*. This combination is referred to as a *load-center unit substation,* since it is the central control for several loads. The rating of these load centers is usually 15,000 volts or less for the high-voltage section, and 600 volts or less for the low-voltage section. Load centers provide flexibility in the electrical power distribution design of industrial plants and commercial buildings.

Metal-enclosed switchgear, or metal-clad switchgear, is a type of equipment that houses all the necessary control devices for the electrical circuits that are connected to them. The control devices contained inside the switchgear include circuit breakers, disconnect switches, interconnecting cables and buses, transformers, and the necessary measuring instruments. Switchgear is used for indoor and outdoor applications at industrial plants, commercial buildings, and substations. The voltage ratings of switchgear are usually from 13.8 to 138 kV, with 1 to 10 MV A power ratings.

THE ELECTRICAL SERVICE ENTRANCE

Electrical power is brought from the overhead power lines, or from the underground cable, into a building by what is called a *service entrance.* A good working knowledge of the *National Electrical Code* (NEC) specifications and definitions is necessary for an understanding of service entrance equipment. The NEC sets the minimum standards that are necessary for *wiring design* inside a building.

The type of *equipment* used for an electrical service entrance of a building may include high-current *conductors* and *insulators,* disconnect *switches, protective* equipment for each load circuit that will be connected to the main power system, and the meters needed to measure power, voltage, current, and/or frequency. It is also necessary to ground the power system at the service entrance location. This is done by a *grounding electrode*, which is a metal rod driven deep into the ground. The grounding conductor is attached securely to this grounding electrode. Then, the grounding conductor is used to make contact with all neutral conductors and safety grounds of the system.

SERVICE ENTRANCE TERMINOLOGY

There are several terms associated with service entrance equipment. The *service entrance conductors* are a set of conductors brought to a building by the local electrical utility company. These conductors must be capable of carrying all of the electrical current that is to be delivered to the various loads inside the building that are to be supplied with power by the power system. Conductors that extend from the service entrance to a power distribution panel or other type of overcurrent protective equipment, are called *feeders*. Feeders are power lines that supply branch circuits. A *branch circuit* is defined as conductors that extend beyond the last overcurrent protective equipment of the power system. Usually, each branch circuit delivers electrical power to a small percentage of the total load of the main power system.

In commercial and industrial installations, *switchboards* and *panelboards* are used to supply power to various loads throughout the power system. A *switchboard* is a large enclosure that has several overcurrent protective devices (fuses or circuit breakers). Each feeder is connected to the proper type of overcurrent device. Often, switchboards contain metering equipment for the power system. *Panelboards* are smaller than switchboards, but are used for a similar purpose. They are enclosures for overcurrent devices for either branch circuits or feeder circuits. A common example of a panelboard is the main power-distribution panel that houses the circuit breakers used for the branch circuits of a home. For more specific definitions of terms, you should refer to the most recent edition of the NEC.

Power Distribution System Components

Several specialized types of power distribution system components are available today and should be reviewed.

Power Distribution Equipment 253

Uninterruptible Power Supply—An uninterruptible power supply (UPS) has computer-controlled diagnostics and monitoring to provide constant on-line power for today's modern equipment. The constant power capability is particularly useful for computer systems used in business and industry.

Power Filters and Conditioners—Power *line filters* ordinarily plug into interior power distribution systems. These filters have power outlets for obtaining filtered AC power. *Power conditioners* are also used to protect power distribution systems from spikes, surges, or other interference that may be damaging to certain types of equipment.

Floor-mounted Raceways—*Floor-mounted raceways*, surface raceways and power outlets, are used in most commercial and industrial facilities.

Conduit Connectors—Several types of *conduit bodies* are used today. These bodies are used to provide a means of connecting conductors, and to allow angular bends in conduit runs throughout a building.

Wire Connectors—Simple but essential components of electrical power distribution systems are *wire end connectors* (sometimes called "wire nuts").

Plastic Components—Flexible *plastic conduit* provides an alternative to *electrical metallic tubing (EMT)* and rigid conduit, in certain distribution systems. Plastic boxes are compatible with flexible conduit and *plastic enclosures* for power distribution systems.

Power Outlets—*Power outlets* have standard configurations that have been established by the National Electrical Manufacturers Association (NEMA). The specific configuration indicates the voltage, current, and phase ratings of the distribution system. *NEMA designs* are shown in Figure 9-5.

International Power Sources—an *international power source* provides a convenient means of converting North American voltage and frequency (120 volt/60 hertz) to international voltages and frequencies. Output power is obtained through the appropriate standard international socket. The system shown has adjustable voltages and frequencies for obtaining power to match that of most countries, and for the use of products purchased in other countries, without modification of power supplies.

254   *Electrical Power Systems Technology*

Note that the following NEMA configurations are not to scale. Canadian, United Kingdom, European, and other required configurations are also available.

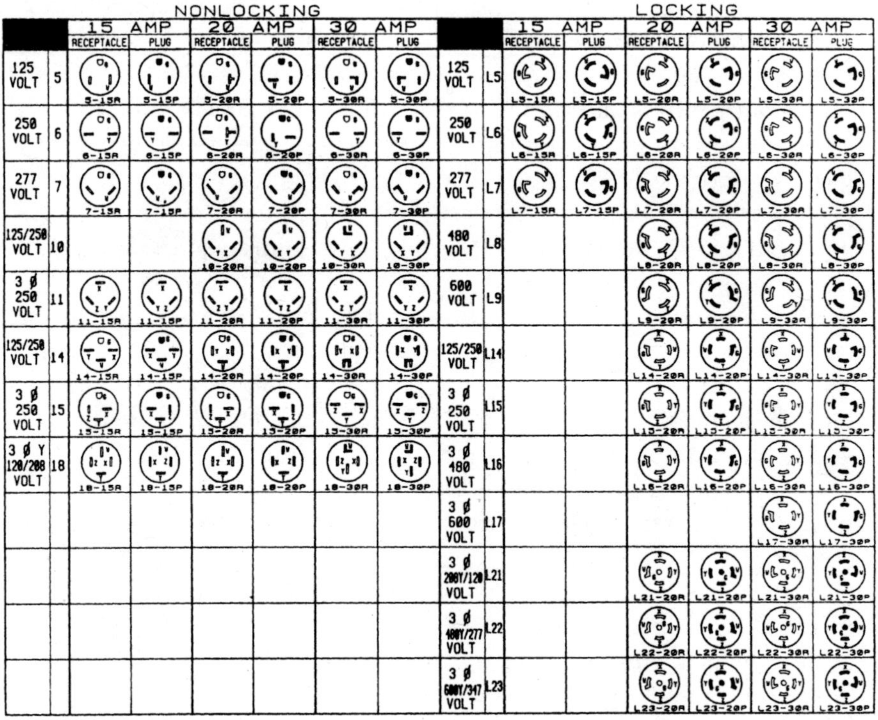

Figure 9-5. Standard NEMA designs for power outlets (*Courtesy Pulizzi Engineering, Inc.*)

Chapter 10

Single-phase and Three-phase Distribution Systems

When electrical power is distributed to its point of utilization, it is normally either in the form of *single-phase* or *three-phase* alternating current (AC) voltage. Single-phase AC voltage is distributed into residences and smaller commercial buildings. Normally, three-phase AC voltage is distributed to industries and larger commercial buildings. Thus, the main types of power distribution systems are *residential* (single-phase) and *industrial or commercial* (three-phase).

An important aspect of both single-phase and three-phase distribution systems is *grounding*. Two grounding methods, *system grounding* and *equipment grounding*, will be discussed in this chapter, along with *ground-fault* protective equipment.

IMPORTANT TERMS

In Chapter 10, single-phase and three-phase power distribution systems are discussed. After studying this chapter, you should have an understanding of the following terms:

Residential Distribution
Commercial Distribution
Industrial Distribution
Single-Phase, Two-Wire Distribution System
Single-Phase, Three-Wire Distribution System
Hot Line
Neutral
System Ground
Equipment Ground

Insulation Color Identification
Three-phase Delta-Delta Transformer Connection
Three-phase Delta-Wye Transformer Connection
Three-phase Wye-Wye Transformer Connection
Three-phase Wye-Delta Transformer Connection
Three-phase Open Delta Transformer Connection
Three-phase, Three-Wire Distribution System
Three-phase, Three-Wire, with Neutral Distribution System
Three-phase, Four-Wire Distribution System
"Wild" Phase
Grounding Electrode
Ground Fault Interrupter (GFI)
Hand-to-Hand Body Resistance
National Electrical Code (NEC)
Electrical Inspection
Voltage Drop of a Branch Circuit
Branch Circuit
Grounding Conductor
Nonmetallic-Sheathed Cable (NMC)
Metal-Clad Cable
Rigid Conduit
Electrical Metallic Tubing (EMT)

SINGLE-PHASE SYSTEMS

Most electrical power, when produced at the power plants, is produced as *three-phase AC* voltage. Electrical power is also transmitted in the form of three-phase voltage over long-distance power-transmission lines. At its destination, three-phase voltage can be changed into three separate *single-phase* voltages for distribution into the residential areas.

Although *single-phase systems* are used mainly for *residential* power distribution systems, there are some industrial and commercial applications of single-phase systems. Single-phase power distribution usually originates from three-phase power lines, so electrical power systems are capable of supplying both three-phase and single-phase loads from the same power lines. Figure 10-1 shows a typical *power distribution system-from* the power station *(source)* to the various single-phase and three-phase *loads* that are connected to the system.

Single-phase and Three-phase Distribution Systems

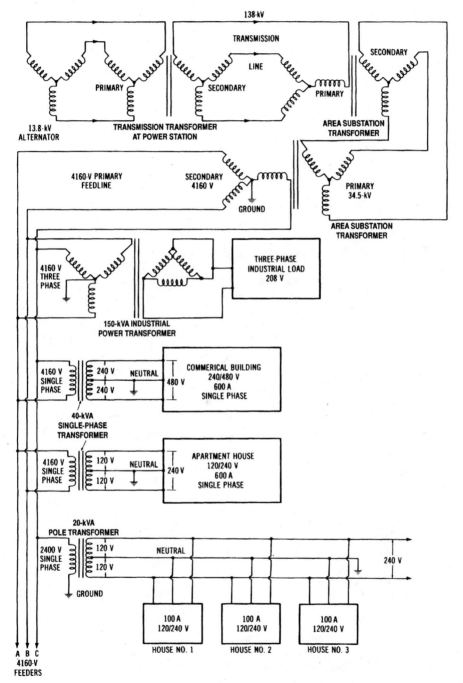

Figure 10-1. A typical power distribution system

Single-phase systems can be of two major types—single-phase two-wire systems or single-phase three-wire systems. A single-phase two-wire system is shown in Figure 10-2A (the top diagram). This system uses a 10-kV A transformer whose secondary produces one single-phase voltage, such as 120 or 240 volts. This system has one hot line and one neutral line. In residential distribution systems, this was the type most used to provide 120-volt service several years ago. However, as appliance power requirements increased, the need for a dual-voltage system was evident.

To meet the demand for more residential power, the single-phase three-wire system is now used. A home service entrance can be supplied with 120/240-volt energy by the methods shown in Figures 10-2B and 10-2C (center and bottom diagrams). Each at these systems is derived from a three-phase power line. The single-phase three-wire system has two hot

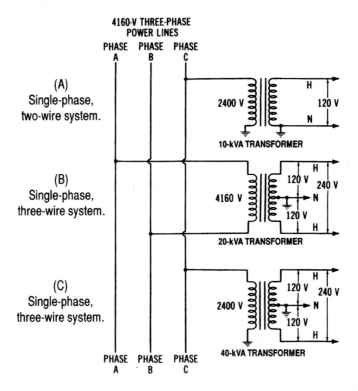

Figure 10-2. Single-phase power distribution systems: (A) Single-phase, two-wire system, (B) Single-phase, three-wire system (taken from two hot lines), (C) Single-phase, three-wire system (taken from one hot line and one grounded neutral)

lines and a neutral line. The hot lines, whose insulation is usually black and red, are connected to the outer terminals of the transformer secondary windings. The neutral line (white insulated wire) is connected to the center tap of the distribution transformer. Thus, from neutral to either hot line, 120 volts for lighting and low-power requirements may be obtained. Across the hot lines, 240 volts is supplied for higher-power requirements. Therefore, the current requirement for large power-consuming equipment is cut in half, since 240 volts rather than 120 volts are used. Either the single-phase two-wire, or the single-phase three-wire system, can be used to supply single-phase power for industrial or commercial use. However, these single-phase systems are mainly for residential power distribution.

THREE-PHASE SYSTEMS

Since industries and commercial buildings use three-phase power predominantly, they rely upon three-phase distribution systems to supply this power. Large three-phase distribution transformers are usually located at substations adjoining the industrial plants or commercial buildings. Their purpose is to supply the proper AC voltages to meet the necessary load requirements. The AC voltages that are transmitted to the distribution substations are high voltages, which must be stepped down by three-phase transformers.

Three-phase Transformer Connections

There are five ways in which the primary and secondary windings of three-phase transformers may be connected. These are the delta-delta, the delta-wye, the wye-wye, the wye-delta, and the open-delta connections. These basic methods are illustrated in Figure 10-3. The delta-delta connection (Figure 10-3A) is used for some lower-voltage applications. The delta-wye method (Figure 10-3B) is commonly used for stepping up voltages, since the voltage characteristic of the wye-connected secondary results in an inherent step-up factor of 1.73 times. The wye-wye connection of Figure 10-3C is ordinarily not used, while the wye-delta method (Figure 10-3D) may be used advantageously to step voltages down. The open-delta connection (Figure 10-3E) is used if one transformer winding becomes damaged, or is taken out of service. The transformer will still deliver three-phase power, but at a lower current and power capacity. This connection may also be desirable when the full capacity of three trans-

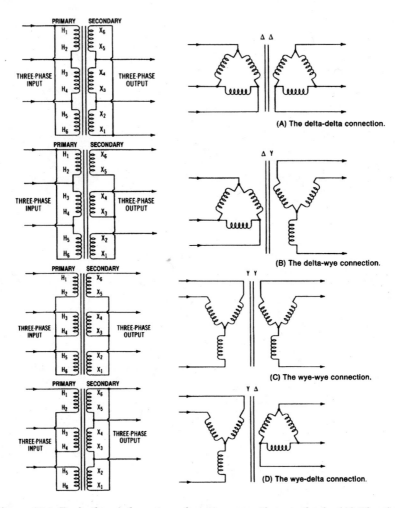

Figure 10-3. Basic three-phase transformer connection methods: (A) The delta-delta connection, (B) The delta-wye connection, (C) The wye-wye connection, (D) The wye-delta connection, and (E) The open delta connection

formers is not needed until a later time. Two identical single-phase transformers can be used to supply power to the load until, at a later time, the third transformer is needed to meet increased load requirements.

Types of Three-phase Systems

Three-phase power distribution systems, which supply industrial and commercial buildings, are classified according to the number of phases and

Single-phase and Three-phase Distribution Systems

number of wires required. These systems, shown in Figure 10-4, are the three-phase three-wire system, the three-phase three-wire system with neutral, and the three-phase four-wire system. The primary winding connection is not considered here. The three-phase three-wire system, shown in Figure 10-4A, can be used to supply motor loads of 240 volts or 480 volts. Its major disadvantage is that it only supplies one voltage, as only three hot lines are supplied to the load. The usual insulation color code for these three hot lines is black, red, or blue, as specified in the NEC.

The disadvantage of the three-phase three-wire system may be partially overcome by add-

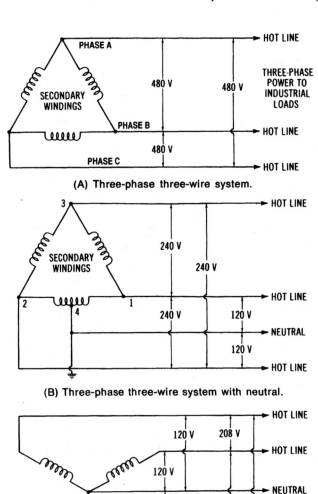

Figure 10-4. Industrial power distribution systems: (A) Three-phase, three-wire system, (B) Three-phase, three-wire system with neutral, (C) Three-phase, four-wire system

ing one center-tapped winding, as shown in the three-phase three-wire with neutral system of Figure 10-4B. This system can be used as a supply for 120/240 volts or 240/480 volts. If we assume that it is used to supply 120/240 volts, the voltage from the hot line at point 1 and the hot line at point 2 to neutral would be 120 volts, because of the center-tapped winding. However, 240 volts would still be available across any two hot lines. The neutral wire is color-coded with a white or gray insulation. The disadvantage of this system is that, when wiring changes are made, it is possible to connect a 120-volt load between the neutral and point 3 (sometimes called the "wild" phase). The voltage present here would be the combination of three-phase voltages between points 1 and 4 and points 1 and 3. This would be a voltage in excess of 300 volts! Although the "wild-phase" situation exists, this system is capable of supplying both high-power loads and low-voltage loads, such as are used for lighting and small equipment.

The most widely used three-phase power distribution system is the three-phase four-wire system. This system, shown in Figure 10-4C, commonly supplies 120/208 volts and 277/480 volts for industrial or commercial load requirements. The 120/208-volt system is illustrated here. From neutral to any hot line, 120 volts for lighting and low-power loads may be obtained. Across any two hot lines, 208 volts is present for supplying motors or other high-power loads. The most popular system for industrial and commercial power distribution is the 277/408-volt system, which is capable of supplying both three-phase and single-phase loads. A 240/416-volt system is sometimes used for industrial loads, while the 120/208-volt system is often used for underground distribution in urban areas. Note that this system is based on the voltage characteristics of the three-phase wye connection, and that the relationship $V_L = V_P \times 1.73$ exists for each application of this system.

GROUNDING OF DISTRIBUTION SYSTEMS

The concept of grounding in an electrical power distribution system is very important. Distribution systems must have continuous uninterrupted grounds. If a grounded conductor is opened, the ground is no longer functional. An open-ground condition can present severe safety problems and cause abnormal system operation.

Distribution systems must be grounded at substations, and at the end of the power lines, before the power is delivered to the load. Grounding is necessary at substations for the safety of the public and the power company's maintenance personnel. Grounding also provides points for transformer neutral connections for equipment grounds. Safety and equipment grounds will be discussed in more detail later.

At substations, all external metal parts must be grounded, and all trans- former, circuit breaker, and switch housings must be grounded. Also, metal fences and any other metal that is part of the substation construction must be grounded. Grounding assures that any person who touches any of the metal parts will not receive a high-voltage shock. Therefore, if a high-voltage line were to come in contact with any of the grounded parts, the system would be opened by protective equipment. Thus, the danger of high voltages at substations is substantially reduced by grounding. The actual ground connection is made by welding, brazing, or bolting a conductor to a metal rod or bar, which is then physically placed in the earth. This rod device is called a grounding electrode. Proper grounding techniques are required for safety, as well as for circuit performance. There are two types of grounding: (1) system grounding, and (2) equipment grounding. Another important grounding factor is ground-fault protective equipment.

SYSTEM GROUNDING

System grounding involves the actual grounding of a current-carrying conductor (usually called the neutral) of a power distribution system. Three-phase systems may be either the wye or delta type. The wye system has an obvious advantage over the delta system, since one side of each phase winding is connected to ground. We will define a ground as a reference point of zero-volt potential, which is usually an actual connection to earth ground. The common terminals of the wye system, when connected to ground, become the neutral conductor of the three-phase four-wire system.

The delta-system does not readily lend itself to grounding, since it does not have a common neutral. The problem of ground faults (line-to-ground shorts) occurring in ungrounded delta systems is much greater than in wye systems. A common method of grounding a delta system is to use a wye-delta transformer connection and ground the common terminals of the wye-connected primary. However, the wye system is now used more often for industrial and commercial distribution, since the secondary is easily grounded, and it provides overvoltage protection from lightning or line-to-ground shorts.

Single-phase 120/240-volt or 240/480-volt systems are grounded in a manner similar to a three-phase ground. The neutral of the single-phase three-wire system is grounded by a metal rod (grounding electrode) driven into the earth at the transformer location. System grounding conductors are insulated with white or gray material for easy identification.

Equipment Grounding

The second type of ground is the equipment ground, which, as the term implies, places operating equipment at ground potential. The conductor that is used for this purpose is either bare wire or a green insulated wire. The NEC describes conditions that require fixed electrical equipment to be grounded. Usually, all fixed electrical equipment located in industrial plants or commercial buildings should be grounded. Types of equipment that should be grounded include enclosures for switching and protective equipment for load control, transformer enclosures, electric motor frames, and fixed electronic test equipment. Industrial plants should use 120-volt, single-phase, duplex receptacles of the grounded type for all portable tools. The grounding of these receptacles may be checked by using a plug-in tester.

GROUND-FAULT PROTECTION

Ground-fault interrupters (GFIs) are used extensively in industrial, commercial, and residential power distribution systems. It is required by the NEC that all 120-volt, single-phase, 15- or 20-ampere receptacle outlets that are installed outdoors or in bathrooms have ground-fault interrupters connected to them. These devices are also called ground-fault circuit interrupters (GFCIs).

GFI Operation

These devices are designed to eliminate electrical shock hazard resulting from individuals coming in contact with a hot AC line (line-to-ground short). The circuit interrupter is designed to sense any change in circuit conditions, such as would occur when a line-to-ground short exists. One type of GFI has control wires that extend through a magnetic toroidal loop (see Figure 10-5). Ordinarily, the AC current flowing through the two conductors inside the loop is equal in magnitude and opposite in direction. Any change in this equal and opposite condition is sensed by the magnetic toroidal loop. When a line-to-ground short occurs, an instantaneous change in circuit conditions occurs. The change causes a magnetic field to be induced into the toroidal loop. The induced current is amplified to a level sufficient to cause the circuit breaker mechanism to open. Thus, any line-to-ground short will cause the ground-fault interrupter to open. The operating speed of the GFI is so fast that the shock hazard to individuals is greatly reduced, since only a minute current opens the circuit.

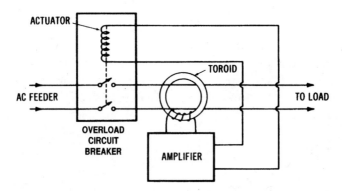

Figure 10-5. Simplified schematic of a ground-fault interrupter

GFI Applications

Construction sites, where temporary wiring is set up, are required to use GFIs for the protection of workers using electrical equipment. Ground-fault protection of individuals and commercial equipment must be provided for wye-connected systems of 150 to 600 volts for each distribution panelboard rated at over 1000 amperes. In this situation, the GFI will open all ungrounded conductors at the panelboard when a line-to-ground short occurs. Now, GFIs are used for all types of residential, commercial, and industrial applications.

Types of Ground-fault Protection Systems

There are four basic types of ground-fault protection systems in use today. They are: hospital applications, residential applications, motor protection applications, and specific electrical power distribution system applications. These ground-fault systems can be classified either by what they are to protect, or by the type of protection they are to provide. Hospital applications and residential applications are designed to protect people from excessive shock. The motor and electrical power applications are designed for protecting electrical equipment.

Another classification method is according to the amount of current required before an alarm system sounds, or the disconnect of an electrical circuit occurs. Typical current values that will cause alarms or disconnects to activate are 0.002 amperes (2 mA) for hospital applications, 0.005 amperes (5 mA) for residential applications, 5 to 100 amperes for motor-protective circuit applications, and 200 to 1200 amperes for electrical power distribution equipment applications.

Need for Ground-fault Protection

In order to understand the need for a ground-fault circuit interrupter (for the protection of people), certain basic facts must first be understood. These facts relate to people as well as to ground faults.

One important fact is that a person's body resistance varies with the amount of moisture present on the skin, the muscular structure of the body, and the voltage to which the body is subjected. Experiments have shown that the body resistance from one hand to the other hand is somewhere between 1000 and 4000 ohms. These estimates are based upon several assumptions concerning moisture and muscular structure. We also know that resistance of the body (hand to hand) is lower for higher voltages. This is because higher voltages are capable of "breaking down" the outer layers of the skin. Thus, higher voltages are more dangerous.

We can use Ohm's law to estimate that the typical current resulting from the average body resistance (from hand to hand) is about 115 mA at 240 volts AC, and about 40 mA at 120 volts AC. The effects of a 60-Hz AC on the human body are generally accepted to be as given in Table 10-1.

Ventricular fibrillation is an abnormal pattern of contraction of the heart. Once ventricular fibrillation occurs, it will continue, and death will occur within a few minutes. Resuscitation techniques, when applied immediately, can save a victim. Deaths caused by electrical shock account for a high percentage of the deaths that occur in the home and in industry. Many of these deaths are due to contact with low-voltage circuits (600 volts and under), mainly 120- and 240-volt systems.

Ground-fault Protection for the Home

Ground-fault interruptors for homes are of three types: (1) circuit breaker, (2) receptacle, and (3) plug-in types. Ground-fault protection de-

Table 10-1. Body Reaction to Alternating Current

Amount of Current	Effect on Body
1 mA or less	No sensation (not felt).
More than 5 mA	Painful shock.
More than 10 mA	Muscle contractions; could cause "freezing" to the electrical circuit for *some* people.
More than 15 mA	Muscle contractions; could cause "freezing" to the electrical circuit for *most* people.
More than 30 mA	Breathing difficult; could cause unconsciousness.
50 to 100 mA	Ventricular fibrillation of the heart is possible.
100 to 200 mA	Ventricular fibrillation of the heart is certain.
Over 200 mA	Severe burns and muscular contractions; the heart is more apt to stop beating than to fibrillate.
1 ampere and above	Permanent damage to body tissues.

vices are constructed according to standards developed by the Underwriters' Laboratories. The GFI circuit breakers combine ground-fault protection and circuit interruption in the same over-current and short-circuit protective equipment as does a standard circuit breaker. A GFI circuit breaker fits the same space required by a standard circuit breaker. It provides the same branch-circuit wiring protection as the standard circuit breaker, as well as ground-fault protection. The GFI sensing system continuously monitors the current balance in the ungrounded (hot) conductor and the grounded (neutral) conductor. The current in the neutral wire becomes less than the current in the hot wire when a ground fault develops. This means that a portion of the circuit current is returning to ground by some means other than the neutral wire. When an imbalance in current occurs, the sensor (a differential current transformer) sends a signal to the solid state circuitry, which activates a trip mechanism. This action opens the hot line. A differential current as low as 5 mA will cause the sensor to send a fault signal and cause the circuit breaker to interrupt the circuit.

Ordinarily, GFI receptacles provide ground-fault protection on 120-, 208-, or 240-volt AC systems. The GFI receptacles come in 15- and 20-ampere designs. The 15-ampere unit has a receptacle configuration for use with 15-ampere plugs only. The 20-ampere device has a receptacle configuration for use with either 15- or 20-ampere plugs. These GFI receptacles have connections for hot, neutral, and ground wires. All GFI receptacles have a two-pole tripping mechanism, which breaks both the hot and the neutral load connections at the time a fault occurs.

The plug-in GFI receptacles provide protection by plugging into a standard wall receptacle. Some manufacturers provide units that will fit either two-wire or three-wire receptacles. The major advantage of this type of unit is that it can be moved from one location to another.

Ground-fault Protection for Power Distribution Equipment

Ground faults can destroy electrical equipment if allowed to continue. Phase-to-phase short circuits and some types of ground faults are usually high current. Normally, they are adequately handled by conventional overcurrent protective equipment. However, some ground faults produce an arcing effect from relatively low currents that are not large enough to trip conventional protective devices. Arcs can severely bum electrical equipment. A 480- or 600-volt system is more susceptible to arcing damage than a 120-, 208-, or 240-volt system, because the higher voltages sustain the arcing effect. High-current faults are quickly detected by conventional overcurrent devices. Low-current values must be detected by GFIs.

Ground faults that cause an arcing effect in the equipment are probably the most frequent faults. They may result from damaged or deteriorated insulation, dirt, moisture, or improper connections. They usually occur between one hot conductor and the grounded equipment enclosure, conduit, or metal housing. The line-to-neutral voltage of the source will cause current to flow in the hot conductor, through the arc path, and back through the ground path. The impedance of the conductor and the ground-return path (enclosure, conduit, or housing) depends on many factors. As a result, the fault-current value cannot be predicted. It can also increase or decrease as the fault condition continues.

It is apparent that many factors influence the magnitude, duration, and effect of an arcing ground fault. Some conditions produce a large amount of fault current, while others limit the fault current to a relatively small amount. Arc-current magnitude and the time that the arc persists can cause very great damage to equipment. Probably the more important factor is the time period of the arcing voltage, since the longer the arcing time, the more chance that the arcs will spread to different areas within the equipment.

Ground-current relaying is one method used to protect equipment from ground faults. Current flows through a load or fault along the hot and neutral conductors and returns to the source on these conductors—and, to some extent, along the ground path. The normal ground path current is very small. Therefore, essentially all the current flowing from the source is also returning on the same hot line and neutral conductors. However, if a ground fault occurs, the ground current will increase to the point where some current will escape through the fault and return via the ground path. As a result, the current returning on the hot and neutral conductors is less than the amount going out. The difference is an indication of the amount of current in the ground path. A relay, which senses this difference in currents, can act as a ground-fault protective device.

Ground-fault Protection for Electric Motors

Motor protective systems offer protection in the 5- to 100-ampere range. This type of ground-fault protective system offers a protection against ground faults in both the single-phase and the three-phase systems. Many insulation system failures begin with a small leakage current, which builds up with time until damage results. These ground-fault systems detect ground leakage currents while they are still small, and thus prevent any extensive damage to the motors.

WIRING DESIGN CONSIDERATIONS FOR DISTRIBUTION SYSTEMS

The wiring design of electrical power distribution systems can be very complex. There are many factors that must be considered in the wiring design of a distribution system installed in a building. Wiring design standards are specified in the National Electrical Code (NEC), which is published by the National Electrical Protection Association (NEP A). The NEC, local wiring standards, and electrical inspection policies should be taken into account when an electrical wiring design is under consideration.

There are several distribution system wiring design considerations that are pointed out specifically in the NEC. In this chapter, we will deal with voltage-drop calculations, branch circuit design, feeder circuit design, and the design for grounding systems.

National Electrical Code (NEC) Use

The NEC sets forth the minimum standards for electrical wiring in the United States. The standards contained in the NEC are enforced by being incorporated into the different city and community ordinances that deal with electrical wiring in residences, industrial plants, and commercial buildings. Therefore, these local ordinances conform to the standards set forth in the NEC.

In most areas of the United States, a license must be obtained by any individual who does electrical wiring. Usually, one must pass a test administered by the city, county, or state, in order to obtain this license. These tests are based on local ordinances and the NEC. The rules for electrical wiring that are established by the local electrical power company are also sometimes incorporated into the license test.

Electrical Inspections

When new buildings are constructed, they must be inspected to see if the electrical wiring meets the standards of the local ordinances, the NEC, and the local power company. The organization that supplies the electrical inspectors varies from one locality to another. Ordinarily, the local power company can advise individuals about whom to contact for information about electrical inspections.

Voltage Drop in Electrical Conductors

Although the resistance of electrical conductors is very low, a long length of wire could cause a substantial voltage drop. This is illustrated in Figure 10-6. Remember that a voltage drop is current times resistance (I × R). Therefore, whenever current flows through a system, a voltage drop is created. Ideally, the voltage drop caused by the resistance of a conductor will be very small.

However, a longer section of electrical conductor has a higher resistance. Therefore, it is sometimes necessary to limit the distance a conductor can extend from the power source to the load that it supplies. Many types of loads do not operate properly when a value less than the full source voltage is available.

You can also see from Figure 10-6 that as the voltage drop (V_D) increases, the voltage applied to the load (V_L) decreases. As current in the system increases, V_D increases, causing V_L to decrease, since the source voltage stays the same.

Voltage Drop Calculations Using Conductor Table

It is important when dealing with electrical wiring design to be able to determine the amount of voltage drop caused by conductor resistance. Table 10-2 is used to make these calculations. The NEC limits the amount

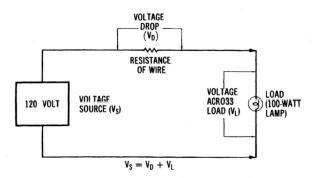

Figure 10-6. Voltage drop in on electrical circuit

Table 10-2. Sizes of Copper and Aluminum Conductors

	Size (AWG or MCM)	$A\text{-}D^2$ Area (cmil)	Number of Wires	Diameter of Each Wire (in.)	DC Resistance (Ω/1000 ft) 25 C	
					Copper	Aluminum
AWE	18	1,620	1	0.0403	6.51	10.7
	16	2,580	1	0.0508	4.10	6.72
	14	4,110	1	0.0641	2.57	4.22
	12	6,530	1	0.0808	1.62	2.66
	10	10,380	1	0.1019	1.018	1.67
	8	16,510	1	0.1285	0.6404	1.05
	6	26,240	7	0.0612	0.410	0.674
	4	41,740	7	0.0772	0.259	0.424
	3	52,620	7	0.0867	0.205	0.336
	2	66,360	7	0.0974	0.162	0.266
	1	83,690	19	0.0664	0.129	0.211
	0	105,600	19	0.0745	0.102	0.168
	00	133,100	19	0.0837	0.0811	0.133
	000	167,800	19	0.0940	0.0642	0.105
	0000	211,600	19	0.1055	0.0509	0.0836
MCM	250	250,000	37	0.0822	0.0431	0.0708
	300	300,000	37	0.09001	0.0360	0.0590
	350	350,000	37	0.0973	0.0308	0.0505
	400	400,000	37	0.1040	0.0270	0.0442
	500	500,000	37	0.1162	0.0216	0.0354
	600	600,000	61	0.0992	0.0180	0.0295
	700	700,000	61	0.1071	0.0154	0.0253
	750	750,000	61	0.1109	0.0144	0.0236
	800	800,000	61	0.1145	0.0135	0.0221
	900	900,000	61	0.1215	0.0120	0.0197
	1000	1,000,000	61	0.1280	0.0108	0.0177

of voltage drop that a system can have. This means that long runs of conductors must ordinarily be avoided. Remember that a conductor with a large cross-sectional area will cause a smaller voltage drop, since its resistance is smaller.

To better understand how to determine the size of conductor required to limit the voltage drop in a system, we will look at a sample problem.

Sample Problem:

Given: a 200-ampere load located 400 feet (121.92 meters) from a 240-volt single-phase source. Limit the voltage drop to 2 percent of the source.

Find: the size of an RH copper conductor needed to limit the voltage drop of the system.

Solution:
1. The allowable voltage drop equals 240 volts times 0.02 (2%). This equals 4.8 volts.
2. Determine the maximum resistance for 800 feet (243.84 meters). This is the equivalent of 400 feet (121.92 meters) × 2, since there are two current-carrying conductors for a single-phase system.

$$R = \frac{V_D}{I}$$

$$= \frac{4.8 \text{ V}}{200 \text{ A}}$$

$$= 0.024 \text{ ohm, resistance for 800 feet}$$

3. Determine the maximum resistance for 1000 feet (304.8 meters) of conductor.

$$\frac{800 \text{ feet}}{1000 \text{ feet}} = \frac{0.024 \text{ ohm}}{R}$$

$$800 \, R = (1000)(0.024)$$

$$R = 0.030 \text{ ohm}$$

Single-phase and Three-phase Distribution Systems   273

4. Use Table 10-2 to find the size of copper conductor that has the nearest direct current (DC) resistance (ohms per 1000 feet) value that is equal to or less than the value calculated in 3, above. The conductor chosen is conductor size 350 MCM, RH Copper.
5. Check this conductor with the proper ampacity table to ensure that it is large enough to carry 200 amperes. Table 10-3 shows that a 350 MCM, RH copper conductor will handle 310 amperes of current; therefore, use 350 MCM conductors. (Always remember to use the largest conductor, if Steps 4 and 5 produce conflicting values.)
6. If the current is larger than listed on the tables, use more than one conductor of the same size for design calculations.

Table 10-3. Ampacities of Conductors in a Raceway or Cable (3 or less)

	Wire Size	Copper With R, T, TW Insulation	Copper With RH, RHW TH, THW Insulation	Aluminum With R, T, TW Insulation	Aluminum With RH, RHW, TH THW Insulation
AWG	14	15	15		
	12	20	20	15	15
	10	30	30	25	25
	8	40	45	30	40
	6	55	65	40	50
	4	70	85	55	65
	3	80	100	65	75
	2	95	115	75	90
	1	110	130	85	100
	0	125	150	100	120
	00	145	175	115	135
	000	165	200	130	155
	0000	195	239	155	180
MCM	250	215	255	170	205
	300	240	285	190	230
	350	260	310	210	250
	400	280	335	225	270
	500	320	380	260	310
	600	355	420	285	340
	700	385	460	310	375
	750	400	475	320	385
	800	410	490	330	395
	900	435	520	355	425
	1000	455	545	375	445

Alternative Method of Voltage Drop Calculation

In some cases, an easier method to determine the conductor size for limiting the voltage drop is to use one of the following formulas to find the cross-sectional (cmil) area of the conductor.

$$\text{cmil} = \frac{p \times I \times 2d}{V_D} \quad \text{(Single-phase Systems)}$$

or,

$$\text{cmil} = \frac{p \times I \times 1.73d}{V_D} \quad \text{Three-phase systems}$$

where:
 p = the resistivity from Table 8-2
 I = the load current in amperes,
 V_D = the allowable voltage drop, and
 d = the distance from source to load, in feet.

The sample problem given for a single-phase system in the preceding section could be set up as follows:

$$\text{cmil} = \frac{p \times I \times 2d}{V_D}$$

$$= \frac{10.4 \times 200 \times 2 \times 400}{240 \times 0.02}$$

$$= \frac{1{,}664{,}000}{4.8}$$

$$= 346{,}666 = 347 \text{ MCM}$$

The next largest size is a 350 MCM conductor.

BRANCH CIRCUIT DESIGN CONSIDERATIONS

A branch circuit is defined as a circuit that extends from the last overcurrent protective device of the power system. Branch circuits, according to the NEC, are either 15, 20, 30, 40, or 50 amperes in capacity. Loads larger than 50 amperes would not be connected to a branch circuit.

There are many rules in the NEC that apply to branch circuit design. The following information is based on the NEC. First, each circuit must be designed so that accidental short circuits or grounds do not cause damage to any part of the system. Then, fuses or circuit breakers are to be used as branch circuit overcurrent protective devices. Should a short circuit or ground condition occur, the protective device should open and interrupt the flow of current in the branch circuit. One important NEC rule is that No. 16 or No. 18 (extension cord) wire may be tapped from No. 12 or No. 14 conductors, but not from conductors larger than No. 12. This means that an extension cord of No. 16 wire should not be plugged into a receptacle that uses No. 10 wire. Damage to smaller wires (due to the heating effect) before the overcurrent device can open is eliminated by applying this rule. Lighting circuits are one of the most common types of branch circuits. They are usually either 15-ampere or 20-ampere circuits.

The maximum rating of an individual load (such as a portable appliance connected to a branch circuit) is 80 percent of the branch circuit current rating. Therefore, a 20-ampere circuit could not have a single load that draws more than 16 amperes. If the load is a permanently connected appliance, its current rating cannot be more than 50 percent of the branch circuit capacity—if portable appliances or lights are connected to the same circuit.

Voltage Drop in Branch Circuits

Branch circuits must be designed so that sufficient voltage is supplied to all parts of the circuit. The distance that a branch circuit can extend from the voltage source or power distribution panel is, therefore, limited. A voltage drop of 3 percent is specified by the NEC as the maximum allowed for branch circuits in electrical wiring design.

The method for calculating the voltage drop in a branch circuit is a step-by-step process that is illustrated by the following problem. Refer to the circuit diagram given in Figure 10-7.

Sample Problem:

Given: a 120-volt 15-ampere branch circuit supplies a load that consists of four lamps. Each lamp draws 3 amperes of current from the source. The lamps are located at 10-foot (3.05-meter) intervals from the power distribution panel.

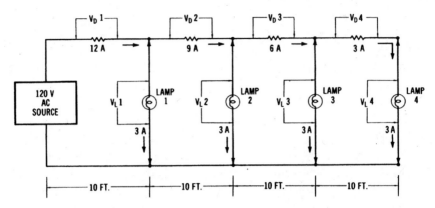

Figure 10-7. Circuit for calculating the voltage drop in a branch circuit

Find: the voltage across lamp number 4.
Solution:

1. Find the resistance for 20 feet (6.1 meters) of conductor (same as for 10-foot conductor × 2). No. 14 copper wire is used for 15-ampere branch circuits. From Table 10-2, we find that the resistance of 1000 feet (304.8 meters) of No. 14 copper wire is 2.57 ohms. Therefore, the resistance of 20 feet of wire is:

$$\frac{20 \text{ feet}}{1000 \text{ feet}} = \frac{R}{2.57 \text{ ohms}}$$

$$1000\ R = (20)(2.57)$$
$$R = 0.0514 \text{ ohm.}$$

2. Calculate voltage drop V_D number 1. (R equals the resistance of 20 feet of wire.)

$$\begin{aligned} V_D \text{ number 1} &= I \times R \\ &= 12 \text{ A} \times 0.0514\ \Omega \\ &= 0.6168 \text{ V AC} \end{aligned}$$

Calculate load V_1, number 1 (the source voltage minus V_D number 1.)

$$\begin{aligned} V_L \text{ number 1} &= 120 \text{ V} - 0.6168 \text{ V} \\ &= 119.383 \text{ volts} \end{aligned}$$

3. Calculate voltage drop and load number 2.

$$V_D \text{ number 2} = I \times R$$
$$= 9 \text{ A} \times 0.0514 \text{ }\Omega$$
$$= 0.4626 \text{ V AC}$$
$$V_L \text{ number 2} = 119.383 \text{ V} - 0.4626 \text{ V}$$
$$= 118.920 \text{ volts}$$

4. Calculate voltage drop and load number 3.

$$V_D \text{ number 3} = I \times R$$
$$= 6 \text{ A} \times 0.0514 \text{ }\Omega$$
$$= 0.384 \text{ V AC}$$
$$V_L \text{ number 3} = 118.920 \text{ V} - 0.3084 \text{ V}$$
$$= 118.612 \text{ volts.}$$

5. Calculate voltage drop and load number 4.

$$V_D \text{ number 4} = I \times R$$
$$= 3 \text{ A} \times 0.0514 \text{ }\Omega$$
$$= 0.1542 \text{ V AC}$$
$$V_L \text{ number 4} = 118.612 \text{ V} - 0.1542 \text{ V}$$
$$= 118.458 \text{ volts}$$

Notice that the voltage across lamp number 4 is substantially reduced from the 120-volt source value because of the voltage drop of the conductors. Also, notice that the resistances used to calculate the *voltage drops* represented both wires (hot and neutral) of the branch circuit. Ordinarily, 120-volt branch circuits do not extend more than 100 feet (30.48 meters) from the power distribution panel. The *preferred* distance is 75 feet (22.86 meters). The *voltage drop* in branch circuit conductors can be reduced by making the circuit shorter in length, or by using larger conductors.

In residential electrical wiring design, the voltage drop in many branch circuits is difficult to calculate, since the lighting and portable appliance receptacles are placed on the same branch circuits. Since portable appliances and "plug-in" lights are not used all of the time, the voltage drop will vary according to the number of lights and appliances in use. This problem is usually not encountered in an industrial or commercial wiring design for lights, since the lighting units are usually larger and are permanently installed on the branch circuits.

Branch Circuit Wiring

A *branch circuit* usually consists of a *nonmetallic-sheathed cable* that is connected into a *power distribution panel.* Each branch circuit that is wired from the power distribution panel is protected by a fuse or circuit breaker. The power panel also has a *main switch* that controls all of the branch circuits that are connected to it.

Single-phase Branch Circuits

A diagram of a *single-phase three-wire (120/240-volt)* power distribution panel is shown in Figure 10-8. Notice that eight *120-volt* branch circuits and one *240-volt* circuit are available from the power panel. This type of system is used in most homes, where several *120-volt* branch circuits and, typically, three or four *240-volt* branch circuits are required. Notice in Figure 10-8 that each hot line has a circuit breaker, while the neutral line

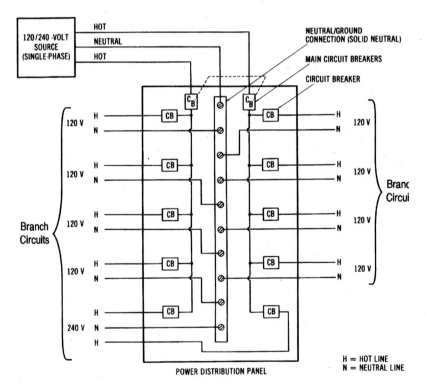

Figure 10-8. Diagram of a power distribution panel for a single-phase, three-wire branch

connects directly to the branch circuits. *Neutrals* should never be opened (fused). This is a safety precaution in electrical wiring design.

Three-phase Branch Circuits

A diagram for a *three-phase, four-wire (120/208-volt)* power distribution panel is shown in Figure 10-9. There are three single-phase *120-volt* branch circuits, and two three-phase *208-volt* branch circuits shown. The single-phase branches are balanced (one *hot* line from each branch). Each hot line has an individual circuit breaker. Three-phase lines should be connected so that an overload in the branch circuit will cause all three lines to open. This is accomplished by using a three-phase circuit breaker, which is arranged internally as shown in Figure 10-9.

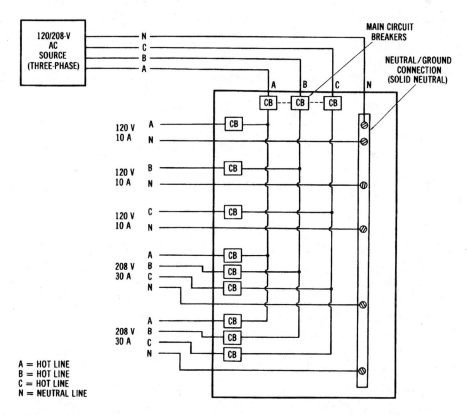

Figure 10-9. Diagram of a power distribution panel for a three-phase, four-wire branch circuit

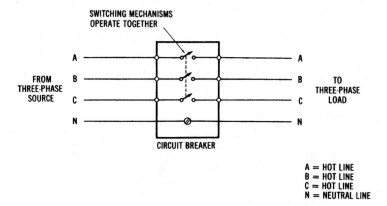

Figure 10-10. Diagram of a three-phase circuit breaker

FEEDER CIRCUIT DESIGN CONSIDERATIONS

Feeder circuits are used to distribute electrical power to power distribution panels. Many feeder circuits extend for very long distances; therefore, voltage drop must be considered in feeder circuit design. In higher voltage feeder circuits, the *voltage drop* is reduced. However, many lower voltage feeder circuits require large-diameter conductors to provide a tolerable level of voltage drop. High-current feeder circuits also present a problem in terms of the massive overload protection that is sometimes required. This protection is usually provided by system switchgear or load centers where the feeder circuits originate.

Determining the Size of Feeder Circuits

The amount of current that a *feeder circuit* must be designed to carry depends upon the actual load demanded by the branch circuit power distribution panels that it supplies. Each power distribution panel will have a separate feeder circuit. Also, each feeder circuit must have its own overload protection.

The following problem is an example of size calculation for a feeder circuit.

Sample Problem:

Given: three 15-kW fluorescent lighting units are connected to a three-phase, four-wire (277/480-volt) system. The lighting units have a power factor of 0.8.

Find: the size of THW aluminum feeder conductors required to supply this load.

Solution:
1. Find the line current:

$$I_L = \frac{P_T}{1.73 \times V_L \times pf}$$

$$= \frac{45,000 \text{ watts}}{1.73 \times 480 \text{ volts} \times 0.8}$$

$$= 67.74 \text{ amperes}$$

2. From Table 10-3, we find that the conductor size that will carry 67.74 amperes of current is a No. 3 AWG THW aluminum conductor.

Voltage Drop Calculation for Feeder Circuits

Feeder circuit design must take the conductor *voltage drop* into consideration. The voltage drop in a feeder circuit must be kept as low as possible so that *maximum* power can be delivered to the loads connected to the feeder system. The NEC allows a maximum 5 percent voltage drop in the combination of a branch and a feeder circuit; however, a 5 percent voltage reduction represents a significant power loss in a circuit. We can calculate power loss due to voltage drop as V^2/R, where V^2 is the voltage drop of the circuit, and R is the resistance of the conductors of the circuit.

The calculation of feeder conductor size is similar to that for a branch circuit voltage drop. The size of the conductors must be large enough to: (1) have the required ampacity, and (2) keep *the voltage drop* below a specified level. If the second requirement is not met, possibly because of a long feeder circuit, the conductors chosen must be larger than the ampacity rating requires. The following problem illustrates the calculation of *feeder conductor size* based upon the voltage drop in a single-phase circuit.

Sample Problem:

Given: ex single-phase 240-volt load in a factory is rated at 85 kilowatts. The feeders (two hot lines) will be 260-foot (79.25 meters) lengths of RHW copper conductor. The maximum conductor voltage drop allowed is 2 percent.

Find: the feeder conductor size required.

Solution:
1. Find the maximum voltage drop of the circuit.

$$V_D = \% \times \text{Load}$$
$$= 0.02 \times 240$$
$$= 4.8 \text{ volts}$$

2. Find the current drawn by the load.

$$I = \frac{\text{Power}}{\text{Voltage}}$$

$$= \frac{85{,}000}{240}$$

$$= 354.2 \text{ amperes}$$

3. Find the minimum circular-mil conductor area required. Use the formula given for finding the cross-sectional area of a conductor in single-phase systems, which was previously given in the "Alternative Method of Voltage Drop Calculation" section.

$$\text{cmil} = \frac{p \times I \times 2d}{V_D}$$

$$= \frac{10.4 \times 354.2 \times 2 \times 260}{4.8}$$

$$= 399{,}065.33 \text{ cmil}$$

4. Determine the *feeder conductor size*. The next larger size conductor in Table 10-2 is also 400 MCM. Check Table 10-3, and you will see that a 400 MCM RHW copper conductor will carry 335 amperes. This is less than the required 354.2 amperes, so use the next larger size, which is a 500 MCM conductor.

The *conductor size* for a three-phase feeder circuit is determined in a similar way. In this problem, the feeder size will be determined on the basis of the circuit voltage drop.

Single-phase and Three-phase Distribution Systems

Sample Problem:

Given: ex 480-volt, three-phase, three-wire (delta) feeder circuit supplies a 45kilowatt balanced load to a commercial building. The load operates at a 0.75 power factor. The feeder circuit (three hot lines) will be a 300-foot (91.44-meter) length of RH copper conductor. The maximum voltage drop is 1 percent.

Find: the feeder size required (based on the voltage drop of the circuit).

Solution:

1. Find the maximum voltage drop of the circuit.

$$V_D = 0.01 \times 480$$
$$= 4.8 \text{ volts}$$

2. Find the line current drawn by the load.

$$I_L = \frac{P}{1.73 \times V \times pf}$$

$$= \frac{45,000 \text{ W}}{1.73 \times 480 \times 0.75}$$

$$= 72.25 \text{ amperes}$$

3. Find the minimum circular-mil conductor area required. Use the formula for finding cmil in three-phase systems, which was given in an earlier section.

$$\text{cmil} = \frac{p \times I \times 1.73 \, d}{V_D}$$

$$= \frac{10.4 \times 72.25 \times 1.73 \times 300}{4.8}$$

$$= 81,245 \text{ cmil}$$

4. Determine the *feeder conductor size*. The closest and next larger conductor size in Table 10-3 is No. 1 AWG. Check Table 10-3, and you will see that a No. 1 AWG RH copper conductor will carry 130 amperes, much more than the required 72.25 amperes. Therefore, use No. 1 AWG RH copper conductors for the feeder circuit.

DETERMINING GROUNDING CONDUCTOR SIZE

Grounding considerations in electrical wiring design were discussed previously. Another necessity of wiring design is to determine the *size* of the grounding conductor required in a circuit. All circuits that operate at 150 volts or less must be grounded; therefore, all residential electrical systems must be grounded. Higher voltage systems used in industrial and commercial buildings have grounding requirements that are specified by the NEC and by local codes. A *ground* at the *service entrance* of a building is usually a metal water pipe that extends, uninterrupted, underground, or a grounding electrode that is driven into the ground near the service entrance.

The *size* of the *grounding conductor* is determined by the current rating of the system. Table 10-4 lists equipment grounding conductor sizes for interior wiring, while Table 10-5 lists the minimum grounding conductor sizes for system grounding of *service entrances*. The sizes of grounding conductors listed in Table 10-4 are for *equipment grounds*, which connect to raceways, enclosures, and metal frames for safety purposes. Note that a No. 12 or a No. 14 wiring cable, such as 12-2 WG NMC, can have a No. 18 equipment ground. The ground is contained in the same cable sheathing as the hot conductors. Table 10-5 is used to find the minimum *size* of grounding conductors needed for service entrances, based upon the size of the hot line conductors used with the system.

PARTS OF INTERIOR ELECTRICAL WIRING SYSTEMS

Some parts of *interior electrical distribution systems* have been discussed previously. Such types of equipment as *transformers, switchgear, conductors, insulators,* and *protective equipment* are parts of interior wiring systems. There are, however, certain parts of interior electrical distribution systems that are unique to the wiring system itself. These parts include the *nonmetallic-sheathed cables (NMC),* the *metal-clad cables,* the rigid conduit, and the *electrical metallic tubing (EMT).*

Nonmetallic-sheathed Cable (NMC)

Nonmetallic-sheathed cable is a common type of electrical cable used for interior wiring. *NMC,* sometimes referred to as *Romex* cable, is used in residential wiring systems almost exclusively. The most common type

Table 10-4. Equipment Grounding Conductor Sizes for Interior Winding

Ampere Rating of Distribution Panel	Grounding Conductor Size (AWG)	
	Copper	Aluminum
15	14	12
20	12	10
30	10	8
40	10	8
60	10	8
100	8	6
200	6	4
400	3	1
600	1	00
800	0	000
1000	00	0000

Table 10-5. System Grounding Conductor Sizes for Service Entrances

Conductor Size (Hot Line)		Grounding Conductor Size	
Copper	Copper Aluminum	Aluminum (AWG)	(AWG)
No. 2 AWG or smaller	No. 0 AWG or smaller	8	6
No. 1 or 0 AWG	No. 00 or 000 AWG	6	4
No. 00 or 000 AWG	No. 0000 AWG or 250 MCM	4	2
No. 000 AWG to 350 MCM	250 MCM to 500 MCM	2	0
350 MCM to 600 MCM	500 MCM to 900 MCM	0	000
600 MCM to 1100 MCM	900 MCM to 1750 MCM	00	0000

used is *No. 12-2 WG*, which is illustrated in Figure 10-11. This type of NMC comes in 250-foot rolls for interior wiring. The cable has a thin plastic outer covering with three conductors inside. The conductors have colored insulation that designates whether the conductor should be used as a *hot*, *neutral*, or *equipment ground* wire. For instance, the conductor connected to the *hot* side of the system has *black* or *red* insulation, while the *neutral* conductor has *white* or *gray* insulation. The equipment grounding conductor has either a *green* insulation or *no insulation* (bare conductor). There are several different sizes of bushings and connectors used for the installation of NMC in buildings.

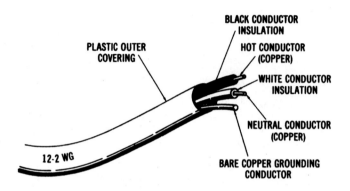

Figure 10-11. Nonmetallic-sheathed cable (MNC)

The designation No. *12-2 WG* means that (1) the copper conductors used are No. 12 AWG, as measured by an *American wire gage (AWG)*, (2) there are two current-carrying conductors, and (3) the cable comes with a ground (WG) wire. A No. *14-3 WG* cable, in comparison, would have three No. 14 conductors and a grounding conductor. NMC ranges in size from No. 14 to No. 1 AWG copper conductors, and from No. 12 to No. 2 AWG aluminum conductors.

Metal-Clad Cable

Metal-clad cable is similar to NMC except that it has a flexible spiral metal covering, rather than a plastic covering. A common type of metal-clad cable is called *BX cable*. Like NMC, BX cable contains two or three conductors. There are also several sizes of connectors and bushings used in the installation of BX cable. The primary advantage of this type of metal-clad cable is that it is contained in a metal enclosure that is flexible, so that it can be bent easily. Other metal enclosures are usually more difficult to bend.

Rigid Conduit

The exterior of *rigid conduit* looks like water pipe. It is used in special locations for enclosing electrical conductors. Rigid conduit comes in *10-*

foot lengths, which must be threaded for joining the pieces together. The conduit is secured to metal wiring boxes by locknuts and bushings. It is bulky to handle and takes a long time to install.

Electrical Metallic Tubing (EMT)

EMT, or *thinwall conduit*, is somewhat like rigid conduit, except that it can be bent with a special conduit-bending tool. EMT is easier to install than rigid conduit, since no threading is required. It also comes in *10-foot* lengths. EMT is installed by using compression couplings to connect the conduit to metal wiring boxes. Interior electrical wiring systems use EMT extensively, since it can be easily bent, can be connected together, and can be connected to metal wiring boxes.

UNIT IV
Electrical Power Conversion Systems

One of the most important aspects of electrical power systems is the conversion of electrical power into some other form of energy. Electrical power is ordinarily converted into light, heat, or mechanical power. The power is converted using either resistive, inductive, or capacitive circuits. The fundamental characteristics of electrical power conversion systems are discussed in Chapter 11.

The basic types of electrical power conversion systems are studied further in the remaining chapters. Heating systems are discussed in Chapter 12. Those systems that convert electrical energy into heat energy use three methods, incorporating resistive, inductive, and capacitive circuits. Basic welding method systems are also included in this chapter, since they are a unique type of electrical load. Chapter 13 deals with the conversion of electrical energy into light energy. Lighting systems include incandescent, fluorescent, and vapor lighting systems. The systems that convert electrical energy into mechanical energy are discussed in Chapter 15. The major mechanical energy conversion takes place in electric motors. Both direct current (DC) and alternating current (AC) motors are studied in that chapter.

Figure IV shows the *electrical power systems model* used in this book, and the major topics of Unit IV—Electrical Power Conversion Systems.

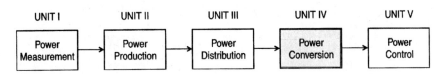

Fundamentals of Electrical Loads (Chapter 11)
Heating Systems (Chapter 12)
Lighting Systems (Chapter 13)
Mechanical Systems (Chapter 14)

Figure IV. Electrical power systems model

UNIT OBJECTIVES

Upon completion of this unit, you should be able to:

1. Define the term "electrical load."
2. List the types and classifications of electrical loads.
3. Explain the differences between resistive, inductive, and capacitive circuits.
4. Calculate load (demand) factor and power factor.
5. Describe power factor correction, using static capacitors or synchronous capacitors.
6. Define true power, apparent power, and reactive power in AC circuits.
7. Calculate power per phase and total power for balanced or unbalanced three-phase loads.
8. Explain the differences between resistive, inductive, and capacitive electric loads.
9. Describe electric resistive and arc-welding systems as electrical loads.
10. Describe electric heating systems and these associated terms:
    BTU
    Design Temperature Difference
    Degree Days
    Thermal Resistance (R)
    Coefficient of Heat Transfer (U)
    Watts of Heat (W)
11. Describe heat pumps and air conditioning systems.
12. Describe the characteristics of light.
13. Define the terms candlepower, lumen, and footcandle as they relate to light.
14. Describe types of street lighting systems.
15. Explain the characteristics of incandescent, fluorescent, and vapor lighting.
16. Describe branch circuits used for controlling electrical lights from one, two, or three locations.
17. Calculate the minimum number of branch circuits and total power requirements for lighting circuits in buildings.
18. Describe the following factors that affect lighting fixture design:
    Luminaire

Unit Objectivees 291

      Coefficient of Utilization (CU)
      Room Ratio
      Depreciation Factor (DF)
19. Calculate light output of a lighting system.
20. Describe the basic principles of electric motor operation.
21. Identify and describe the following types of electric motors:
    DC Motors
    Single-Phase AC Motors
    Three-Phase AC Motors
22. Explain the operation of synchro/servo systems and DC stepping motors.
23. Calculate the following, as each relates to electric motor operation:
    Horsepower
    Speed Regulation
    Starting Current
    Synchronous Speed
    Slip
    Rotor Frequency
    Efficiency

Chapter 11

Fundamentals of Electrical Loads

The electrical load devices that are used in industry, in our homes, and in commercial buildings are very important parts of electrical power systems. The load of any system performs a function that involves power conversion. A load converts one form of energy to another form. An electrical load converts electrical energy to some other form of energy, such as heat, light, or mechanical energy. Electrical loads may be classified according to the function that they perform (lighting, heating, mechanical), or by the electrical characteristics that they exhibit (resistive, inductive, capacitive).

IMPORTANT TERMS

Chapter 11 deals with fundamentals of electrical loads. After studying this chapter, you should have an understanding of the following terms:

Load
Resistive Load
Capacitive Load
Inductive Load
Load (Demand) Factor
Power Factor
Power Factor Correction
True Power
Apparent Power
Reactive Power
Static Capacitor

293

Synchronous Capacitor
Balanced Three-phase Load
Unbalanced Three-phase Load
Line Voltage (V_L)
Phase Voltage (V_P)
Line Current (I_L)
Phase Current (I_P)
Power per Phase (P_P)
Total Three-phase Power (P_T)

LOAD CHARACTERISTICS

In order to plan for power system load requirements, we must understand the electrical characteristics of all the loads connected to the power system. The types of power supplies and distribution systems that a building uses are determined by the load characteristics. All loads may be considered as either resistive, inductive, capacitive, or a combination of these. We should be aware of the effects that various types of loads will have on the power system. The nature of AC results in certain specific electrical circuit properties.

You should review that portion of Chapter 2 that deals with resistive, inductive, and capacitive effects in an electrical circuit. One primary factor that affects the electrical power system is the presence of inductive loads. These are mainly electric motors. To counteract the inductive effects, utility companies use power factor corrective capacitors as part of the power system design. Capacitor units are located at substations to improve the power factor of the system. The inductive effect, therefore, increases the cost of a power system and reduces the actual amount of power that is converted to another form of energy.

Load (Demand) Factor

One electrical load relationship that is important to understand is the *load* (or *demand*) factor. Load factor expresses the ratio between the average power requirement and the peak power requirement, or:

$$\text{load demand factor} = \frac{\text{average demand (kW)}}{\text{peak demand (kW)}}$$

Sample Problem:

Given: a factory has a peak demand of 12 MW and an average power demand of 9.86 MX.

Find: the load (demand) factor for the factory.

Solution:

$$DF = \frac{\text{Avg Demand}}{\text{Peak Demand}}$$

$$= \frac{9.86 \text{ MW}}{12 \text{MW}}$$

$$DF = 0.82$$

The average demand of an industry or commercial building is the average electrical power used over a specific time period. The peak demand is the maximum amount of power (kW) used during that time period. The load profile shown in Figure 11-1 shows a typical industrial demand-versus-time curve for a working day. Demand peaks that far exceed the average demand cause a decrease in the load factor ratio. Low load factors result in an additional billing charge by the utility company.

Utility companies must design power distribution systems that take peak demand time into account, and ensure that their generating capacity will be able to meet this peak power demand. Therefore, it is inefficient electrical design for an industry to operate at a low-load factor, since this represents a significant difference between peak-power demand and average-power demand. Every industry should attempt to raise its load factor to the maximum level it can. By minimizing the peak demands of indus-

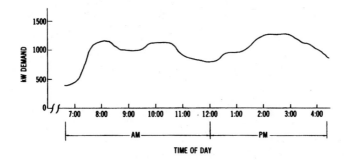

Figure 11-1. Load profile for an industrial plant

trial plants, power demand control systems and procedures can help increase the efficiency of our nation's electrical power systems.

Be careful not to confuse the load factor of a power system with the power factor. The power factor is the ratio of power converted (true power), to the power delivered to a system (apparent power). If necessary, you should review power factor (see Chapter 2).

Most industries use a large number of electric motors; therefore, industrial plants represent highly inductive loads. This means that industrial power systems operate at a power factor of less than unity (1.0). However, it is undesirable for an industry to operate at a low-power factor, since the electrical power system will have to supply more power to the industry than is actually used.

A given value of volt-amperes (voltage × current) is supplied to an industry by the electrical power system. If the power factor (pf) of the industry is low, the current must be higher, since the power converted by the total industrial load equals VA × pf. The value of the power factor decreases as the reactive power (unused power) drawn by the industry increases. This is shown in Figure 11-2. We will assume a constant value of true power, in order to see the effect of increases in reactive power drawn

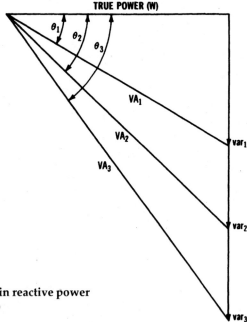

Figure 11-2. Effect of increases in reactive power (VAR) on apparent power (VA)

by a load. The smallest reactive power shown (VAR$_1$) results in the volt-ampere value of VA$_1$. As reactive power is increased, as shown by the VAR$_2$ and VAR$_3$ values, more volt-amperes (VA$_2$ and VA$_3$) must be drawn from the source. This is true since the voltage component of the supplied volt-amperes remains constant. This example represents the same effect as a decrease in the power factor, since pf = W/VA, and, as VA increases, the pf will decrease if W remains constant.

Utility companies usually charge industries for operating at power factors below a specified level. It is desirable for industries to "correct" their power factor to avoid such charges and to make more economical use of electrical energy. Two methods may be used to cause the power factor to increase: (1) power-factor-corrective capacitors, and (2) three-phase synchronous motors. Since the effect of capacitive reactance is opposite to that of inductive reactance, their reactive effects will counteract one another. Either power-factor-corrective capacitors, or three-phase synchronous motors, may be used to add the effect of capacitance to an AC power line.

An example of power factor correction is shown in Figure 11-3. We will assume from the example that both true power and inductive reactive power remain constant at values of 10 kW and 10 kVAR. In Figure 11-3A, the formulas show that the power factor equals 70 percent. However, if 5-kVAR capacitive reactive power is introduced into the electrical power system, the net reactive power becomes 5 kVAR (10-kVAR inductive minus 5-kVAR capacitive), as shown in Figure 11-3B. With the addition of 5-kVAR capacitive to the system, the power factor is increased to 89 percent. Now, in Figure 11-3C, if 10-kVAR capacitive is added to the power system, the total reactive power (kVAR) becomes zero. The true power is now equal to the apparent power; therefore, the power factor is 1.0, or 100 percent, which is characteristic of a purely resistive circuit. The effect of the increased capacitive reactive power in the system is to increase or "correct" the power factor and, thus, to reduce the current drawn from the power distribution lines that supply the loads. In many cases, it is beneficial for industries to invest in either power-factor-corrective capacitors, or three-phase synchronous motors, to correct their power factor. Calculations may be simplified by using the chart of Table 11-1.

Utility companies also attempt to correct the power factor of the power distribution system. A certain quantity of inductance is present in most of the power distribution system, including the generator windings,

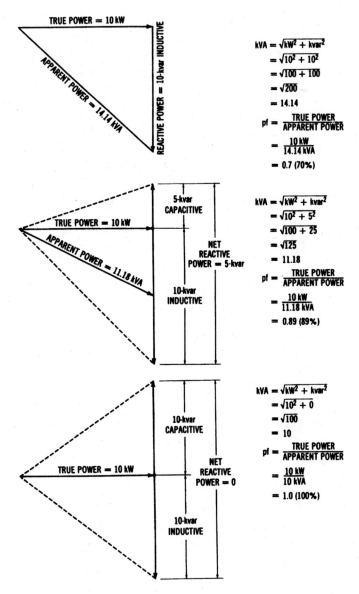

Figure 11-3. Illustration of the effect of capacitive reactance on an inductive circuit: (A) Reactive power = 10 kVAR inductive, (B) Reactive power = 10 kVAR inductive, and 5 kVAR capacitive, (C) Reactive power = 10 kVAR inductive, and 10 kVAR capacitive

the transformer windings, and the power lines. To counteract the inductive effects, utilities use power-factor-corrective capacitors.

Capacitors for Power Factor Correction

Static capacitors are used for power factor correction in the system. They are constructed similarly to the smaller capacitors used in electrical equipment, which have metal-foil plates separated by paper insulation. Ordinarily, static capacitors are housed in metal tanks, so that the plates can be immersed in an insulating oil to improve high-voltage operation. The usual operating voltages of static capacitors are from 230 volts to 13.8 kilovolts. These units are connected in parallel with power lines, usually at the industrial plants, to increase the system power factor. Their primary disadvantage is that their capacitance cannot be adjusted to compensate for changing power factors.

Power factor correction can also be accomplished by using *synchronous capacitors* connected across the power lines. (Three-phase synchronous motors are also called synchronous capacitors; see Chapter 14 for a discussion of three-phase synchronous motors.) The advantage of synchronous capacitors over static capacitors is that their capacitive effect can be adjusted as the system power factor increases or decreases. The capacitive effect of a synchronous capacitor is easily changed by varying the DC excitation voltage applied to the rotor of the machine. Industries considering the installation of either static or synchronous capacitors should first compare the initial equipment cost and the operating cost against the savings brought about by an increased system power factor.

THREE-PHASE LOAD CHARACTERISTICS

In Chapter 2 single-phase circuit fundamentals were discussed. However, the major type of load in electrical power systems is the three-phase load. Therefore, you should become familiar with the characteristics of three-phase loads.

Balanced Three-phase Loads

A balanced three-phase load means that the resistance or impedances connected across each phase of the system are equal. Three-phase motors are one type of balanced load. The following relationships exist in a balanced, three-phase delta system. The line voltages (V_L) are equal to the

Table 11-1. Kilowatt (kW) Multipliers for Determining Capacitor Kilovars (kVAR)

	\multicolumn{21}{c}{Desired Power Factor in Percentage}
	80	81	82	83	84	85	86	87	88	89	90	91	92	93	94	95	96	97	98	99	1.0
50	0.982	1.008	1.034	1.060	1.086	1.112	1.139	1.165	1.192	1.220	1.248	1.276	1.306	1.337	1.369	1.403	1.440	1.481	1.529	1.589	1.732
51	0.937	0.962	0.989	1.015	1.041	1.067	1.094	1.120	1.147	1.175	1.203	1.231	1.261	1.292	1.324	1.358	1.395	1.436	1.484	1.544	1.686
52	0.893	0.919	0.945	0.971	0.997	1.023	1.050	1.076	1.103	1.131	1.159	1.187	1.217	1.248	1.280	1.314	1.351	1.392	1.440	1.500	1.643
53	0.850	0.876	0.902	0.928	0.954	0.980	1.007	1.033	1.060	1.088	1.116	1.144	1.174	1.205	1.237	1.271	1.308	1.349	1.397	1.457	1.600
54	0.809	0.835	0.861	0.887	0.913	0.939	0.966	0.992	1.019	1.047	1.075	1.103	1.133	1.164	1.196	1.230	1.267	1.308	1.356	1.416	1.559
55	0.769	0.795	0.821	0.847	0.873	0.899	0.926	0.952	0.979	1.007	1.035	1.063	1.093	1.124	1.156	1.190	1.227	1.268	1.316	1.376	1.519
56	0.730	0.756	0.782	0.808	0.834	0.860	0.887	0.913	0.940	0.968	0.996	1.024	1.054	1.085	1.117	1.151	1.188	1.229	1.277	1.337	1.480
57	0.692	0.718	0.744	0.770	0.796	0.822	0.849	0.875	0.902	0.930	0.958	0.986	1.016	1.047	1.079	1.113	1.150	1.191	1.239	1.299	1.442
58	0.655	0.681	0.707	0.733	0.759	0.785	0.812	0.838	0.865	0.893	0.921	0.949	0.979	1.010	1.042	1.076	1.113	1.154	1.202	1.262	1.405
59	0.619	0.645	0.671	0.697	0.723	0.749	0.776	0.802	0.829	0.857	0.885	0.913	0.943	0.974	1.006	1.040	1.077	1.118	1.166	1.226	1.369
60	0.583	0.609	0.635	0.661	0.687	0.713	0.740	0.766	0.793	0.821	0.849	0.877	0.907	0.938	0.970	1.004	1.041	1.082	1.130	1.190	1.333
61	0.549	0.575	0.601	0.627	0.653	0.679	0.706	0.732	0.759	0.787	0.815	0.843	0.873	0.904	0.936	0.970	1.007	1.048	1.096	1.156	1.299
62	0.516	0.542	0.568	0.594	0.620	0.646	0.673	0.699	0.725	0.754	0.782	0.810	0.840	0.871	0.903	0.937	0.974	1.015	1.063	1.123	1.266
63	0.483	0.509	0.535	0.561	0.587	0.613	0.640	0.666	0.693	0.721	0.749	0.777	0.807	0.838	0.870	0.904	0.941	0.982	1.030	1.090	1.233
64	0.451	0.474	0.503	0.529	0.555	0.581	0.608	0.634	0.661	0.689	0.717	0.745	0.775	0.806	0.838	0.872	0.909	0.950	0.998	1.068	1.201
65	0.419	0.445	0.471	0.497	0.523	0.549	0.576	0.602	0.629	0.657	0.685	0.713	0.743	0.774	0.805	0.840	0.877	0.918	0.966	1.026	1.169
66	0.388	0.414	0.440	0.466	0.492	0.518	0.545	0.571	0.598	0.626	0.654	0.682	0.712	0.743	0.775	0.809	0.846	0.887	0.935	0.995	1.138
67	0.358	0.384	0.410	0.436	0.462	0.488	0.515	0.541	0.568	0.596	0.624	0.652	0.682	0.713	0.745	0.779	0.816	0.857	0.905	0.965	1.108
68	0.328	0.354	0.380	0.406	0.432	0.458	0.485	0.511	0.538	0.566	0.594	0.622	0.652	0.683	0.715	0.749	0.786	0.827	0.875	0.935	1.078
69	0.299	0.325	0.351	0.377	0.403	0.429	0.456	0.482	0.509	0.537	0.565	0.593	0.623	0.654	0.686	0.720	0.757	0.798	0.846	0.906	1.049
70	0.270	0.296	0.322	0.348	0.374	0.400	0.427	0.453	0.480	0.508	0.536	0.564	0.594	0.625	0.657	0.691	0.728	0.769	0.817	0.877	1.020
71	0.242	0.268	0.294	0.320	0.346	0.372	0.399	0.425	0.452	0.480	0.508	0.536	0.566	0.597	0.629	0.663	0.700	0.741	0.789	0.849	0.992
72	0.214	0.240	0.266	0.292	0.318	0.344	0.371	0.397	0.424	0.452	0.480	0.508	0.538	0.569	0.601	0.635	0.672	0.713	0.761	0.821	0.964
73	0.186	0.212	0.238	0.264	0.290	0.316	0.343	0.369	0.396	0.424	0.452	0.480	0.510	0.541	0.573	0.607	0.644	0.685	0.733	0.793	0.936

Original Power Factor in Percentage

Electrical Power Conversion Systems

74	0.159	0.185	0.211	0.237	0.263	0.289	0.316	0.342	0.369	0.397	0.425	0.453	0.483	0.514	0.546	0.580	0.617	0.658	0.706	0.766	0.909
75	0.132	0.158	0.184	0.210	0.236	0.262	0.289	0.315	0.342	0.370	0.398	0.426	0.456	0.487	0.519	0.553	0.590	0.631	0.679	0.739	0.882
76	0.105	0.131	0.157	0.183	0.209	0.235	0.262	0.288	0.315	0.343	0.371	0.399	0.429	0.460	0.492	0.526	0.563	0.604	0.652	0.712	0.855
77	0.079	0.105	0.131	0.157	0.183	0.209	0.236	0.262	0.289	0.317	0.345	0.373	0.403	0.434	0.466	0.500	0.537	0.578	0.626	0.686	0.829
78	0.052	0.078	0.104	0.130	0.156	0.182	0.209	0.235	0.262	0.290	0.318	0.346	0.376	0.407	0.439	0.473	0.510	0.554	0.599	0.659	0.802
79	0.026	0.052	0.078	0.104	0.130	0.156	0.183	0.209	0.236	0.264	0.292	0.320	0.350	0.381	0.413	0.447	0.484	0.525	0.573	0.633	0.776
80	0.000	0.026	0.052	0.078	0.104	0.130	0.157	0.183	0.210	0.238	0.266	0.294	0.324	0.355	0.387	0.421	0.458	0.499	0.547	0.609	0.750
81		0.000	0.026	0.052	0.078	0.104	0.131	0.157	0.184	0.212	0.240	0.268	0.298	0.329	0.361	0.395	0.432	0.473	0.521	0.581	0.724
82			0.000	0.026	0.052	0.078	0.105	0.131	0.158	0.186	0.214	0.242	0.272	0.303	0.335	0.369	0.406	0.447	0.495	0.555	0.698
83				0.000	0.026	0.052	0.079	0.105	0.132	0.160	0.188	0.216	0.246	0.277	0.309	0.343	0.380	0.421	0.469	0.529	0.672
84					0.000	0.026	0.053	0.079	0.106	0.134	0.162	0.190	0.220	0.251	0.283	0.317	0.354	0.395	0.443	0.503	0.646
85						0.000	0.027	0.053	0.080	0.108	0.136	0.164	0.194	0.225	0.257	0.291	0.328	0.369	0.417	0.477	0.620
86							0.000	0.026	0.053	0.081	0.109	0.137	0.167	0.198	0.230	0.264	0.301	0.342	0.390	0.450	0.593
87								0.000	0.027	0.055	0.083	0.111	0.141	0.172	0.204	0.238	0.275	0.316	0.364	0.424	0.567
88									0.000	0.028	0.056	0.084	0.114	0.145	0.177	0.211	0.248	0.289	0.337	0.397	0.540
89										0.000	0.028	0.056	0.086	0.117	0.149	0.183	0.220	0.261	0.309	0.369	0.512
90											0.000	0.028	0.058	0.089	0.121	0.155	0.192	0.233	0.281	0.341	0.484
91												0.000	0.030	0.061	0.093	0.127	0.164	0.205	0.253	0.313	0.456
92													0.000	0.031	0.063	0.097	0.134	0.175	0.223	0.283	0.426
93														0.000	0.032	0.066	0.103	0.144	0.192	0.252	0.395
94															0.000	0.034	0.071	0.112	0.160	0.220	0.363
95																0.000	0.037	0.079	0.126	0.186	0.329
96																	0.000	0.041	0.089	0.149	0.292
97																		0.000	0.048	0.108	0.254
98																			0.000	0.060	0.203
99																				0.000	0.143

301

phase voltages (V_P). The line currents (I_L) are equal to the phase currents (I_P) multiplied by 1.73. Thus:

$$V_L = V_P$$
$$I_L = I_P \times 1.73$$

For a balanced three-phase wye system, the method used to find the voltages and currents is similar. The voltage across the AC lines (V_L) is equal to the square root of 3 (1.73) multiplied by the voltage across the phase windings (V_P), or

$$V_L = V_P \times 1.73.$$

The line currents (I_L) are equal to the phase currents (I_P), or

$$I_L = I_P$$

The power developed in each phase (P_P), for either a wye or a delta circuit, is expressed as:

$$P_P = V_P \times I_P \times pf$$

where:
pf is the power factor (phase angle between voltage and current) of the load.

The total power developed by all three phases of a three-phase generator (P_T) is expressed as:

$$\begin{aligned} P_T &= 3 \times P_P \\ &= 3 \times V_P \times I_P \times pf \\ &= 1.73 \times V_L \times I_L \times pf \end{aligned}$$

As an example, we can calculate the phase current (I_P), line current (I_L), and total power (P_T) for a 240-volt, three-phase, delta-connected system with 1000-watt load resistances connected across each power line. The phase current is:

$$I_P = \frac{P_P}{V_P}$$

$$= \frac{1000 \text{ watts}}{240 \text{ volts}}$$

$$= 4.167 \text{ amperes}$$

The line current is calculated as:

$$\begin{aligned} I_L &= I_P \times 1.73 \\ &= 4.167 \times 1.73 \\ &= 7.21 \text{ amperes} \end{aligned}$$

Thus, the total power is determined as:

$$\begin{aligned} P_T &= 3 \times P_P \\ &= 3 \times 1000 \text{ watts} \\ &= 3000 \text{ watts} \end{aligned}$$

Another way to calculate the total power is:

$$\begin{aligned} P_T &= 1.73 \times V_L \times I_L \times pf \\ &= 1.73 \times 240 \text{ volts} \times 7.21 \text{ amperes} \times 1 \\ &= 2993.59 \text{ watts} \end{aligned}$$

(Note that your answer depends upon the number of decimal places to which you carry your calculations. Round off I_L at 7.2 amperes, and your answer for the last calculation is 2989.44 watts, instead of 2993.59 watts.)

The following is an example of the power calculation for a three-phase wye system. For the problem, "find the line current (I_L) and power per phase (P_P) of a balanced, 20,000-watt, 277/480-volt, three-phase wye system operating at a 0.75 power factor," we use the formula:

$$P_T = V_L \times I_L \times 1.73 \times pf$$

Transposing and substituting, we have:

$$I_L = \frac{P_T}{V_L \times P_T \times pf}$$

$$= \frac{20{,}000 \text{ watts}}{480 \text{ volts} \times 1.73 \times 0.75}$$

$$= 32.11 \text{ amperes}$$

Then, for the power per phase (P_P), divide the 20,000 watts of the system by 3 (the number of phases), for a value of 6,666.7 watts (6.66 kW).

Unbalanced Three-phase Loads

Often, three-phase systems are used to supply power to both three-phase and single-phase loads. If three, identical, single-phase loads were connected across each set of power lines, the three-phase system would still be balanced. However, this situation is usually difficult to accomplish, particularly when the loads are lights. Unbalanced loads exist when the individual power lines supply loads that are not of equal resistances or impedances.

The total power converted by the loads of an unbalanced system must be calculated by looking at each phase individually. Total power of a three-phase unbalanced system is:

$$P_T = P_{P\text{-}A} + P_{P\text{-}B} + P_{P\text{-}C}$$

where the power-per-phase (P_P) values are added. Power per phase is found in the same way as when dealing with balanced loads:

$$P_P = V_P \times I_P \times \text{power factor}$$

The current flow in each phase may be found if we know the power per phase and the phase voltage of the system. The phase currents are found in the following manner:

Sample Problem:

Given: the following 120-volt single-phase loads are connected to a 120/208-volt wye system. Phase A has 2000 watts at a 0.75 power factor, Phase B has 1000 watts at a 0.85 power factor, and Phase C has 3000 watts at a 1.0 power factor.

Find: the total power of the three-phase system, and the current flow through each line.

Solution: To find the phase currents, use the formula $P_P = V_P \times I_P \times$

power factor, and transpose. Thus, we have $I_P = P_P/V_P \times pf$. Substitution of the values for each leg of the system gives us:

1.
$$*I_{P-A} = \frac{P_{P-A}}{V_P \times pf}$$

$$= \frac{2000 \text{ watts}}{120 \text{ volts} \times 0.75}$$

$$= 22.22 \text{ amperes}$$

*This notation means phase current (I_P) of "A" power line.

2.
$$*I_{P-B} = \frac{P_{P-B}}{V_P \times pf}$$

$$= \frac{1000 \text{ watts}}{120 \text{ volts} \times 0.85}$$

$$= 9.8 \text{ amperes}$$

3.
$$*I_{P-C} = \frac{P_{P-C}}{V_P \times pf}$$

$$= \frac{3000 \text{ watts}}{120 \text{ volts} \times 1.0}$$

$$= 25.0 \text{ amperes}$$

To determine the total power, use the total power formula for a three-phase unbalanced system, which was just given:

$$\begin{aligned} P_T &= P_{P-A} + P_{P-B} + P_{P-C} \\ &= 2000 + 1000 + 3000 \\ &= 6000 \text{ watts (6 kW)} \end{aligned}$$

Chapter 12

Heating Systems

Power conversion systems are commonly referred to by the function they perform. Power conversion takes place in the *load* of an electrical power system. Common types of power system loads are those that convert electrical energy into heat, light, or mechanical energy. There are various types of lighting, heating, and mechanical loads used in industry, commercial buildings, and homes. Several of these power conversion systems will be discussed in Chapters 12, 13, and 14.

IMPORTANT TERMS

Chapter 12 deals with heating systems. After studying this chapter, you should have an understanding of the following terms:

Resistive Heating
Induction Heating
Dielectric Heating
Electric Welding
Resistance Welding
Arc Welding
Induction Welding
SCR Contactors
Electric Heating Systems
British Thermal Units (Btus)
Design Temperature Difference
Degree Days
Thermal Resistance (R)
Coefficient of Heat Transfer (U)
Heat Pump
Air Conditioning System

BASIC HEATING LOADS

Most loads that are connected to electrical power systems produce a certain amount of heat, mainly as the result of current flow through resistive devices. In many instances, heat represents a power loss in the circuit, since heat energy is not the type of energy that the system was intended to produce. Lights, for instance, produce heat energy as well as light energy. The conversion of electrical energy to heat energy in a light-producing load reduces the efficiency of that load device, since not all of the available source energy is converted to light energy. There are, however, several types of power conversion systems that are mainly heating loads. Their primary function is to convert electrical energy into heat energy. Some basic systems include resistance heating, inductive heating, and dielectric (capacitive) heating.

Resistance Heating

Heat energy is produced when an electrical current flows through a resistive material. In many instances, the heat energy produced by an electrical current is undesirable; however, certain applications require controlled resistance heating. Useful heat may be transferred from a resistive element to a point of utilization by the common methods of heat transfer—convection, radiation, or conduction. A heating-element enclosure is needed to control the transfer of heat by convection and radiation. For heat transfer by conduction, the heating element is in direct contact with the material to be heated. Actual heat transfer usually involves a combination of these methods.

Figure 12-1 illustrates the principle of resistance heating. The self-contained heating element uses a coiled resistance wire, which is placed inside a heat-conducting material and enclosed in a metal sheath. This principle may be used to heat water, oil, the surrounding atmosphere, or various other media. This type of heater may be employed in the open air or immersed in the media to be heated. The useful life of the resistance elements depends mainly upon the operating temperature. As the temperature increases, the heat output also increases. Basically, the heat energy produced is dependent upon the current flow and the resistance of the element; it can be calculated as current squared times resistance (I^2R).

Induction Heating

The principle of induction heating is illustrated in Figure 12-2. Heat is produced in magnetic materials when they are exposed to an alternating current (AC) field. In the example shown, current is induced in the material heated by electromagnetic induction. This is brought about by the application of an AC to the heating coil. The material to be heated must be a *conductor* in order for current to be induced. Ordinarily, a high-frequency AC source in the range of 100-500 kHz is used to produce a high heat output. This high heat output is due to greater amounts of induced voltage.

As the magnetic field created by the high-frequency AC source moves across the material to be heated, the induced voltage causes *eddy currents* (circulating currents) to flow in the material. Heat results because of the resistance of the material to the flow of the eddy currents. The heat is produced rapidly by this method, which is an advantage.

The major application of the induction-heating process is in metalworking industries, for such processes as hardening, soldering, melting, and annealing of metals. Compared to that of other methods of heating

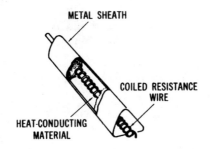

Figure 12-1. Resistance heating principle

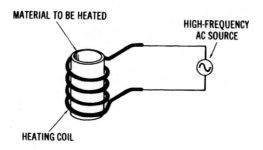

Figure 12-2. Induction heating principle

the heat production of this process is extremely rapid. The area of the metal that is actually heated can be controlled by the size and position of the heating coils of the induction heater. This type of control is difficult to accomplish by other methods. Induction furnaces use the induction-heating principle.

By varying the frequency of the voltage applied to the induction heater windings, it is possible to vary the depth of heat penetration into the heated metal. At higher frequencies the heat produced by the induced current from the heating coils will not penetrate as deeply, because of the so-called "skin effect." Thus, heat will penetrate more deeply at lower frequencies. When heat must be localized onto the surface of a material only—for example, for surface hardening of a metal—higher frequencies are used. The cost of higher-frequency induction heaters is greater, because more complex oscillator circuits are required to produce these frequencies.

Dielectric (Capacitive) Heating

Induction heating can only be used with conductive materials. Therefore, some other method must be used to heat nonconductive materials. Such a method is illustrated in Figure 12-3 and is referred to as *dielectric* or *capacitive* heating. Nonconductors may be heated by placing them in an electrostatic field, created between two metal electrodes that are supplied by a high-frequency AC source. The material to be heated becomes the dielectric or insulation of a capacitive device. The metal electrodes constitute the two plates.

When high-frequency AC is applied to a dielectric heating assembly, the changing nature of the applied AC causes the internal atomic structure of the dielectric material to become distorted. As the frequency of the AC increases, the amount of internal atomic distortion also increases.

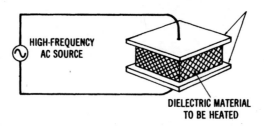

Figure 12-3. Capacitive heating principle

This internal friction produces a large amount of heat in the nonconductive material. Frequencies in the 50-Mhz range may be used for dielectric heating. Dielectric heating produces rapid heating, which is spread evenly throughout the heated material. Common applications of this heating method are the gluing of plywood and the bonding together of plastic sheets.

ELECTRICAL WELDING LOADS

Electrical welding is another common type of a heat-producing power conversion system. The types of electrical welding systems include resistance welding, electric arc welding, and induction welding.

Resistance Welding

Several familiar welding methods, such as spot welding, seam welding, and butt welding, are resistance welding processes. All of these processes rely upon the resistance heating principle. Spot welding, illustrated in Figure 12-4A, is performed on overlapping sheets of metal, which are usually less than 1/4-inch thick. The metal sheets are clamped between two electrodes, and an electrical current is passed through the electrodes and metal sheets. The current causes the metals to fuse together. The instantaneous current through the electrodes is usually in excess of 5000 amperes, while the voltage between the electrodes is less than 2 volts.

Seam welding, shown in Figure 12-4B, is accomplished by passing sheets of metal between two pressure rollers, while a continuously interrupted current is passed through the electrodes. The operational principle of seam welding is the same as that of spot welding. Several other similar methods, which are referred to as butt welding, edge welding, and projection welding, are also commonly used.

Electric Arc Welding

While resistance welding utilizes pressure on the materials to be welded, electric arc welding produces welded metals by localized heating without pressure, as shown in Figure 12-5. An electric arc is created when the electrode of the welder is brought in contact with the metal to be welded. Carbon electrodes are used for DC or AC arc welding of nonferrous metals and alloys. Not all metals can be welded by the arc welding process. When metals are welded together, part of the metals to be welded

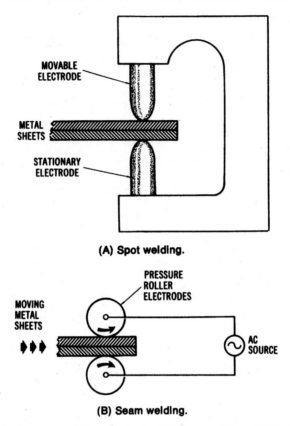

Figure 12-4. Resistance welding methods: (A) Spot welding, (B) Seam welding

are melted, creating a metal pool that is added to, when necessary, by the use of a filler rod. The puddle (molten metal pool) then fills in the gap (arc crater) that was created by the arc of the electrode. Various types and various current-voltage ratings of electric arc welders are available.

A smaller amount of current is required for arc welding than for resistance welding. The currents may range from 50 to 200 amperes, or higher for some applications. Voltages typically range from 10 to 50 volts. An electric arc welder may be powered by a portable generator, a storage-battery unit, a step-down transformer, or a rectification unit.

Induction Welding

The induction welding process uses the principle of induction heating to fuse metals together. High-frequency AC is applied to a heating coil,

Heating Systems

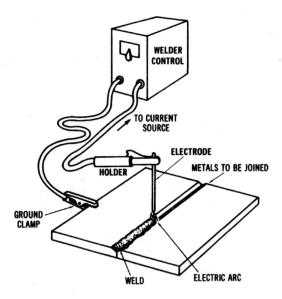

Figure 12-5. Electric arc welding

into which the materials to be welded are placed. Tubular metal is often welded in this way.

POWER CONSIDERATIONS FOR ELECTRIC WELDERS

Electric welders are rather specialized types of equipment, since they use very high amounts of current at low voltage levels. They have a peculiar effect on the power system operation. They draw large amounts of current for short periods of time. Silicon-controlled rectifiers (SCRs) are commonly used to control the starting and stopping of the large currents associated with electric welders. The current rating of these devices must be very high, sometimes in the range of 1000 to 100,000 amperes, and the power distribution equipment must be able to handle these high currents. SCRs are discussed in Chapter 17.

Figure 12-6 illustrates a typical electric welding system. The AC power supplied from the branch circuit of the power system is either stepped down by a transformer to deliver AC voltage to the welder, or rectified to produce DC voltage for DC welders. In either type of machine, an SCR contactor may be used to control the on and off time of the welder.

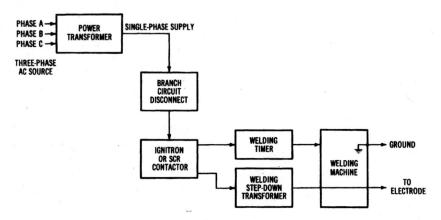

Figure 12-6. Block diagram of a typical electric welding system

SCR Contactors

SCR contactors are electronic control devices designed to handle large amounts of current. SCRs are triggered or turned on by pulses supplied by the timing or sequencing circuits of the welder. The SCRs are usually cooled by circulating water. The operating principles of SCRs are discussed in Chapter 17.

ELECTRIC HEATING AND AIR CONDITIONING SYSTEMS

A very important type of electrical power conversion is that which takes place in the heating and air conditioning systems of homes, industries, and commercial buildings. These loads convert a high percentage of the total amount of electrical power that is furnished. Although natural gas and fuel oil heating systems are still used, electrical heating is becoming more prevalent each year. Air conditioning systems are also becoming more commonly used to cool buildings. The use of electric heating and air conditioning systems has made us even more dependent upon electrical power. We now use electrical power to maintain a comfortable environment inside buildings.

Basics of Electric Heating

There are several important factors that you must understand in order to have a knowledge of electric heating. Heat is measured in British

Heating Systems

thermal units (Btus). One Btu is the amount of heat required to raise the temperature of one pound of water one degree Fahrenheit. Heat energy in the amount of 3.4 Btus per hour is equivalent to one watt of electrical energy.

Another basic factor to be considered in the study of electric heating is *design temperature difference*. This is the difference between inside and outside temperatures in degrees Fahrenheit. The outside temperature is considered to be the lowest temperature that is expected to occur several times a year. The inside temperature is the desired temperature (thermostat setting).

A factor used in conjunction with design temperature difference is called *degree days*. The degree-day factor is used to determine the average number of degrees that the mean temperature is below 65°F. These data are averaged over seasonal periods for consideration in insulating buildings.

Importance of Insulation

The insulation of a building is a very important consideration in electric heating systems. Insulation is used to oppose the escape of heat. The quality of insulation is expressed by a thermal resistance factor (R). The total thermal resistance of a building is found by considering the thermal resistance of the entire structure (wood, concrete, insulation, et cetera). The inverse of thermal resistance is called the coefficient of heat transfer (U), which is an expression of the amount of heat flow through an area, expressed in Btus per square foot per hour per degree Fahrenheit. The following formulas are used in the conversion of either U or R to electrical units (watts):

$$\text{thermal resistance} = \frac{1}{\text{coefficient of heat transfer}}$$

$$\text{watts} = \frac{\text{coefficient of heat transfer}}{3.4}$$

or:

$$\text{watts} = 0.29 \times U$$

The manufacturers of insulation can supply various tables that can be used to estimate the heat loss that can occur in buildings of various types of construction. Heat loss occurs particularly through the windows and doors of buildings. The periodic opening of doors also has a considerable effect on heat loss. A building must have sufficient insulation to reduce heat loss; otherwise, electrical heating and air conditioning systems will be very inefficient. The heat loss of a building depends primarily upon the building construction, and upon the design temperature difference factor in the area where the building is located. Buildings made of concrete have a different amount of heat loss than those of a wood-frame construction. Heat loss will occur through the walls, floors, windows, and ceilings. Each of these must be considered when estimating the heat loss of a building.

The following sample problem will help you to understand the importance of adding insulation to a building.

Sample Problem:

Given: a building constructed to provide the following thermal resistance (R) factors:
- (a) Exterior shingles are R = 0.80
- (b) Plywood sheathing is R = 0.75
- (c) Building paper used is R = 0.04
- (d) Wall structure is R = 0.85
- (e) Wall plaster is R = 0.35
- (f) Insulation is R = 11.0

Find: the total thermal resistance (R), the coefficient of heat transfer (U), and the watts (W) of heat loss both with and without the insulation.

Solution (without insulation):

$$R = a + b + c + d + e$$

$$= 0.80 + 0.75 + 0.04 + 0.85 + 0.35 = 2.79$$

$$U = \frac{1}{R}$$

$$= \frac{1}{2.79}$$

$$= 0.3584 \text{ Btu per ft}^2 \text{ per hour per } °F$$

Heating Systems

$$W = \frac{U}{3.4}$$

$$= \frac{0.3584}{3.4}$$

$$= 0.1054 \text{ watts heat loss}$$

Solution (with insulation):

$$R = a + b + c + d + e + f$$
$$= 0.80 + 0.75 + 0.04 + 0.85 + 0.35 + 11.0$$
$$= 13.79$$

$$U = \frac{1}{R}$$

$$U = \frac{1}{13.79}$$

$$= 0.0725 \text{ Btu per ft}^2 \text{ per hour per }°F$$

$$W = \frac{U}{3.4}$$

$$= \frac{0.0725}{3.4}$$

$$= 0.0213 \text{ watts heat loss}$$

You can see from the results of this problem that the adding of insulation into the walls of a building has a great effect upon heat loss. The insulation has a much greater effect in controlling heat loss than do the construction materials used for the building.

Electric Heating and Cooling Systems

Several types of electric heating systems are used today. Some common types are baseboard heaters, wall- or ceiling-mounted heaters, and heat pumps. Most of these systems use forced air to circulate the heat. Some electric heaters have individual thermostats, while others are connected to one central thermostat that controls the temperature in an entire building. The possibility of having temperature control in each room is an advantage of electric heating systems.

Heat Pumps

In recent years, the heat pump has become very popular as a combination heating and cooling unit for buildings. The heat pump is a heat-transfer unit. When the outside temperature is warm, the heat pump acts as an air conditioning unit and transfers the indoor heat to the outside of the building. This operational cycle is reversed during cool outside temperatures. In the winter, the outdoor heat is transferred to the inside of the building. This process can take place during cold temperatures, since there is always a certain amount of heat in the outside air, even at subzero temperatures. However, at the colder temperatures, there is less heat in the outside air.

Thus, heat pumps transfer heat rather than produce it. Since heat pumps do not produce heat, as resistive-heating units do, they are more economical in terms of energy conservation. Heating and cooling are reversible processes in the heat-pump unit; thus, the unit is self-contained. The reversible feature of heat pumps reduces the space requirement for separate heating and cooling units. Another advantage is that the change-over from heating to cooling can be made automatically. This feature might be desirable during the spring and autumn seasons, in the many areas where temperatures are very variable. In extremely cold areas, the heat pump can be supplemented by an auxiliary resistance-heating unit. This auxiliary unit will operate when the outside temperatures are very cold, and will be useful in maintaining the inside temperatures at a comfortable level. Air is circulated past these heating elements into the heat vents of the building.

Heat pumps are used for residential as well as commercial and industrial applications, and they are being used more extensively each year. Figure 12-7 shows a simplified circuit arrangement of a heat pump, in which a compressor takes a refrigerant from a low-temperature, low-pressure evaporator and converts it to a high temperature and a high pressure. The refrigerant is then delivered to a condenser, in much the same way as in a refrigerator.

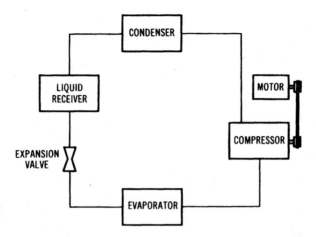

Figure 12-7. A simplified heat-pump circuit diagram

Air Conditioning Systems

The increased use of air conditioning systems provides greater comfort in homes, industrial plants, and commercial facilities. Most air conditioning units are used for the purpose of controlling the inside temperature of buildings, so as to make working and living conditions more comfortable. However, many units are used to cool the insides of various types of equipment. Both air temperature and relative humidity are changed by air conditioning units. In the design of air conditioning systems, all heat-producing items in the immediate environment must be considered. Body heat, electrical appliances, and lights represent some common sources of heat. The diffusion of heat takes place through floors, walls, ceilings, and the windows of buildings. A simplified diagram of a room air conditioning unit is shown in Figure 12-8.

Considerations for Heating Loads

Heating systems used for residential, commercial, and industrial applications are usually referred to as HVAC systems. This means heating ventilation and air conditioning system. The electrical power requirement for HVAC systems is a major concern for electrical system design for a building. Electrical HVAC systems provide individual thermostatic temperature control, have long equipment life, and are safe to use. A well-insulated building is necessary to reduce heat loss for economical use of HVAC systems.

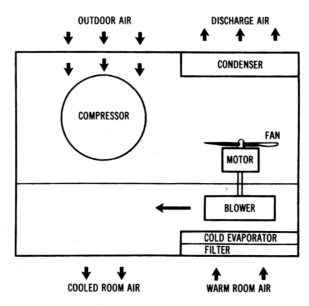

Figure 12-8. A simplified diagram of a room air conditioning unit

Electrically energized comfort heating systems are widely used today to produce heat for commercial, industrial, and residential buildings. Electrical energy is readily available at nearly any building site, and it has a number of advantages over fuel-burning methods of producing heat. Ecologically, any fuel needed to produce electricity is burned or consumed at a power plant, which is usually located some distance from the building where it is being used. With this method of heating, there is less pollution than there would be if fuel were burned at each building. Electric heat is also clean to use, easy to control, and highly efficient.

Electric heating is important today because of its high level of efficiency. Theoretically, when electrical energy is applied to a system, virtually all of it is transformed into heat energy. Essentially, this means that, when a specified amount of electricity is applied, it produces an equivalent Btu output. One thousand watts or 1 kW of electricity, when converted to heat, produces 3412 Btus of heat energy.

Heating can be achieved in a variety of ways through the use of electricity. Comfort heating systems contain an energy source, transmission path, control load device, and the possibility of one or more optional indicators. The primary difference between heat-pump and resistance electrical systems is in the production of heat energy. Resistance heating is accomplished by passing an electrical current through wires or conductors to the load device. By comparison, the heat pump operates by circulating a gas or liquid through pipes that connect an inside coil to an outside coil. Electricity is needed in both cases as an energy source to make the systems operational.

Resistance Heating in Buildings

When an electric current flows through a conductive material, it encounters a type of opposition called resistance. In most circuits this opposition is unavoidable, to some extent, because of the material of the conductor, its length, its cross-sectional area, and its temperature. The conductor wires of a heating system are purposely kept low in resistance to minimize heat production between the source and the load device. Heavy-gauge insulated copper wire is used for this part of the system.

The load device of a resistance heating system is primarily responsible for the generation of heat energy. The amount of heat developed by the load depends upon the value of current that passes through the resistive element. Element resistance is purposely designed to be quite high compared to that of the connecting wires of the system. An alloy of nickel and chromium called Nichrome is commonly used for the heating elements.

Resistive elements may be placed under windows, or at strategic locations throughout the building. In this type of installation, the elements are enclosed in a housing that provides electrical safety and efficient use of the available heat. Air entering at the bottom of the unit circulates around the fins to gain heat, than exits at the top. Different configurations may be selected according to the method of circulation desired, unit length, and heat-density production.

Resistive elements are also used as a heat source in forced-air central heating systems. In this application, the element is mounted directly in the main airstream of the system. The number of elements selected for a particular installation is based upon the desired heat output production. Individual elements are generally positioned in a staggered configuration to provide uniform heat transfer and to eliminate hot spots. The element has spring-coil construction supported by ceramic insulators. Units of this type provide an auxiliary source of heat when the outside temperature becomes quite cold. Air circulating around the element is warmed and forced into the duct network for distribution throughout the building.

Heat Pump Systems in Buildings

A heat pump is defined as a reversible air conditioning system that transfers heat either into or away from an area that is being conditioned. When the outside temperature is warm, the heat pump takes indoor heat and moves it outside, thus acting as an air conditioning unit. Operation during cold weather causes it to take outdoor heat and move it indoors, functioning as a heating unit. Heating can be performed even during cold temperatures, because there is always a certain amount of heat in the outside air. At 0°F (–22°C), for example, the air will have approximately 89 percent of the heat that it has at 100°F (38°C). Even at subzero temperatures, it is possible to develop some heat from the outside air. However, it is more difficult to develop heat when the temperature drops below 20°F (–6°C). For installations that encounter temperatures colder than this, heat pumps are equipped with resistance heating coils to supplement the system.

A heat pump, like an air conditioner, consists of a compressor, an outdoor coil, an expansion device, and an indoor coil. The compressor is responsible for pumping a refrigerant between the indoor and outdoor coils. The refrigerant is alternately changed between liquid and gaseous states, depending upon its location in the system. Electric fans or blowers are used to force air across the respective coils, and to circulate cool or warm air throughout the building.

A majority of the heat pumps in operation today consist of indoor and outdoor units that are connected together by insulated pipes or tubes. The indoor unit houses the supplemental electric heat elements, the blower and motor assembly, the electronic air cleaner, the humidifier, the control panel, and the indoor coil. The outdoor unit is covered with a heavy-gauge steel cabinet that encloses the outdoor coil, the blower fan assembly, the compressor, the expansion device, and the cycle-reversing valve. Both units are designed for maximum performance, high operational efficiency, and low electrical power consumption.

The Heating Cycle of a Heat Pump

If a unit air conditioner were turned around in a window during its operational cycle, it would be extracting heat from the outside air and pumping it into the building. This condition, which is the operational basis of the heat pump, is often called the reverse-flow air conditioner principle. The heat pump is essentially "turned around" from its cooling cycle by a special valve that reverses the flow of refrigerant through the system.

Heating Systems

When the heating cycle occurs, the indoor coil, outdoor coil, and fans are reversed. The outdoor coil is now responsible for extracting heat from the outside air and passing it along the indoor coil, where it is released into the duct network for distribution.

During the heating cycle, any refrigerant that is circulating in the outside coil is changed into a low-temperature gas. It is purposely made to be substantially colder than the outside air. Since heat energy always moves from hot to cold, there is a transfer of heat from the outside air to the cold refrigerant. In a sense, we can say that the heat of the cold outside air is absorbed by the much colder refrigerant gas.

The compressor of the system is responsible for squeezing together the heat-laden gas that has passed through the outside coil. This action is designed to cause an increase in the pressure of the gas that is pumped to the indoor coil. As air is blown over the indoor coil, the high-pressure gas gives up its heat to the air. Warm air is then circulated through the duct network to the respective rooms of the system.

When the refrigerant gas of the indoor coil gives up its heat, it cools and condenses into a liquid. It is then pumped back to the outside coil by compressor action. Once again it is changed into a cool gaseous state and is applied to the outside coil to repeat the cycle. If the outside temperature drops too low, the refrigerant may not be able to collect enough heat to satisfy the system. When this occurs, electric-resistance heaters are energized to supplement the heating process. The place where electric heat is supplied to the system is called the *balance point*.

Figure 12-9 shows an illustration of heat-pump operation during its heating cycle. At (1), the heat is absorbed from the cold outside air by the pressurized, low-temperature refrigerant circulating through the outside coil. As (2), the refrigerant is applied to the compressor and compressed into a high-temperature, high-pressure gas. At (3), the heated gas is transferred to the indoor coil and released as heat. At (4), warm air is circulated through the duct network. Note that the supplemental resistance heat element is placed in this part of the system. At (5), the refrigerant is returned to the compressor and then to an expansion device, where the liquid refrigerant is condensed and then returned to the outdoor coil. The cycle repeats itself from this point.

The Cooling Cycle of a Heat Pump

A heat pump is designed to respond as an air conditioning unit during the summer months. For this to occur, the reversing valve must be

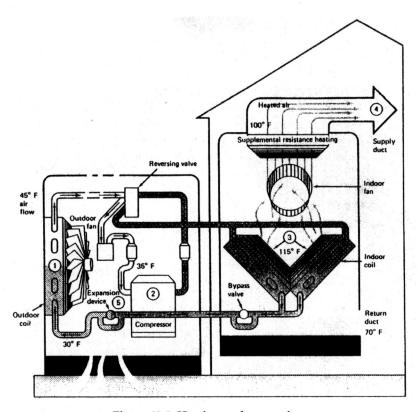

Figure 12-9. Heating cycle operation

placed in the cooling-cycle position. In some systems this is accomplished by a manual changeover switch, whereas in others it is achieved automatically, according to the thermostat setting. The operating position of the valve simply directs the flow path of the refrigerant.

When the cooling cycle is placed in operation, it first causes the refrigerant to flow from the compressor into the indoor coil. During this part of the cycle the refrigerant Is in a low-pressure gaseous state and is quite cool. As the circulation process continues, the indoor coil begins to absorb heat from the inside air of the building. Air passing over the indoor coil is cooled and circulated into the duct network for distribution throughout the building.

After leaving the indoor coil, the refrigerant must pass through the reversing valve and into the compressor. The compressor is responsible for increasing the pressure of the refrigerant and circulating it into the outdoor coil. At this point of the cycle, the refrigerant gives up its heat to

Heating Systems

the outside air, is cooled, and is changed into a liquid state. It then returns to the compressor, where it is pumped through an expansion device and returned to the indoor coil. The process then repeats itself

Figure 12-10 shows an illustration of the heat pump during its air conditioning cycle. At (1), heat is absorbed from the inside air and cool air is transferred into the building. At (2), the pressure of the heat-laden refrigerant is increased by the compressor and cycled into the outside coil for transfer to the air. At (3), cool, dehumidified air is circulated through the duct network as a result of passing through the cooled indoor coil. At (4), the refrigerant condenses back into a liquid as it circulates through the outdoor coil. At (5), the liquid refrigerant flows through the compressor and expansion device, where it is vaporized and returned to the indoor coil to complete the cycle.

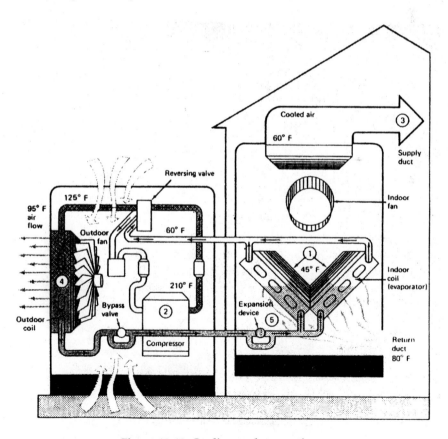

Figure 12-10. Cooling cycle operation

Chapter 13

Lighting Systems

Electrical lighting systems are designed to create a comfortable and safe home or working environment. There are several types of lighting systems in use today. Among the most popular are incandescent, fluorescent, and vapor lights, as well as several special-purpose types of lighting. Lighting systems are one type of electrical *load*.

The three types of lighting systems discussed in this chapter constitute most of the lighting loads placed on electrical power systems. Incandescent, fluorescent, and vapor lights are used for many lighting applications. However, there are a few other specialized types of lighting that are also used. These include zirconium and xenon arc lights, glow lights, black lights, and infrared lights. The planning involved in obtaining proper lighting is quite complex and may involve several types of lights.

IMPORTANT TERMS

Chapter 13 deals with lighting systems. After studying this chapter, you should have an understanding of the following terms:

Visible Light
Electromagnetic Spectrum
Candlepower
Lumen
Footcandle
Reflection Factor Footlambert
Street Lighting
Incandescent Lighting
Fluorescent Lighting
Vapor Lighting
Lighting Circuits
Three-Way Switch

Four-Way Switch
Lighting Branch Circuits
Lighting Fixture
Luminaire
Coefficient of Utilization (CU)
Room Ratio
Depreciation Factor (DF)
Light Output

CHARACTERISTICS OF LIGHT

In order to better understand lighting systems, you should know something about the basic characteristics of light. Light is a visible form of radiation that is actually a narrow band of frequencies along the vast electromagnetic spectrum. The electromagnetic spectrum, shown in Figure 13-1, includes bands of frequencies for radio, television, radar, infrared radiation, visible light, ultraviolet light, x-rays, gamma rays, and various other frequencies. The different types of radiation, such as light, heat, radio waves, and x-rays, differ only with respect to their frequencies, or wavelengths.

The human eye responds to electromagnetic waves in the *visible light* band of frequencies. Each color of light has a different frequency, or wavelength. In order of increasing frequency (or decreasing wavelengths), the colors are red, orange, yellow, green, blue, indigo, and violet. The wavelengths of visible light are in the 400-millimicrometer (violet) to 700-millimicrometer (red) range. A micrometer (mm), which is also called a micron (μm), is one millionth of a meter, and a nanometer (nm) is 1×10^{-3} micrometer. Angstrom units (Å) are also used for light measurement. An angstrom unit is one-tenth of a nanometer. In order to avoid confusion, use the conversion chart given in Table 13-1.

Visible light ranges from 4000 Å to 7000 Å. The response of the human eye to visible light exhibits a frequency selective characteristic, as shown in Figure 13-2. The greatest sensitivity is near 5500 Å. The poorest sensitivity is around 4000 Å on the lower wavelengths, and 7000 Å on the higher wavelengths. Our eyes perceive various degrees of brightness, depending on their response to the wavelengths of light. The normal human eye cannot see a wavelength of less than 4000 Å (<400 nm), or more than 7000 Å (>700 nm).

Lighting Systems

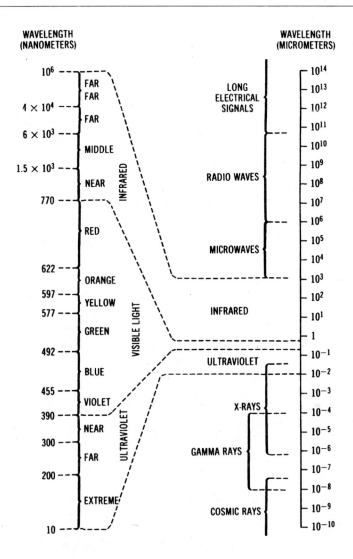

Figure 13-1. The electromagnetic spectrum

When dealing with light, there are several characteristic terms that you should understand. The unit of luminous intensity is a standard light source called a *candela* or *candlepower*. The intensity of light is expressed in one of these units. The amount of light falling on a unit surface, all points of which are a unit distance from a uniform light source of one candela, is one *lumen*. The illumination of a surface is the number of lumens falling

Table 13-1. Conversion Chart for Electromagnetic Wavelengths

Known Quantity	Multiply by	Quantity To Find
Angstrom (Å)	10	Nanometer (nm)
	10^4	Micron (μm)
	10^7	Millimeter (mm)
	10^8	Centimeter (cm)
	10^{10}	Meter (m)
	10^{13}	Kilometer (km)
Micron	10^{-4}	Angstrom (Å)
	10^{-3}	Nanometer (nm)
	10^3	Millimeter (mm)
	10^4	Centimeter (cm)
	10^6	Meter (m)
Nanometer	10^{-1}	Angstrom (Å)
	10^3	Micron (μm)
	10^6	Millimeter (mm)
	10^7	Centimeter (cm)
	10^9	Meter (m)

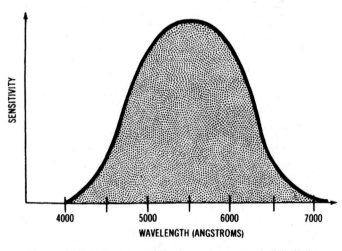

Figure 13-2. Response of the human eye to visible light

Lighting Systems

on it per unit area. The unit of illumination is the *lux* (lumens per square meter) or *footcandle* (lumens per square foot).

We see only light that is reflected. Reflected light is measured in candelas per square meter of surface (or footlamberts). The *reflection factor* is the percent of light reflected from a surface expressed as a decimal. Therefore, the light reflected from a surface is equal to the illumination of the surface times the reflection factor. The brightness of a surface that reflects one lumen per square foot is one *footlambert*.

Units of light measurement are given in both English and International Systems of Units (metric) units. This can become somewhat confusing. Table 13-2 summarizes the most common terms used for light measurement.

Table 13-2. Lighting Terms

Lighting Quantity	Definition	Symbol	Unit SI	Unit English	Definition of Unit
Luminous intensity (candlepower)	Ability of source to produce light in a given direction	I		Candela (cd)	Approximately equal to the luminous intensity produced by a standard candle
Luminous flux	Total amount of light	φ		Lumen (lm)	Luminous flux emitted by a 1 candela uniform point source
Illumination	Amount of light received on a unit area of surface (density)	E	Lux (lx)	Footcandle (fc)	One lumen equally distributed over one unit area of surface
Luminance (brightness)	Intensity of light per unit of area reflected or transmitted from	L	cd/m^2	$cd/in.^2$ (foot-lambert)	A surface reflecting or emitting light at the rate of 1 candela per unit of projected area.

Incandescent Lighting

Incandescent lighting is a widely used method of lighting, used for many different applications. The construction of an incandescent lamp is shown in Figure 13-3. This lamp is simple to install and maintain. The initial cost is low; however, incandescent lamps have a relatively low efficiency and a short life span.

Incandescent lamps usually have a thin tungsten filament, which is in the shape of a coiled wire. This filament is connected through the lamp base to a voltage source (usually 120 volts AC). When an electric current is passed through the filament, the temperature of the filament rises to between 3000° to 5000° Fahrenheit. At this temperature range, the tungsten produces a high-intensity white light. When the incandescent lamp is manufactured, the air is removed from the glass envelope to prevent the filament from burning; also, an inert gas is added.

Incandescent lamps are purely resistive devices, and thus have a power factor of 1.0. As they deteriorate, their light output is reduced. Typically, at the time that an incandescent lamp burns out, its light output is less than 85 percent of its original output. A decrease in the voltage of the power system also reduces the light output. A 1 percent decrease in voltage will cause a 3 percent decrease in light output.

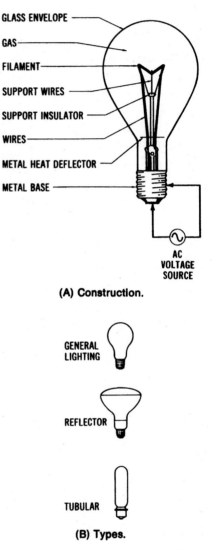

(A) Construction.

(B) Types.

Figure 13-3. The incandescent lamp: (A) Construction, (B) Types

Fluorescent Lighting

Fluorescent lighting is used extensively today, particularly in industrial and commercial buildings. Fluorescent lamps are tubular bulbs with a filament at each end. There is no electrical connection between the two filaments. The operating principle of the fluorescent lamp is shown in Figure 13-4. The tube is filled with mercury vapor; thus, when an electrical current flows through the two filaments, a continuous arc is formed between them by the mercury vapor. High-velocity electrons passing between the filaments collide with the mercury atoms, producing an ultraviolet radiation. The inside of the tube has a phosphor coating that reacts with the ultraviolet radiation to produce visible light.

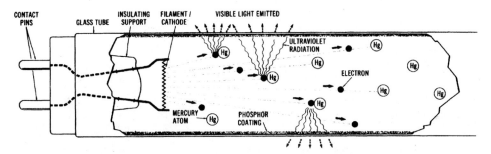

Figure 13-4. Operating principle of a fluorescent lamp

One circuit for a fluorescent light is shown in Figure 13-5. Note that a thermal starter, which is basically a bimetallic strip and a heater, is connected in series with the filaments. The bimetallic strip remains closed long enough for the filaments to heat and vaporize the mercury in the tube. The bimetallic switch will then bend and open as the result of the

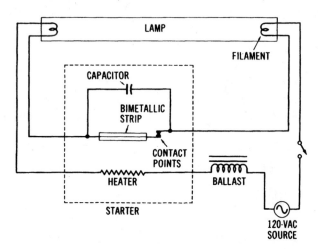

Figure 13-5. Circuit diagram of a fluorescent lamp

heat produced by current flow through the heater. The filament circuit is now opened. A capacitor is connected across the bimetallic switch to reduce contact sparking. Once the contacts of the starter open, a high voltage momentarily occurs between the filaments of the lamp, because of the action of the inductive ballast coil. The ballast coil has many turns of small-diameter wire, and, thus, it produces a high, counter-electromotive force when the contacts of the starter separate. This effect is sometimes called "inductive kickback." The high voltage across the filaments causes the mercury to ionize and initiates a flow of current through the tube. There are several other methods used to start fluorescent lights; however, this method illustrates the basic operating principle.

Fluorescent lights produce more light per watt than incandescent lights; therefore, they are cheaper to operate. Since the illumination is produced by a long tube, there is also less glare. The light produced by fluorescent bulbs is very similar to natural daylight. The light is whiter, and the operating temperature is much lower with fluorescent than with incandescent lights. Various sizes and shapes of fluorescent lights are available. The bulb sizes are expressed in eighths of an inch, with the common sizes being T-12 and T-8. (A T-12 bulb is 1-1/2 inches.) Common lengths are 24, 48, 72, and 96 inches.

Vapor Lighting

Another popular form of lighting is the vapor type. The mercury-vapor light is one of the most common types of vapor light. Another common type is the sodium-vapor light. These lights are filled with a gas that produces a characteristic color. For instance, mercury vapor produces a greenish-blue light, and argon a bluish white light. Gases may be mixed to produce various color combinations for vapor lighting. This is often done with signs used for advertising.

A mercury-vapor lamp is shown in Figure 13-6. It consists of two tubes with an arc tube placed inside an outer bulb. The inner tube contains mercury. When a voltage is applied between the starting probe and an electrode, an arc is started between them. The arc current is limited by a series resistor; however, the current is sufficient to cause the mercury in the inner tube to ionize. Once the mercury has ionized, an intense greenish-blue light is produced. Mercury-vapor lights are compact, long-lasting, and easy to maintain. They are used to provide a high-intensity light output. At low voltages, mercury is slow to vaporize, so these lamps require a long starting time (sometimes 4 to 8 minutes). Mercury-vapor

Lighting Systems

lights can also be used for outdoor lighting.

The sodium-vapor light, shown in Figure 13-7, is popular for outdoor lighting and for highway lighting. This lamp contains some low-pressure neon gas and some sodium. When an electric current is passed through the heater, electrons are given off. The ionizing circuit causes a positive charge to be placed on the electrodes. As electrons pass from the heater to the positive electrodes, the neon gas is ionized. The ionization of the neon gas produces enough heat to cause the sodium to ionize. A yellowish light is produced by the sodium vapor. The sodium-vapor light can produce about three times the candlepower per watt as an incandescent light.

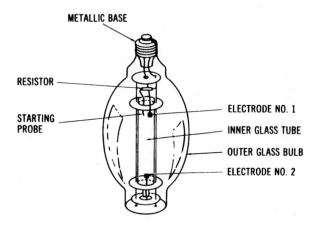

Figure 13-6. A mercury-vapor lamp

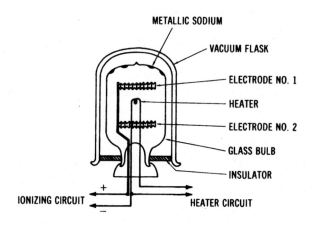

Figure 13-7. A sodium-vapor lamp.

Another type of vapor lamp is called a *metal halide* type. This light source is a high intensity mercury-vapor lamp in which metallic substances, called metal halides, are added to the bulb. The addition of these substances improves the efficacy of the lamps. The efficacy of a metal halide lamp is typically 75 lumens/watt, compared to approximately 50 lumens/watt for mercury-vapor lamps.

Another type of vapor lamp is the high-pressure sodium (HPS) lamp. Sodium is the primary element used to fill the lamp's tube when it is manufactured. These lamps have very high efficacies-approximately 110 lumens/watt.

Street Lighting

The lighting systems of today are highly reliable, compared to the systems that were installed many years ago. Earlier systems were turned on and off either manually, or by timing devices that were regulated by the time of day, rather than by the natural light intensity. Some were controlled by electrical impulses that were transmitted on the power lines. Now, most systems are controlled by automatic photoelectric circuits. The lights that are now used to illuminate streets and highways have photoelectric controls. They operate during periods of darkness and are automatically turned off when natural light is present. These street lights are usually connected to existing 120/240-volt power distribution systems.

There are many types of street lights in use. The earliest types of electric street lights used were 200- to 1000-watt incandescent lamps. Now, mercury-vapor and sodium-vapor lamps are the primary types used. (Mercury lamps produce a white light, and sodium lamps have a yellowish color.) Several different lamp designs and mounting fixture designs are used.

Several years ago, street lights were converted from incandescent lamps to mercury-vapor lamps. Now the trend in street lighting seems to be toward the use of sodium-vapor lights, and several areas of the country have converted to the use of sodium-vapor lamps. Sodium-vapor lights produce more illumination than a similar mercury-vapor light. They also require less electrical power to produce a specific amount of illumination. Thus, the ability of sodium-vapor lights to deliver more light with less power consumption makes them more economically attractive than mercury-vapor lights.

Comparison of Light Sources

The purpose of a light source is to convert electrical energy into light energy. A measure of how well this is done is called efficacy. Efficacy is the lumens of light produced per watt of electrical power converted. A comparison of some different types of light sources is shown below:

Lamp Type	Efficacy (Lumens/watt)
200 Watt incandescent lamp	20
400 Watt mercury lamp	50
40 Watt fluorescent lamp	70
400 Watt metal halide lamp	75
400 Watt high-pressure sodium lamp	110

ELECTRICAL LIGHTING CIRCUITS

There are several types of electrical circuits that are used for electrical lighting control. We will study some of the circuits used for incandescent and fluorescent lighting systems. Electrical lighting circuits for different areas of a building must be wired properly. Some lighting fixtures are controlled from one point by one switch, while other fixtures may be controlled from two or more points by a switch at each point.

Incandescent lighting circuits are a typical type of branch circuit. The most common type of incandescent lighting circuit is a 120-volt branch that extends from a power distribution panel to a light fixture or fixtures in some other area of a building. The path for the electrical distribution is controlled by one or more switches that are usually placed in small metal or plastic enclosures inside a wall. These switches are then covered by rectangular plastic plates to prevent possible shock hazards.

Switches are always placed in a wiring circuit so that they can open or close a hot wire. This hot wire distributes electrical power to the lighting fixtures that the switches control. These switches are referred to as "T-rated" switches. They are designed to handle the high, instantaneous current drawn by lights when they are turned on. A switch that accomplishes control from a single location is a simple, single-pole, single-throw (spst) switch, such as that shown in Figure 13-8. A pictorial view of the same circuit is shown in Figure 13-9.

Control from two locations, such as near the kitchen door and inside the garage door, is accomplished by two 3-way switches. This circuit is

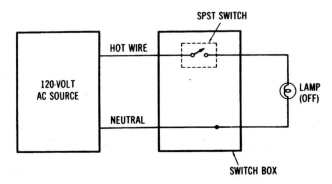

Figure 13-8. Schematic diagram for the control of a light from one location

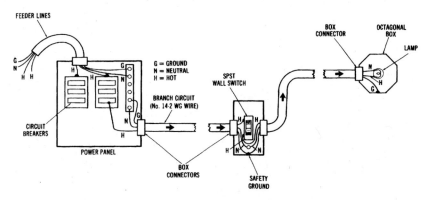

Figure 13-9. Pictorial view of the circuit shown in Figure 13-8

shown in Figure 13-10A, while Figure 13-11 presents a pictorial view of the same circuit. When control of a lighting fixture from more than two points is desired, two 3-way switches and one or more 4-way switches are used. For instance, control of one light from five points could be accomplished by using a combination of two 3-way switches and three 4-way switches. Figure 13-12 and 13-13 show a circuit for controlling a light from three locations. The 3-way switches are always connected to the power panel and to the light fixture, with the 4-way switch between them. The use of 3-way and 4-way switch combinations makes it possible to achieve control of a light from any number of points.

Each lighting-control circuit requires an entirely different type of switching combination to adequately accomplish control of a lighting fixture. These circuits are usually wired into buildings during construction,

Lighting Systems   339

in order to provide the type of lighting control desired for each room. You should study the diagrams of Figures 13-8 through 13-13 to fully understand how these lighting circuits operate.

BRANCH CIRCUIT DESIGN

The design of lighting branch circuits involves the calculation of the maximum current that can be drawn by the lights that are connected to the branch circuit. Many times, particularly in homes, a lighting branch circuit also has duplex receptacles for portable appliances. This makes the exact current calculation more difficult, since not all the lights or the appliances will be in use at the same time.

The National Electrical Code (NEC) specifies that there should be at least one branch circuit for each 500 square feet of lighting area. The NEC further specifies a minimum requirement for lighting for various types of buildings. Table 13-3 lists the number of watts of light per square foot required in some buildings.

The solution of a typical branch circuit lighting problem follows.

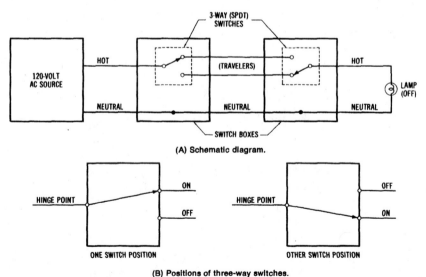

Figure 13-10. Circuit for controlling a light from two locations: (A) Schematic diagram, (B) Positions of three-way switches

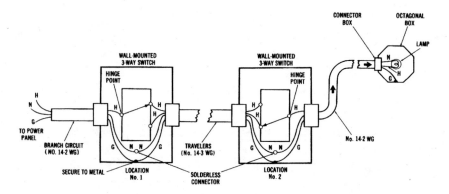

Figure 13-11. Pictorial view of the circuit of Figure 13-10 (A)

Table 13-3. Building Lighting Requirements
in Wattage per Square Foot

Type of Building	Watts
Armories and Auditoriums	1
Banks	2
Churches	1
Dwellings (other than hotels)	3
Garages (for commercial storage)	1/2
Hospitals	2
Hotels and Motels	2
Industrial or Commercial Buildings	2
Office Buildings	5
Restaurants	2
Schools	3
Stores (grocery)	3
Warehouses (storage)	1/4

Sample Problem:

Given: a room in a commercial building is 70 feet × 120 feet (21.34 × 36.58 meters). It will use 125 fluorescent lighting units that draw 3 amperes each. Also, 20-ampere branch circuits will be used. The power factor of the units is 0.75. The operating voltage is 120 volts.

Lighting Systems 341

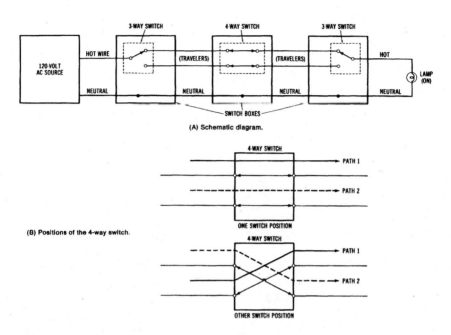

Figure 13-12. Circuit for controlling a light from three locations: (A) Schematic diagram, (B) Positions of the four-way switch

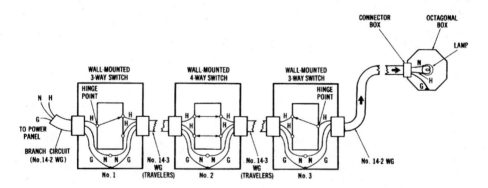

Figure 13-13. Pictorial view of the circuit shown in Figure 13-12(A)

Find: the minimum number of branch circuits needed, and the total power requirement for the lighting units.

Solution for the branch circuits:

1. Find the number of units that should be installed on each branch circuit.

$$\text{units} = \frac{\text{branch current}}{\text{load current}}$$

$$= \frac{20 \text{ amperes}}{3 \text{ amperes}}$$

$$= 6.67 \text{ units (round off at 6)}$$

2. Find the number of branch circuits needed.

$$\text{circuits} = \frac{\text{lighting units}}{\text{No. units per branch circuit}}$$

$$= \frac{125}{6}$$

$$= 20.8 \text{ (round up to 21)}$$

Twenty-one 20-ampere branch circuits will be needed.

Solution for total power:
1. Find the total current drawn by the lights.

$$I = 125 \text{ units} \times 3 \text{ amperes}$$
$$= 375 \text{ amperes}$$

2. Calculate the total power.

$$p = V \times I \times pf$$
$$= 120 \times 375 \times 0.75$$
$$= 33{,}750 \text{ watts (33.75 kW)}$$

An even more common type of branch circuit lighting problem is given in the following discussion. This problem begins with the minimum number of lights required in a room. The number of branch circuits that are required must be determined.

Lighting Systems

Sample Problem:

Given: a large room in an industrial building needs to have 40,000 watts of incandescent lighting. It is decided that 120-volt 20-ampere branch circuits will be used. The lights will use 200-watt bulbs.

Find: the number of branch circuits required for the lighting.

Solution:

1. Find the amount of current each lamp will draw.

$$I = \frac{P}{V}$$

$$= \frac{200 \text{ watts}}{120 \text{ volts}}$$

$$= 1.67 \text{ amperes}$$

2. Find the maximum number of lamps in each branch circuit.

$$\text{lamps} = \frac{\text{amperage of circuit}}{\text{amperage of lamps}}$$

$$= \frac{20 \text{ amperes}}{1.67 \text{ amperes}}$$

$$= 11.97 \text{ (round down to 11)}$$

3. Find the total number of bulbs (lamps needed.

$$\text{bulb required} = \frac{\text{wattage of room}}{\text{wattage of bulbs}}$$

$$= \frac{40,000 \text{ watts}}{200 \text{ watts}}$$

$$= 200 \text{ bulbs required}$$

4. Find the minimum number of branch circuits needed.

$$\text{branch circuits} = \frac{\text{total bulbs required}}{\text{number of bulbs per circuit}}$$

$$= \frac{200}{11}.$$

$$= 18.18 \text{ (round up to 19)}$$

Nineteen branch circuits will be needed.

It should be pointed out that, in these problems, we determined the minimum number of branch circuits. But room for flexibility should be provided for each branch circuit. This can be done by designing the circuit to handle only 16 amperes (80 percent) rather than 20 amperes. Also, room should be provided on the power distribution panel to allow for connecting some additional lighting branch circuits. This will allow additional loads to be connected at a later time.

LIGHTING FIXTURE DESIGN

Lighting fixtures are the devices that are used to hold lamps in place. They are commonly referred to as luminaires. Luminaires are used to efficiently transfer light from its source to a work surface. The proper design of luminaires allows a more efficient transfer of light. It is important to keep in mind that light intensity varies inversely as the square of the distance from the light source. Thus, if the distance is doubled, the light intensity will be reduced four times.

Many factors must be considered in determining the amount of light that is transferred from a light bulb to a work surface. Some light is absorbed by the walls and by the light fixture itself; thus, not all light is efficiently transferred. The manufacturers of lighting systems develop charts that are used to predict the amount of light that will be transferred to a work surface. These charts take into consideration the necessary variables for making a prediction of the quantity of light falling onto a surface.

Each luminaire has a rating, which is referred to as its *coefficient of utilization*. The coefficient of utilization is a factor that expresses the percentage of light output that will be transferred from a lamp to a work area. The coefficient of a luminaire is determined by laboratory tests made by the manufacturer. These coefficiency charts also take into consideration the light absorption characteristics of walls, ceilings, and floors, when the coefficient of utilization of a luminaire is to be determined.

Another factor used for determining the coefficient of utilization is the *room ratio*. Room ratio is very simply determined by the formula:

$$\text{room ratio} = \frac{W \times L}{H(W+L)}$$

where:
 W = the room width in feet,
 L = the room length in feet, and
 H = the distance in feet from the light source to the work surface.

Note: Work surfaces are considered to be 2.5 feet from the floor, unless otherwise specified.

FACTORS IN DETERMINING LIGHT OUTPUT

There are several other factors that must be considered in order to determine the amount of light transferred from a light source to a work surface. We know, for instance, that the age of a lamp has an effect on its light output. Lamp manufacturers determine a *depreciation factor* or *maintenance factor* for luminaires. This factor expresses the percentage of light output available from a light source. A depreciation factor of 0.75 means that in the daily use of a light source, only 75 percent of the actual light output is available for transfer to the work surface. The depreciation factor is an average value. It takes into consideration the reduction of light output with age, and the accumulation of dust and dirt on the luminaires. Some collect dust more easily than others.

The following problem shows the effect of the coefficient of utilization (CD) and the depreciation factor (OF) on the light output transferred to a work surface.

Sample Problem:

Given: a lighting system for a building has 16 luminaries. Each luminaire has two fluorescent lamps. Each lamp has a light output of 3000 lumens. The CD and OF found in the chart developed by the manufacturer are 0.45 and 0.90, respectively.

Find: the total light output from the lighting system.
Solution:
1. Find the total light output from the lamps.

$$3000 \text{ lumens per lamp} \times 16 \text{ luminaires} \times 2 \text{ lamps per luminaire} = 96{,}000 \text{ lumens}$$

2. Find the total light output:

$$\begin{aligned}\text{light output} &= \text{total lumens} \times CD \times DF \\ &= 96{,}000 \times 0.45 \times 0.90 \\ &= 38{,}880 \text{ lumens}\end{aligned}$$

The distribution of light output onto work surfaces must also be considered. The CD and the DF are used to find the total light output of a lighting system. A greater light output is required for larger areas. The light output to a work surface is expressed as lumens per square foot, or footcandles. A light meter is used to measure the quantity of light that reaches a surface.

The following two formulas are useful for finding the required lumens for a work area, or the light output available from a particular lighting system:

$$\text{total lumens} = \frac{\text{desired footcandles} \times \text{room area in ft.}}{CD \times DF}$$

and

$$\text{footcandles available} = \frac{\text{total lumens} \times CD \times DF}{\text{room area in feet}}$$

Sample Problem:
Given: a lighting system with 12,000 total lumen output, luminaire CD = 0.83 and DF = 0.85, is installed in a 600 sq. ft. room.
Find: the footcandles of light available on the work surface.
Solution:

$$FC = \frac{\text{lumens} \times CD \times DF}{\text{room area}}$$

$$= \frac{12{,}000 \times 0.83 \times 0.85}{600}$$

FC = 14.11 footcandles

Considerations for Electrical Lighting Loads

Incandescent lamps produce light by the passage of electrical current through a tungsten filament. The electrical current heats the filament to the point of incandescence, which causes the lamp to produce light. The primary advantage of incandescent lights is their low initial cost. However, they have a very low efficacy (lumen/watt) rating. They also have a very high operating temperature and a short life expectancy. Incandescent lights are usually not good choices of light sources for commercial, industrial, or outdoor lighting applications; their primary applications are for residential use.

Fluorescent light sources have a higher efficacy (lumens/watt) than incandescent lights. They have a much longer life, and lower brightness and operating temperature. Fluorescent lights are used for residential (120-volt applications) and for general-purpose commercial and industrial lighting (120-volt and 277-volt systems). Disadvantages of fluorescent lights include the necessity for a ballast and a rather large luminaire. They also have a higher initial cost than incandescent lamps. It is estimated that fluorescent light sources provide approximately 70 percent of the lighting in the United States.

Vapor light sources also have very good efficacy ratings and long life expectancies. They have a high light output for their compact size. They are typically used for industrial, commercial, and outdoor applications, since they can be operated economically on higher-voltage systems. The initial cost of vapor lights is high, they require a ballast, and they are a very bright light source. Their very high efficacy ratings have led to increased use of vapor light sources.

Chapter 14

Mechanical Systems

Another broad category of electrical loads includes those devices that convert electrical energy into mechanical energy. Electric motors fall into this category of load devices. They are mechanical power-conversion systems. There are many types of motors used today. The electrical motor load is the major power-consuming load of electrical power systems. Motors of various sizes are used for purposes that range from large industrial machine operation, to operating power blenders and mixers in the home.

IMPORTANT TERMS

Chapter 14 deals with mechanical systems. After studying this chapter, you should have an understanding of the following terms:

Motor
Rotor
Stator
Brush/Commutator Assembly
Torque
Load
Speed
Counterelectromotive Force (CEMF)
Armature Current
Horsepower
Speed Regulation
Direct Current (DC) Motors
    Permanent-Magnet DC Motor
    Series-wound DC Motor
    Shunt-wound DC Motor
    Compound-wound DC Motor

Motor Reversal
Dynamotor
Brushless DC Motor
DC Stepping Motor
Single-Phase Alternating Current (AC) Motors
    Universal Motors
    Split-Phase Induction Motors
    Capacitor Motors
    Shaded Pole Motor
    Repulsion Motors
    Synchronous Motors
Synchronous Speed
Slip
Rotor Frequency
Three-Phase AC Motors
Three-Phase AC Induction Motor
Three-Phase AC Synchronous Motor
Three-Phase Wound-Rotor Induction Motor
Damper Windings
Auxiliary Starting Machine
Synchro System
Servo System
Efficiency

BASIC MOTOR PRINCIPLES

The function of a motor is to convert electrical energy into mechanical energy in the form of a rotary motion. To produce a rotary motion, a motor must have an electrical power input. Generator action is brought about by a magnetic field, a set of conductors within the magnetic field, and relative motion between the two. Motion is similarly produced in a motor through the interaction of a magnetic field and a set of conductors.

All motors, regardless of whether they operate from an AC or a DC power line, have several basic characteristics in common. Their basic parts include (1) a stator, which is the frame and other stationary components, (2) a rotor, which is the rotating shaft and its associated parts, and (3) auxiliary equipment, such as a brush/commutator assembly for DC motors, or a starting circuit for single-phase AC motors. The basic parts of a DC motor are shown in Figure 14-1. A simple DC motor is constructed in the same way as a DC generator. Their basic parts are the same. DC generators were discussed in Chapter 7.

The motor principle is illustrated in Figure 14-2. In Figure 14-2A, no current is flowing through the conductors because of the position of the brushes in relation to the commutator. In this state, no motion is produced. When current flows through the conductor, a circular magnetic field develops around the conductor. The direction of the current flow determines the direction of the circular magnetic fields, as shown in the cross-sectional diagram of Figure 14-2C.

When current flows through the conductors within the main magnetic field, this secondary field interacts with the main field. The interaction of these two magnetic fields results in the production of motion. The circular magnetic field around the conductors causes a compression of the main magnetic flux at points A and B in Figure 14-2B. This compression causes the magnetic field to produce a reaction in the direction opposite to that of the compression. Therefore, motion is produced away from points A and B. In actual motor operation, a rotary motion in a clockwise direction would be produced. If we wished to change the direction of rotation, we would merely have to reverse the direction of the current flow through the conductors.

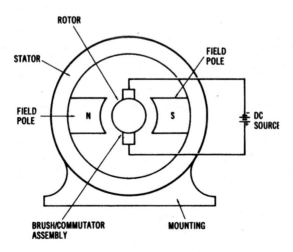

Figure 14-1. Basic parts of a DC motor

Figure 14-2. Illustration of the basic motor principle: (A) Condition with no current flowing through the conductors, (B) Condition with current flowing through the conductors, (e) Direction of current flow through the conductors determines the direction of the magnetic field around the conductors

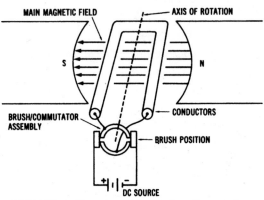

Sample Problem: Force Acting on a Conductor

When a current-carrying conductor is contained within a magnetic field, the force produced is called Lorentz force. This force is basic to the operation of electric motors. The force is greatest when the conductor is perpendicular to the magnetic field, and minimum (0) when it is parallel. The maximum Lorentz force is expressed as:

$$F = B \times l \times I$$

where:
- F = force acting on a conductor in newtons,
- B = flux density in teslas,
- ℓ = length of the conductor in meters, and
- I = current through the coil in amperes.

Given: a conductor 1.8 meters long has a current flow of 50 A through it. It is contained within a magnetic field of 0.25 tesla density.

Find: the maximum Lorentz force acting on the conductor.

Solution:

$$F = B \times I \times \ell$$
$$= 0.25 \times 1.8 \times 50$$
$$F = 22.5 \text{ newtons}$$

The rotating effect produced by the interaction of two magnetic fields is called *torque* or *motor action*. The torque produced by a motor depends on the strength of the main magnetic field and the amount of current flowing through the conductors. As the magnetic field strength or the current through the conductors increases, the amount of torque or rotary motion will increase also.

DC MOTORS

Motors that operate from DC power sources are often used in industry when speed control is desirable. DC motors are almost identical in construction to DC generators. They are also classified in a similar manner as series, shunt, or compound machines, depending on the method of connecting the armature and field windings. The permanent-magnet DC motor is another type of motor that is used for certain applications.

DC Motor Characteristics

The general operational characteristics common to all DC motors are shown in Figure 14-3. Most electric motors exhibit characteristics similar to those shown in the block diagram. In order to discuss DC motor characteristics, you should be familiar with the terms *load, speed, counterelectromotive force* (cemf), *armature current*, and *torque*. The amount of mechanical load applied to the shaft of a motor determines its operational characteristics. As the mechanical load is increased, the speed of a motor tends to decrease. As speed decreases, the voltage induced into the conductors of the motor through generator action (cemf) decreases. The generated voltage or counterelectromotive force depends upon the number of rotating conductors and the speed of rotation. Therefore, as speed of rotation decreases, so does the cemf.

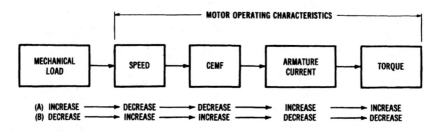

Figure 14-3. Operational characteristics of DC motors

The counterelectromotive force generated by a DC motor is in opposition to the supply voltage. Therefore, the actual working voltage of a DC motor may be expressed as:

$$V_T = V_C + I_A R_A$$

where:
    V_T = the terminal voltage of the motor in volts,
    V_C = the cemf generated by the motor in volts, and
    $I_A R_A$ = the voltage drop across the armature of the motor in volts.

Since the cemf is in opposition to the supply voltage, the actual working voltage of a motor will increase as the cemf decreases. As the result of an increase in working voltage, more current will flow through the armature conductors that are connected to the DC power supply. Since torque is directly proportional to armature current, the torque will increase as the armature current increases.

To briefly discuss the opposite situation, if the mechanical load connected to the shaft of a motor decreases, the speed of the motor will tend to increase. An increase in speed causes an increase in generated voltage. Since cemf is in opposition to the supply voltage, as cemf increases, the armature current decreases. A decrease in armature current causes a decrease in torque. We can see that torque varies with changes in load, but we need to consider each of the steps involved, in order to understand DC motor operation. As the load on a motor is increased, its torque also increases as the motor tries to meet the increased load requirement. However, the current drawn by a motor also increases when load is increased.

The presence of a cemf to oppose armature current is very important in motor operation. The lack of any cemf when a motor is being started explains why motors draw a very large initial starting current, compared to the running current they draw once full speed is reached. Maximum armature current flows when there is no cemf. As cemf increases, armature current decreases. Thus, resistances in series with the armature circuit are often used to compensate for the lack of cemf, and to reduce the starting current of a motor. After a motor has reached full speed, these resistances may be bypassed by automatic or manual switching systems, in order to allow the motor to produce maximum torque. Keep in mind that the armature current, which directly affects torque, can be expressed as:

$$I_A = \frac{V_T - V_C}{R_A}$$

where:
I_A = the armature current in amperes,
V_T = the terminal voltage of the motor in volts,
V_C = the cemf generated by the motor in volts, and
R_A = the armature resistance in ohms.

In determining the functional characteristics of a motor, the torque developed can be expressed as:

$$T = K \Phi I_A$$

where:
T = the torque in foot-pounds,
K = a constant based on physical characteristics (conductor size, frame size, etc.),
Φ = the quantity of magnetic flux between poles, and
I_A = the armature current in amperes.

Torque can be measured by several types of motor analysis equipment. The horsepower rating of a motor is based on the amount of torque produced at the rated full-load values. Horsepower, which is the usual method of rating motors, can be expressed mathematically as:

$$hp = \frac{2\pi NT}{33,000}$$

$$= \frac{NT}{5252}$$

where:
    HP = the horsepower rating,
    2π = a constant,
    N = the speed of the motor in revolutions per minute (rpm), and
    T = the torque developed by the motor in foot-pounds.

Sample Problem: Power of a Motor

The mechanical power output of a motor is dependent upon the speed of rotation and the torque produced. Power output is expressed as:

$$P = \frac{n \times T}{9.55}$$

where:
    P = mechanical power in Watts,
    n = speed of rotation in revolutions per minute (r/min),
    T = torque in newtons per meter (N/m), and
    9.55 = a constant equal to 30/pi.

Given: a motor rotating at 3,600 r/min produces a torque of 5N/m.
Find: the power output developed by the motor.
Solution:

$$P = \frac{n \times T}{9.55}$$

$$= \frac{3,500 \times 5}{9.55}$$

$$P = 1,885 \text{ watts}$$

$$\text{power in HP} = \frac{W}{746}$$

$$= \frac{1{,}885 \text{ W}}{746}$$

$$\text{horsepower} = 2.53 \text{ HP}$$

Another DC motor characteristic that we should discuss is armature reaction. Armature reaction was discussed in relation to DC generators in Chapter 7. This effect distorts the main magnetic field in DC motors, as well as in DC generators. Similarly, the brushes of a DC motor can be shifted to counteract the effect of armature reaction. However, interpoles are ordinarily used to control armature reaction in DC motors.

The most desirable characteristic of DC motors is their speed-control capability. Varying the applied DC voltage with a rheostat allows speed to be varied from zero to the maximum rpm of the motor. Some types of DC motors have more desirable speed characteristics than others. For this reason, we can determine the comparative speed regulation for different types of motors. Speed regulation is expressed as:

$$\%SR = \frac{S_{NL} - S_{FL} \times 100}{S_{FL}}$$

where:
 $\%SR$ = the percentage of speed regulation,
 S_{NL} = the no-load speed in rpm, and
 S_{FL} = the rated, full-load speed in rpm.

Sample Problem:

Given: the no-load speed of a universal motor is 6,250 r/min, and its full-load speed is 5,000 r/min.

Find: the motor's speed regulation.

Solution:

$$\%SR = \frac{S_{NL} - S_{FL} \times 100}{S_{FL}}$$

$$= \frac{6250 - 5000 \times 100}{5000}$$

$$= \%SR = 25\%$$

Good speed regulation (low % SR) results when a motor has nearly constant speeds under varying load situations.

Types of DC Motors

The types of commercially available DC motors basically fall into four categories: (1) permanent-magnet DC motors, (2) series-wound DC motors, (3) shunt-wound DC motors, and (4) compound-wound DC motors. Each of these motors has different characteristics that are due to its basic circuit arrangement and physical properties.

Permanent-Magnet DC Motors—The permanent-magnet DC motor, shown in Figure 14-4, is constructed in the same manner as its DC generator counterpart, which was discussed in Chapter 7. The permanent-magnet motor is used for low-torque applications. When this type of motor is used, the DC power supply is connected directly to the armature conductors through the brush/commutator assembly. The magnetic field is produced by permanent magnets mounted on the stator.

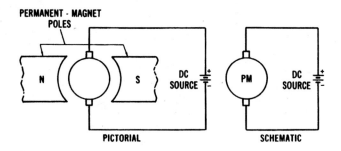

Figure 14-4. Permanent-magnet DC motor

This type of motor ordinarily uses either alnico or ceramic permanent magnets, rather than field coils. The alnico magnets are used with high-horsepower applications. Ceramic magnets are ordinarily used for low-horsepower, slow-speed motors. Ceramic magnets are highly resistant to demagnetization, yet they are relatively low in magnetic-flux level. The magnets are usually mounted in the motor frame, and then magnetized prior to the insertion of the armature.

The permanent-magnet motor has several advantages over conventional types of DC motors. One advantage is a reduced operational cost. The speed characteristics of the permanent-magnet motor are similar to those of the shunt-wound DC motor. The direction of rotation of a permanent-magnet motor can be reversed by reversing the two power lines.

Series-wound DC Motors—The manner in which the armature and field circuits of a DC motor are connected determines its basic characteristics. Each type of DC motor is similar in construction to the type of DC generator that corresponds to it. The only difference, in most cases, is that the generator acts as a voltage source, while the motor functions as a mechanical power-conversion device.

The series-wound motor, shown in Figure 14-5, has the armature and field circuits connected in a series arrangement. There is only one path for current to flow from the DC voltage source. Therefore, the field is wound of relatively few turns of large-diameter wire, giving the field a low resistance. Changes in load applied to the motor shaft cause changes in the current through the field. If the mechanical load increases, the current also increases. The increased current creates a stronger magnetic field. The speed of a series motor varies from very fast at no load, to very slow at heavy loads. Since large currents may flow through the low resistance field, the series motor produces a high torque output. Series motors are used when heavy loads must be moved, and speed regulation is not important. A typical application is automobile starter motors.

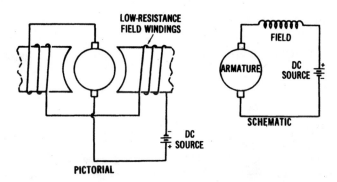

Figure 14-5. Series-wound DC motor

Shunt-wound DC Motors—Shunt-wound DC motors are more commonly used than any other type of DC motor. As shown in Figure 14-6, the shunt-wound DC motor has field coils connected in parallel with its armature. This type of DC motor has field coils that are wound of many turns of small-diameter wire and have a relatively high resistance. Since the field is a high-resistant parallel path of the circuit of the shunt motor, a small amount of current flows through the field. A strong electromagnetic field is produced because of the many turns of wire that form the field windings.

A large majority (about 95 percent) of the current drawn by the shunt motor flows in the armature circuit. Since the field current has little effect on the strength of the field, motor speed is not affected appreciably by variations in load current. The relationship of the currents that flow through a DC shunt motor is as follows:

$$I_T = I_A + I_F$$

where:
I_T = the total current drawn from the power source,
I_A = the armature current, and IF = the field current.

The field current may be varied by placing a variable resistance in series with the field windings. Since the current in the field circuit is low, a low-wattage rheostat may be used to vary the speed of the motor in accordance with the variation in field resistance. As field resistance increases, field current will decrease. A decrease in field current reduces the strength of the electromagnetic field. When the field flux is decreased, the armature will rotate faster, because of reduced magnetic-field interaction. Thus, the speed of a DC shunt motor may be easily varied by using a field rheostat.

The shunt-wound DC motor has very good speed regulation. The speed does decrease slightly when the load increases, as the result of the increase in voltage drop across the armature. Because of its good speed regulation, and its ease of speed control, the DC shunt motor is commonly used for industrial applications. Many types of variable-speed machine tools are driven by DC shunt motors.

Compound-wound DC Motors—The compound-wound DC motor, shown in Figure 14-7, has two sets of field windings, one in series with the armature and one in parallel. This motor combines the desirable characteristics of the series- and shunt-wound motors. It has high torque similar to that of a series-wound motor, along with good speed regulation similar to that of a shunt motor. Therefore, when good torque and good speed regulation are needed, the compound-wound DC motor can be used. A major disadvantage of a compound-wound motor is its expense.

Comparison of DC Motor Characteristics—The characteristics of DC motors should be considered when motors for particular applications are selected. Figure 14-8 shows comparative graphs that illustrate the relative

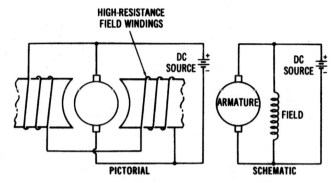

Figure 14-6. Shunt-wound DC motor

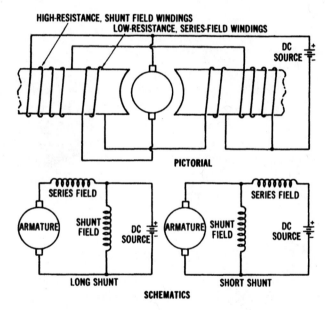

Figure 14-7. Compound-wound DC motor

torque and speed characteristics of DC motors.

A DC motor is designed so that its shaft will rotate in either direction. It is a very simple process to reverse the direction of rotation of any DC motor. By reversing the relationship between the connections of the armature winding and field windings, reversal of rotation is achieved. Usually, this is done by changing the terminal connections at the point

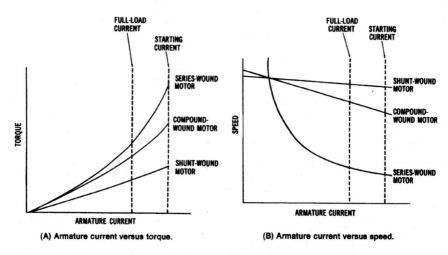

Figure 14-8. Torque and speed characteristics of DC motors: (A) Armature current versus torque, (B) Armature current versus speed

where the power source is connected to the motor. Four terminals are ordinarily used for interconnection purposes. They may be labeled A1 and A2 for the armature connections, and F1 and F2 for the field connections. If either the armature connections or the field connections are reversed, the rotation of the motor will reverse. However, if both are reversed, the motor shaft will rotate in its original direction, since the relationship between the armature and field windings will be the same.

SPECIALIZED DC MOTORS

There are several specialized types of motors that operate on direct current. Among these are dynamotors, brushless DC motors, and DC stepping motors.

Dynamotors

One specialized type of DC motor is called a dynamotor. This motor, depicted in Figure 14-9, converts DC voltage of one value to DC voltage of another value. It is actually a motor-generator housed in one unit. The armature has two separate windings. One winding is connected to the commutator of the motor section, and the other winding is connected to the commutator of a generator unit. A magnetic field, developed by either permanent magnets or electromagnetic windings, surrounds the armature assembly. Since the magnetic field remains relatively constant, the generator voltage output depends upon the ratio of the number of motor windings to the number of generator windings. For instance, if there are twice as many generator windings as motor windings, the generated DC-voltage output will be twice the value of the DC voltage that is input to the motor section of the dynamotor.

Brushless DC Motors

The use of transistors has resulted in the development of brushless DC motors, which have neither brushes nor commutator assemblies. Instead, they make use of solid state switching circuits. The major problem with most DC motors is the low reliability of the commutator/brush assembly. The brushes have a limited life and cause the commutator to wear. This wearing produces brush dust, which can cause other maintenance problems.

Although some brushless DC motors use other methods, the transistor-switched motor is the most common (see Figure 14-10). The motor itself is actually a single-phase, AC, permanent-capacitor, induction motor, with a center-tapped main winding. Transistors, operated by an oscillator circuit, conduct alternately through the paths of the main winding. The oscillator circuit requires a feedback winding wound into the stator slots, in order to generate a control voltage to determine the frequency. A capacitor (C2) is placed across the main winding to reduce voltage peaks and to keep the frequency of the circuit at a constant value.

The main disadvantage of this motor is its inability to develop a very high-starting torque. As a result, it is suitable only for driving very low-torque loads. When used in a low-voltage system, this motor is not very efficient. Also, since only half of the main winding is in use at any instant,

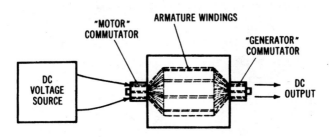

Figure 14-9. Dynamotor construction

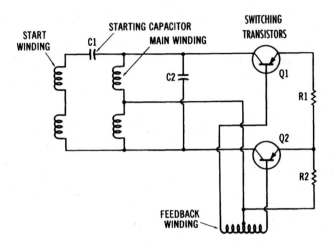

Figure 14-10. A brush less DC motor circuit

copper losses are relatively high. However, the advantages outweigh this disadvantage for certain applications. Since there are no brushes and commutator, motor life is limited mainly by the bearing. With proper lubrication, a brushless DC motor can be used for an indefinite period. Also, the motor frequency, and thus the speed, can be adjusted by varying the oscillator circuit.

DC Stepping Motors

DC stepping motors are unique DC motors that are used to control automatic industrial processing equipment. DC motors of this type are found in numerically controlled machines and robotic systems used by industry. They are very efficient and develop a high torque The stepping motor is used primarily to change electrical pulses into a rotary motion that can be used to produce mechanical movements.

The shaft of a DC stepping motor rotates a specific number of mechanical degrees with each incoming pulse of electrical energy. The amount of rotary movement or angular displacement produced by each pulse can be repeated precisely with each succeeding pulse from the drive source. The resulting output of this device is used to accurately locate or position automatic process machinery.

The velocity, distance, and direction of movement of a specific machine can be controlled by DC stepping motors. The movement error of this device is generally less than 5 percent per step. This error factor is not cumulative, regardless of the distance moved, or the number of steps taken. Motors of this type are energized by a DC drive amplifier that is controlled by digital logic circuits. The drive-amplifier circuitry is a key factor in the overall performance of this motor. The stator construction and coil layout are shown in Figure 14-11. The rotor of a stepping motor is of a permanent-magnet type of construction.

Figure 14-11. Stator and coil layout of a DC stepping motor (*Courtesy Superior Electric Co.*)

A very important principle that applies to the operation of a DC stepping motor is that like magnetic poles repel, and unlike magnetic poles attract. If a permanent-magnet rotor is placed between two series-connected stator coils, it produces the situation shown in Figure 14-12. With power applied to the stator, the rotor can be repelled in either direction. The direction of rotation in this case is unpredictable. Adding two more stator coils to a simple motor, as indicated in Figure 14-12B, will make the direction of rotation predictable. With the stator polarities indicated, the rotor will align itself midway between the two pairs of stator coils. The direction of rotation can now be predicted, and is determined by the polarities

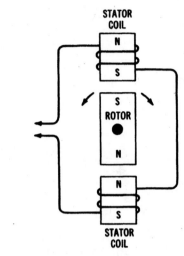

(A) With one set of stator coils, direction of rotor rotation is unpredictable.

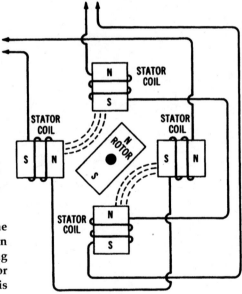

Figure 14-12. Illustration of the magnetic principle involved in the operation of a DC stepping motor: (A) With one set of stator coils, direction of rotor rotation is unpredictable, (B) With two sets of stator coils and polarities as shown, rotor will align itself as shown.

(B) With two sets of stator coils and polarities as shown, rotor will align itself as shown.

of the stator-coil sets. Adding more stator-coil pairs to a motor of this type improves its rotation and makes the stepping action very accurate.

Figure 14-13 shows an electrical diagram of a DC stepping motor. The stator coils of this motor are wound in a special type of construction called bifilar. Two separate wires are wound into the coil slots at the same time. The two wires are small in size, which permits twice as many turns as can be achieved with a larger-sized wire. Construction of this type simplifies the control circuitry and DC energy-source requirements.

Operation of the stepping motor illustrated in Figure 14-13 is achieved in a four-step switching sequence. Any of the four combinations of switches 1 or 2 will produce an appropriate rotor position location. After the four switch combinations have been achieved, the switching cycle repeats itself. Each switching combination causes the motor to move one-fourth of a step.

A rotor, similar to the one shown in Figure 14-13 normally has 50 teeth. Using a 50-tooth rotor in the circuit of Figure 14-13 would permit four steps per tooth, or 200 steps per revolution. The amount of displacement, or step angle, of this motor is, therefore, determined by the number of teeth on the rotor, and by the switching sequence.

A stepping motor that takes 200 steps to produce one revolution will

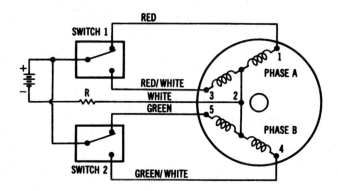

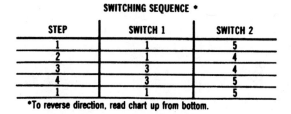

Figure 14-13. Circuit diagram and switching sequence of a DC stepping motor (*Courtesy Superior Electric Co.*)

move 360 degrees every 200 steps, or 1.8 degrees per step. It is not unusual for stepping motors to use eight switching combinations to achieve one step. In this case, each switching combination could be used to produce 0.9 degree of displacement. Motors and switching circuits of this type permit a very precise type of controlled movement.

SINGLE-PHASE AC MOTORS

Another broad classification of mechanical power-conversion equipment includes single-phase AC motors. These motors are common for industrial as well as commercial and residential usage. They operate from a single-phase AC power source. There are three basic types of single-phase AC motors-universal motors, induction motors, and synchronous motors.

Universal Motors

Universal motors may be powered by either AC or DC power sources. The universal motor, shown in Figure 14-14 is constructed in the same way as a series-wound DC motor. However, it is designed to operate with either AC or DC applied. The series-wound motor is the only type of DC motor that will operate with AC applied. The windings of shunt-wound motors have inductance values that are too high to allow the motor to function with AC applied. However, series-wound motors have windings that have low inductances (few turns of large diameter wire), and they therefore offer a low impedance to the flow of AC. The universal motor is one type of AC motor that has concentrated or salient field windings. These field windings are similar to those of all DC motors.

The operating principle of the universal motor, with AC applied, involves the instantaneous change of both field and armature polarities. Since the field windings have low inductance, the reversals of field polarity brought about by the changing nature of the applied AC also create reversals of current direction through the armature conductors at the proper time intervals. The universal motor operates in the same manner as a series-wound DC motor, except that the field polarity and the direction of armature current change at a rate of 120 times per second when connected to a 60-Hz AC source. The speed and torque characteristics of universal motors are similar to those of DC series-wound motors. Universal motors are used mainly for portable tools and small, motor-driven equipment, such as mixers and blenders.

Mechanical Systems 369

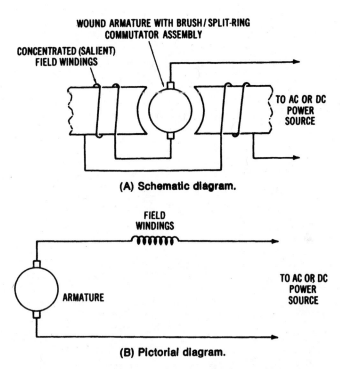

Figure 14-14. The universal motor: (A) Schematic diagram, (B) Pictorial diagram

Induction Motors

Another popular type of single-phase AC motor operates on the induction principle. This principle is illustrated in Figure 14-15. The coil symbols along the stator represent the field coils of an induction motor. These coils are energized by an AC source; therefore, their instantaneous polarity changes 120 times per second when 60-Hz AC is applied.

Induction motors have a solid rotor, which is referred to as a squirrel-cage rotor. This type of rotor, which is illustrated in Figure 14-15B, has large-diameter copper conductors that are soldered at each end to a circular connecting plate. This plate actually short circuits the individual conductors of the rotor. When current flows in the stator windings, a current is induced in the rotor. This current is developed by means of "transformer action" between the stator and rotor. The stator, which has AC voltage applied, acts as a transformer primary. The rotor is similar to a transformer secondary, since current is induced into it.

Since the stator polarity changes in step with the applied AC, it develops a rotating magnetic field. The rotor becomes instantaneously polarized, as the result of the induced current flow through the short-circuited copper conductors. The rotor will, therefore, tend to rotate in step with the revolving magnetic field of the stator. If some method of initially starting rotation is used, the rotor will continue to rotate. However, because of inertia, a rotor must be put into motion initially by some auxiliary method.

It should be pointed out that the speed of an AC induction motor is dependent on the speed of the rotating magnetic field and the number of stator poles that the motor has. The speed of the rotor will never be as high as the speed of the rotating stator field. If the two speeds were equal, there

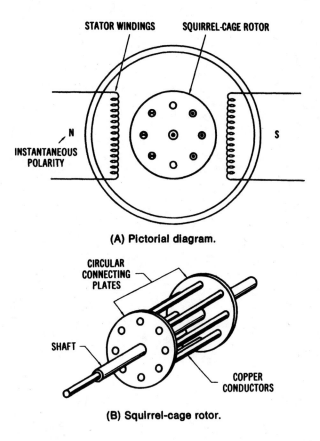

(A) Pictorial diagram.

(B) Squirrel-cage rotor.

Figure 14-15. Illustration of the induction principle: (A) Pictorial diagram, (8) Squirrel-cage rotor

would be no relative motion between the rotor and stator, and, therefore, no induced rotor current and torque would develop. The rotor speed (operating speed) of an induction motor is always somewhat less than the rotating stator field developed by the applied AC voltage.

The speed of the rotating stator field may be expressed as:

$$S = \frac{f \times 120}{n}$$

where:
- S = the speed of the rotating stator field in rpm,
- f = the frequency of the applied AC voltage in hertz,
- n = the number of poles in the stator windings, and
- 120 = a conversion constant.

A two-pole motor operating from a 60-Hz source would have a stator speed of 3600 revolutions per minute. The stator speed is also referred to as the synchronous speed of a motor. The difference between the revolving stator speed of an induction motor and the rotor speed is called slip. The rotor speed must lag behind the revolving stator speed in order to develop torque. The more the rotor speed lags behind, the more torque is developed. Slip is expressed mathematically as:

$$\% \text{ slip} = \frac{Ss - Sr \times 100}{Ss}$$

where:
- Ss = the synchronous (stator) speed in rpm, and
- Sr = the rotor speed in rpm.

As the rotor speed becomes closer to the stator speed, the percentage of slip becomes smaller.

Another factor, referred to as rotor frequency, affects the operational characteristics of an induction motor under load. As the load on the shaft of the motor increases, the rotor speed tends to decrease. The stator speed, however, is unaffected. When a two-pole induction motor connected to a 60-Hz source operates at 10 percent slip, the slip will equal 360 rpm (3600 rpm × 10%). Functionally, this means that a revolving stator field sweeps across a rotor conductor 360 times per minute. Current is induced into a rotor conductor each time the stator field revolves past the conductor. As slip is increased, more current is induced into the rotor, causing more torque to be developed. The rotor frequency depends on the amount of slip, and can be expressed as:

$$fr = fs \times slip$$

where:
    fr = the frequency of the rotor current,
    fs = the frequency of the stator current, and
    slip is expressed as a decimal.

Rotor frequency affects the operational characteristics of induction motors.

Single-phase AC induction motors are classified according to the method they use for starting. Some common types of single-phase AC induction motors include split-phase motors, capacitor motors, shaded-pole motors, and repulsion motors.

Split-phase Induction Motors—The split-phase AC induction motor, shown in Figure 14-16 has two sets of stator windings. One set, called the run windings, is connected directly across the AC line. The other set, called the start windings, is also connected across the AC line. However, the start winding is connected in series with a centrifugal switch that is mounted on the shaft of the motor. The centrifugal switch is in the closed position when the motor is not rotating.

Before discussing the functional principle of the split-phase AC motor, we should understand how rotation is developed by an AC motor. Refer to Figure 14-17. In Figure 14-17, we have a two-pole stator with single-phase AC applied. For the purposes of our discussion, a permanent magnet is placed within the stator to represent the squirrel-cage rotor of an induction motor. At time t_0 of the AC sine wave, no stator field is developed. Time interval t_1 will cause a stator field to be produced. Assume a north polarity on the right pole of the stator, and a south polarity on the left pole. These polarities will cause the rotor to align itself horizontally, in accordance with the laws of magnetic attraction. At time t_2, the poles will become demagnetized, and then begin to magnetize in the opposite direction. At time interval t_3, the stator poles will be magnetized in the opposite direction. The rotor will now align itself horizontally, as before, but in the opposite direction. This effect will continue at a rate of 120 polarity changes per second, if 60-Hz AC is applied to the stator. The rotor will not start unless it is positioned initially to be drawn toward a pole piece.

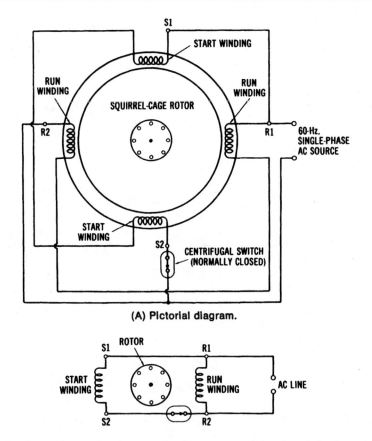

Figure 14-16. The split-phase induction motor: (A) Pictorial diagram, (8) Schematic diagram

Therefore, some starting method must be used for single-phase AC motors, since they are not self-starting.

Assume that we have a two-phase situation, as shown in Figure 14-17B. We now have two sets of stator windings with one phase connected to each set. Two-phase voltage is, of course, not produced by power companies in the United States; however, this example will show the operational principle of a split-phase motor. As shown in the two-phase voltage-curve diagram, when one phase is at minimum value, the other is at maximum.

At time interval t_1, phase 1 is maximum while phase 2 is minimum. Assume that the right stator pole becomes a north polarity, and the left

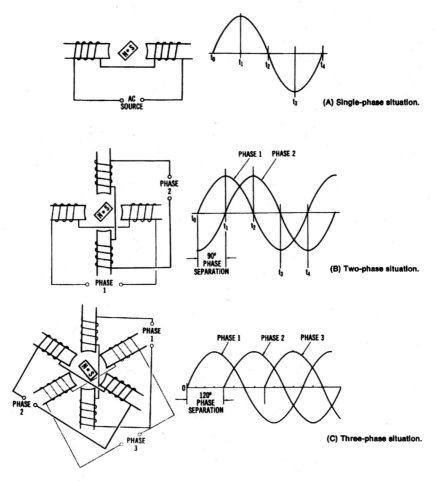

Figure 14-17. Illustration of how AC motors develop rotation: (A) Single-phase situation, (8) Two-phase situation, (C) Three-phase situation

pole becomes a south polarity. The rotor will align itself horizontally, since no polarity is developed in the vertical poles at this time. Now, as we progress to time t_2, phase 1 is minimum and phase 2 is maximum. Assume that the upper stator pole becomes a south polarity, and the bottom stator pole becomes a north polarity. The rotor will now align itself vertically, by moving 90°. At time t_3, phase 1 becomes maximum in the opposite direction, and phase 2 is minimum. This time interval results in a north pole on the left and a south pole on the right. Thus, the rotor moves 90° further in a clockwise direction. This effect will continue as two-phase voltage is applied to the stator poles. We can see from the two-phase situation that a direction of rotation is established by the relationship of the phase 1 and the phase 2 curves, and that an AC motor with two-phase voltage applied will be self-starting.

The same is true for a three-phase situation, as illustrated in Figure 14-16. Rotation of the rotor will result because of the 120° phase separation of the three-phase voltage applied to the stator poles. Three-phase induction motors are, therefore, self-starting, with no auxiliary starting method required.

Going back to the split-phase motor of Figure 14-17, we can see that the purpose of the two sets of windings is to establish a simulated two-phase condition, in order to start the motor. The single-phase voltage applied to this motor is said to be "split" into a two-phase current. A rotating or revolving magnetic field is created by phase splitting. The start winding of the split-phase motor is made of relatively few turns of small diameter wire, giving it a high resistance and a low inductance. The run winding is wound with many turns of large diameter wire, causing it to have a lower resistance and a higher inductance. We know that inductance in an AC circuit causes the current to lag the applied AC voltage. The more inductance present, the greater is the lag in current.

When single-phase AC is applied to the stator of a split-phase induction motor, the situation illustrated in Figure 14-18 will result. Notice that the current in the start winding lags the applied voltage because of its inductance. However, the current in the run windings lags by a greater amount, because of its higher inductance. The phase separation of the currents in the start and run windings creates a two-phase situation. The phase displacement, however, is usually around 30° or less, which gives the motor a low starting torque, since this phase separation does not nearly approach the 90° separation of two-phase voltage. When the split-phase AC induction motor reaches about 80 percent of its normal operating speed, needed. The removal of the start winding minimizes energy losses in the machine and prevents the winding from overheating. When the motor is turned off and its speed reduced, the centrifugal switch closes, in order to connect the start winding back into the circuit.

Split-phase motors are fairly inexpensive, compared to other types of single-phase motors. They are used when low torque is required to drive mechanical loads, such as in small machinery.

Capacitor Motors—Capacitor motors are an improvement over the split-phase AC motor. Notice that, in this cutaway illustration, some of the

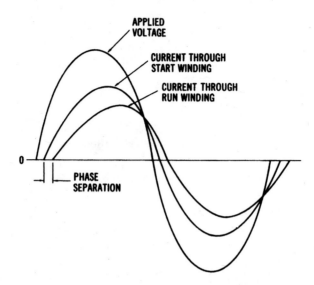

Figure 14-18. Voltage/current relationships in a split-phase induction motor

internal features of capacitor motors are similar to split-phase induction motors. All induction motors have squirrel-cage rotors that look similar to the one shown in the illustration. You can also clearly see that capacitor motors have a centrifugal switch assembly, start windings, and run windings. Also, notice the location of the starting capacitor. The wiring diagram of a capacitor-start, single-phase induction motor is shown in Figure 14-19. Notice that, except for a capacitor placed in series with the start winding, this diagram is the same as for the split-phase motor. The purpose of the capacitor is to cause the current in the start winding to lead (rather than lag) the applied voltage. This situation is illustrated by the voltage/current curves shown in Figure 14-20. The current in the start winding now leads the applied voltage, because of the high value of capacitance in the circuit. Since the run winding is highly inductive, the current through it lags the applied voltage. Note that the amount of phase separation now approaches 90°, or an actual two-phase situation. The starting torque produced by a capacitor-start induction motor is much greater than that of a split-phase motor. Thus, this type of motor can be used for applications requiring greater initial torque. However, they are somewhat more expensive than split-phase AC motors. Most capacitor motors, as well as split-phase motors, are used in fractional-horsepower sizes (less than one hp).

Another type of capacitor motor is called a capacitor-start, capaci-

Mechanical Systems 377

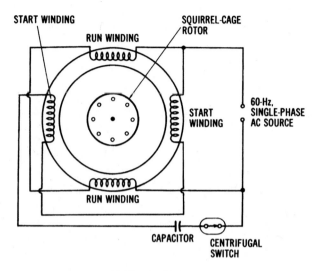

(A) Pictorial diagram.

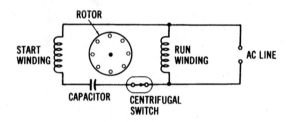

(B) Schematic diagram.

Figure 14-19. The capacitor-start, single-phase induction motor: (A) Pictorial diagram, (B) Schematic diagram

tor-run (or two-value capacitor) motor. Its circuit is shown in Figure 14-21. This motor employs two capacitors. One, of low value, is in series with the start winding and remains in the circuit during operation. The other, of higher value, is in series with the start winding and a centrifugal switch. The larger capacitor is used only to increase starting torque, and is removed from the circuit during normal operation by the centrifugal switch. The smaller capacitor, and the entire start winding, are part of the operational circuit of the motor. The smaller capacitor helps to produce a more constant-running torque, as well as quieter operation and an improved power factor.

Still another type of capacitor motor is one that is called a perma-

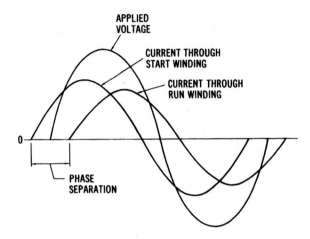

Figure 14-20. Voltage/current relationships in a capacitor-start, single-phase induction motor

nent capacitor motor. Its circuit is shown in Figure 14-21. This motor has no centrifugal switch, so its capacitor is permanently connected into the circuit. These motors are only used for very low-torque requirements, and are made in small fractional-horsepower-size units.

Both split-phase motors and capacitor motors may have their direction of rotation reversed easily. Simply change the relationship of the start winding and the run winding. When either the start winding connections or the run winding connections (but not both) are reversed, the rotational direction will be reversed.

Shaded-pole Induction Motors—Another method of producing torque by a simulated two-phase method is called pole shading. These motors are used for very low-torque applications, such as fans and blower units. They are low-cost, rugged, and reliable motors that ordinarily come in low horsepower ratings, from 1/3000 to 1/30 hp, with some exceptions.

The operational principle of a shaded-pole motor is shown in Figure 14-22. The single-phase alternation shown is for discussion purposes only. The dotted lines represent induced voltage into the shaded section of the field poles. Note the shading coils in the upper right and lower left of the two poles. The shaded pole is encircled by a heavy copper conductor and is actually a part of the main field pole. This closed-loop conductor will cause current to be induced into the shaded pole when AC is applied to the field.

When an AC voltage is applied to the stator windings, the magnetic

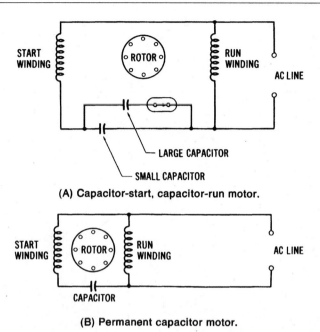

Figure 14-21. Other types of capacitor motors: (A) Capacitor-start, capacitor-run motor, (B) Permanent capacitor motor

flux in the main poles induces a voltage into the shaded sections of the poles. Since the shaded section acts like a transformer secondary, its voltage is out of phase with the main field voltage, as shown in the waveform diagram of Figure 14-22. Note the four time intervals that are shown in sequence in Figure 14-22. The voltage induced in the shaded pole from the main pole field causes movement of the rotor to continue. Study Figure 14-22 carefully to understand the basic operating principle of the shaded-pole AC induction motor more fully.

The shaded-pole motor is inexpensive, since it uses a squirrel-cage rotor and has no auxiliary starting winding or centrifugal mechanism. Application is limited mainly to small fans and blowers, and other low-torque applications.

Repulsion Motors—Another type of AC induction motor is the repulsion-start induction motor. This motor was once used for many applications, but is now being replaced by other types of single-phase motors. The principle of operation of the repulsion motor provides an interesting contrast to other induction motors.

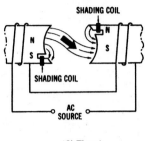

(C) Time t_2.

Figure 14-22. Illustration of the operational principle of the shaded-pole AC induction motor: (A) A single-phase AC alternation, (B) Time t_1 (C) Time t_2, (D) Time t_3, (E) Time t_4

Figure 14-23 shows the operational principle of the repulsion motor. This motor has a wound rotor that functions similarly to a squirrel-cage rotor. It also has a commutator/brush assembly. The brushes are shorted together to produce an effect similar to the shorted conductors of a squirrel-cage rotor. The position of the brush axis determines the amount of torque developed and the direction of rotation of the repulsion motor.

In position 1, Figure 14-23A, the brush axis is horizontally aligned with the stator poles. Equal and opposite currents are now induced into both halves of the rotor. Thus, no torque is developed with the brushes in this position. In position 2 (Figure 14-23B), the brushes are placed at a 90° angle to the stator field poles. The voltages induced into the rotor again counteract one another, and no torque is developed. In position 3 (Figure 14-23C), the brush axis is shifted about 60° from the stator poles. The current flow in the armature now causes a magnetic field around the rotor. The rotor field will now follow the revolving stator field in a clockwise direction. As might be expected, if we shift the brush axis in the opposite direction, as shown in position 4 (Figure 14-23D), rotation reversal will result. Thus, magnetic repulsion between the stator field and the induced rotor field causes the rotor to turn in the direction of the brush-shift.

Repulsion-start induction motors, and some similar types of modified repulsion motors, have very high starting torque. Their speed may be varied by varying the position of the brush axis. However, the mechanical problems inherent with this type of motor have caused it to become obsolete.

Mechanical Systems

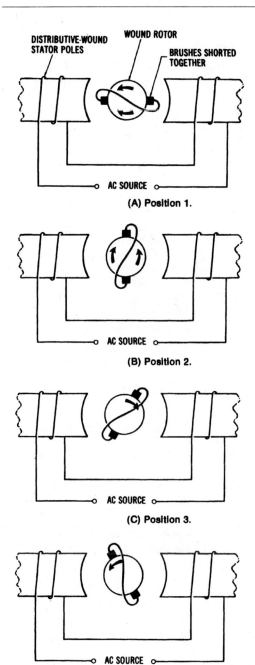

Figure 14-23. Operational principle of the repulsion motor: (A) Position 1, (B) Position 2, (C) Position 3, (D) Position 4

Single-phase Synchronous Motors

It is often desirable, in timing or clock applications, to use a constant-speed drive motor. Such a motor, which operates from a single-phase AC line, is called a synchronous motor. The single-phase synchronous motor has stator windings that are connected across the AC line. Its rotor is made of a permanent-magnetic material. Once the rotor is started, it will rotate in synch with the revolving stator field, since it does not rely upon the induction principle. The calculation of the speeds of synchronous motors is based on the speed formula. This formula states that

$$\text{revolutions per minute} = \frac{\text{frequency} \times 60}{\text{number of pairs of poles}}$$

Therefore, for 60-Hz operation, the following synchronous speeds would be obtained:

1. Two-pole   3600 rpm.
2. Four-pole   1800 rpm.
3. Six-pole   1200 rpm.
4. Eight-pole   900 rpm.
5. Ten-pole   720 rpm.
6. Twelve-pole   600 rpm.

Small synchronous motors are used in single-phase applications for low-torque applications. Such applications include clocks, drives, and timing devices that require constant speeds.

The construction of an AC synchronous motor is quite simple. It contains no brushes, commutators, slip rings, or centrifugal-force switches. It is simply made up of a rotor and a stator assembly. There is no direct physical contact between the rotor and stator. A carefully maintained air gap is always present between the rotor and stator. As a result of this construction, the motor has a long operating life and is highly reliable.

The speed of a synchronous motor is directly proportional to the frequency of the applied AC, and inversely proportional to the number of pairs of stator poles. Since the number of stator poles cannot be effectively altered after the motor has been manufactured, frequency is the most significant speed factor. Speeds of 28, 72, and 200 rpm are typical, with 72 rpm being a common industrial numerical control standard.

The stator layout of a two-phase synchronous motor with four poles per phase is the same as that of the DC stepping motor shown in Figure 14-11. Refer now to Figure 14-12. In the diagram shown, poles N1-53 and N5-57 represent one phase, while poles N2-54 and N6-58 represent the second phase. There are places for 48 teeth around the inside of the stator. One tooth per pole has been eliminated, however, to provide a space for the windings. Five teeth per pole, or a total of 40 teeth, are formed on the stator. The four coils of each phase are connected in series to achieve the correct polarity.

The rotor of the synchronous motor is an axially magnetized permanent magnet. There are 50 teeth cast into its form. The front section of the rotor has one polarity, while the back section has the opposite polarity. The physical difference in the number of stator teeth (40) and rotor teeth (50) means that only two teeth of each part can be properly aligned simultaneously. With one section of the rotor being a north pole and the other section being a south pole, the rotor has the ability to stop very quickly. It can also produce complete direction reversals without hesitation, because of this gear-like construction.

A circuit diagram of a single-phase synchronous motor is shown in Figure 14-24. The resistor and capacitor of this circuit are used to produce a 90° phase shift in one winding. As a result, the two windings are always out of phase, regardless of whether the switch is in the clockwise (cw) or counterclockwise (ccw) position. When power is applied, the four coils of one phase produce an electromagnetic field. The rotor is attracted and aligns itself to these stator coils. Then, 90° later, the four coils of the second phase produce a corresponding field. The stator is again attracted to this position. As a result of this action, the rotor "sees" a moving force across first one phase and then the other. This force gives the rotor the

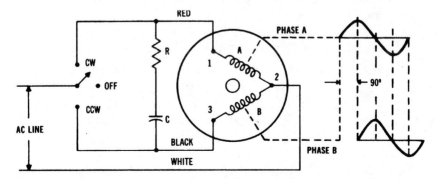

Figure 14-24. Circuit diagram of a single-phase synchronous motor

needed torque that causes it to start and continue rotation when power is applied.

The synchronous motor just described has the capability of starting in one and one-half cycles of the applied line-voltage frequency. In addition to this, it can be stopped in five mechanical degrees of rotation. These two characteristics are primarily due to the geared rotor and stator construction. Synchronous motors of this type have one other important characteristic—they draw the same amount of line current when stalled as they do when operating. This characteristic is very important in automatic machine-tool applications, where overloads occur frequently.

THREE-PHASE AC MOTORS

Three-phase AC motors are often called the "workhorses of industry." Most motors used in industry, and several types used in commercial buildings, are operated from three-phase power sources. There are three basic types of three-phase motors: (1) induction motors, (2) synchronous motors, and (3) wound-rotor induction motors (wrim).

Induction Motors

A pictorial diagram of the construction of a three-phase induction motor is given in Figure 14-25. Note that the construction of this motor is very simple. It has only a distributive-wound stator, which is connected in either a wye or a delta configuration, and a squirrel-cage rotor. Since three-phase voltage is applied to the stator, phase separation is already established (Figure 14-26). No external starting mechanisms are needed. Three-phase induction motors come in a variety of integral horsepower sizes, and have good starting and running torque characteristics.

The direction of rotation of a three-phase motor of any type can be changed very easily. If any two power lines coming into the stator windings are reversed, the direction of rotation of the shaft will change. Three-phase induction motors are used for many applications, such as mechanical-energy sources for machine tools, pumps, elevators, hoists, conveyors, and other systems that use large amounts of power.

Synchronous Motors

The three-phase synchronous motor is a unique and very specialized motor. It is considered a constant-speed motor, and it can be used to

Mechanical Systems

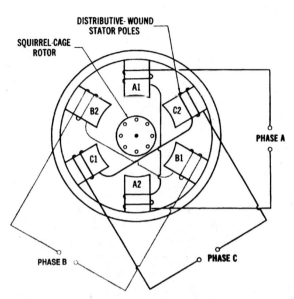

RIGHT: Figure 14-25. Pictorial diagram of the construction of a three-phase induction motor

BELOW: Figure 14-26. Waveforms and the resulting stator fields that show the operating principle of a three-phase induction motor

NOTE: Stator windings may be connected in either a wye or a delta configuration.

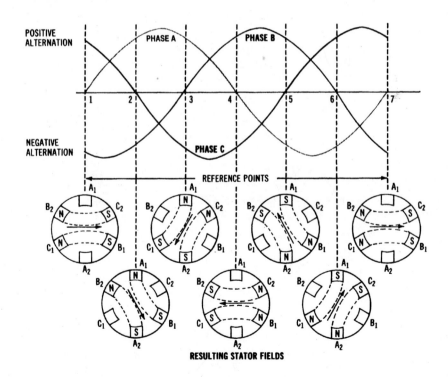

correct the power factor of three-phase systems. Synchronous motors are usually very large in size and horsepower rating.

Figure 14-27 shows a pictorial diagram of the construction of a three-phase synchronous motor. Physically, this motor is constructed like a three-phase alternator (see Chapter 6). DC is applied to the rotor to produce a rotating electromagnetic field; the stator windings are connected in either a wye or delta configuration. The only difference is that three-phase AC power is applied to the synchronous motor, while three-phase power is extracted from the alternator. Thus, the motor acts as an electrical load, while the alternator functions as a source of three-phase power. This relationship should be kept in mind during the following discussion.

The three-phase synchronous motor differs from the three-phase induction motor in that the rotor is wound and is connected through a slip ring/brush assembly to a DC power source. Three-phase synchronous motors, in their pure form, have no starting torque. Some external means must be used to initially start the motor. Synchronous motors are constructed so that they will rotate at the same speed as the revolving stator field. We can say that at synchronous speed, rotor speed equals stator

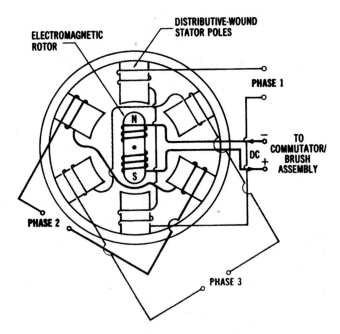

Figure 14-27. Pictorial diagram of three-phase synchronous motor construction

speed, and the motor has zero slip. Thus, we can determine the speed of a synchronous motor by using the following formula:

$$S = \frac{f \times 120}{n/3}$$

where:
    S = the speed of a synchronous motor in r/min,
    f = the frequency of the applied AC voltage in hertz,
    $n/3$ = the number of stator poles per phase, and
    120 = a conversion constant.

Note that this is the same as the formula used to determine the stator speed of a single-phase motor, except that the number of poles must be divided by three (the number of phases). A three-phase motor with twelve actual poles will have four poles per phase. Therefore, its stator speed will be 1800 rpm. Synchronous motors have operating speeds that are based on the number of stator poles they have.

Three-phase synchronous motors usually are employed in very large horse power ratings. One method of starting a large synchronous motor is to use a smaller auxiliary DC machine connected to the shaft of the synchronous motor, as illustrated in Figure 14-28. The method of starting is as follows:

Step 1. DC power is applied to the auxiliary motor, causing it to increase in speed. Three-phase AC power is applied to the stator.
Step 2. When the speed of rotation reaches a value near the synchronous speed of the motor, the DC power circuit is opened and, at the same time, the terminals of the auxiliary machine are connected across the slip ring/brush assembly of the rotor.
Step 3. The auxiliary machine now converts to generator operation and supplies exciter current to the rotor of the synchronous motor, using the motor as its prime mover.
Step 4. Once the rotor is magnetized, it will "lock" in step, or synchronize, with the revolving stator field.
Step 5. The speed of rotation will remain constant under changes in load condition.

Another starting method is shown in Figure 14-28. This method utilizes damper windings, which are similar to the conductors of a squirrel-

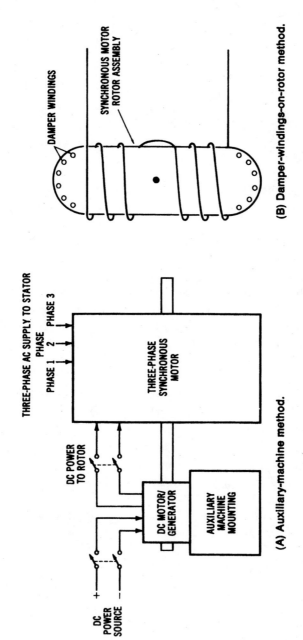

Figure 14-28. Three-phase synchronous motor starting methods: (A) Auxiliary machine starting method, (B) Damper windings placed in the rotor as a method of starting

cage rotor. These windings are placed within the laminated iron of the rotor assembly. No auxiliary machine is required when damper windings are used. The starting method used is as follows:

Step 1. Three-phase AC power is applied to the stator windings.
Step 2. The motor will operate as an induction motor, because of the "transformer action" of the damper windings.
Step 3. The motor speed will build up, so that the rotor speed is somewhat less than the speed of the revolving stator field.
Step 4. DC power from a rotating DC machine, or more commonly from a rectification system, is applied to the slip ring/brush assembly of the rotor.
Step 5. The rotor becomes magnetized and builds up speed until rotor speed is equal to stator speed.
Step 6. The speed of rotation remains constant regardless of the load placed on the shaft of the motor.

An outstanding advantage of the three-phase synchronous motor is that it can be connected to a three-phase power system to increase the overall power factor of the system. Power factor correction was discussed previously. Three-phase synchronous motors are sometimes used only to correct the system power factor. If no load is to be connected to the shaft of a three-phase synchronous motor, it is called a synchronous capacitor. It is designed to act only as a power factor corrective machine. Of course, it might be beneficial to use this motor as a constant-speed drive connected to a load, as well as for power factor correction.

We know from previous discussions that a low power factor cannot be tolerated by an electrical power system. Thus, the expense of installing three-phase synchronous machines can be justified in industrial use, as a means of appreciably increasing the system power factor. To understand how a three-phase synchronous machine operates as a power factor corrective machine, refer to the curves of Figure 14-29. We know that the synchronous motor operates at a constant speed. Variation in rotor DC excitation current has no effect on speed. The excitation level will change the power factor at which the machine operates. Three operational conditions may exist, depending on the amount of DC excitation applied to the rotor. These conditions are:

1. Normal excitation—operates at a power factor of 1.0.

2. Under excitation—operates at a lagging power factor (inductive effect).
3. Over excitation—operates at a leading power factor (capacitive effect).

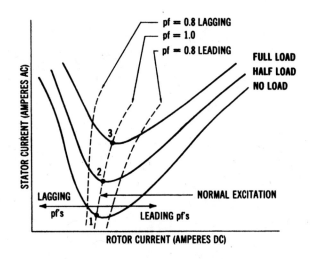

(A) Rotor current versus stator current.

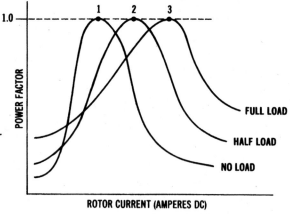

Figure 14-29. Power relationships of a three-phase synchronous motor: (A) Rotor current versus stator current—"V curves," (B) Rotor current versus power factor

(B) Rotor current versus power factor.

Note the variation of stator current drawn by the synchronous motor as the rotor current varies. You should also see that stator current is minimum when the power factor equals 1.0, or 100%. The situations shown on the graph in Figure 14-31A indicate stator and rotor currents under no-load, half-load, and full-load conditions. Current values when the power factors are equal to 1.0, 0.8-leading, and 0.8—lagging conditions are also shown. These curves are sometimes referred to as V-curves for a synchronous machine. The graph in Figure 14-29B shows the variation of power factor with changes in rotor current under three different load conditions. Thus, a three-phase synchronous motor, when over-excited, can improve the overall power factor of a three-phase system.

As the load increases, the angle between the stator pole and the corresponding rotor pole on the synchronous machine increases. The stator current will also increase. However, the motor will remain synchronized unless the load causes "pull-out" to take place. The motor would then stop rotating because of the excessive torque required to rotate the load. Most synchronous motors are rated greater than 100 horsepower and are used for many industrial applications requiring constant-speed drives.

Wound-Rotor Induction Motor

The wound-rotor induction motor (wrim), shown in Figure 14-30 is a specialized type of three-phase motor. This motor may be controlled externally by placing resistances in series with its rotor circuit. The starting torque of a wrim motor can be varied by the value of external resistance. The advantages of this type of motor are a lower starting current, a high starting torque, smooth acceleration, and ease of control. The major disadvantage of this type of motor is that it costs a great deal more than an equivalent three-phase induction motor using a squirrel-cage rotor. Thus, they are not used as extensively as other three-phase motors.

SPECIALIZED MECHANICAL POWER SYSTEMS

There is a need for specialized mechanical systems that can produce a rotary motion that is somewhat different from that produced by most electric motors. This type of system employs rotary motion to control the angular position of a shaft that is used to position the shaft of a second device. Synchro systems and servomechanisms are used to achieve this basic operation. With these devices, it becomes possible to transmit a rotary mo-

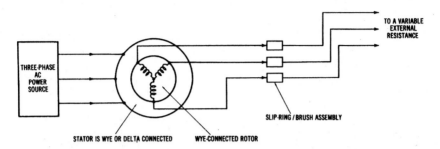

Figure 14-30. Diagram of a three-phase, wound-rotor induction motor

tion from one location to another without any direct mechanical linkage.

Synchro systems are classified as two or more motor/generator units connected together in a way that permits the transmission of angular shaft positions by means of electromagnetic field changes. When an operator turns the generator shaft of a unit to a certain position, it causes the motor shaft at a remote location to automatically rotate to an equivalent position. With this type of system, it is possible to achieve accurate control of devices over long distances. Computers often employ these units to determine the physical changes that take place in automated operations.

In synchro systems that require increased rotational torque or precise movements of a control device, servomechanisms are employed. A servomechanism is ordinarily a special type of AC or DC motor that drives a precision piece of equipment in specific increments. Systems that include servomechanisms generally require amplifiers and error-detecting devices to control the angular displacement of a shaft.

Synchro System Operation

A synchro system contains two or more electromagnetic devices that are similar in appearance to small electric motors. These devices are connected together in such a way that the angular position of the generator shaft can easily be transmitted to the motor or receiver unit. Figure 14-31 shows the schematic diagram of a basic synchro unit. As a general rule, the generator and motor units are identical electrically. Physically, the motor unit has a metal flywheel attached to its shaft to prevent shaft oscillations or vibrations when it is powered. The letter G or M inside of the electrical symbol denotes generator or motor functions.

Figure 14-32 shows the circuit diagram of a basic synchro system. Single-phase AC line voltage is used to power this system. Note that the

Mechanical Systems

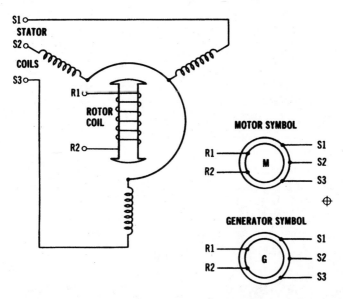

Figure 14-31. Schematic diagram and symbols for a basic synchro unit

line voltage is applied to the rotors of both the generator and motor. The stationary coils, or stator windings, are connected together as indicated. When power is initially applied to the system, the motor will position itself according to the location of the generator shaft. No physical change will take place after the motor unit aligns itself with the generator position. Both units will remain in a stationary condition until some further action takes place. Turning the generator shaft a certain number of degrees

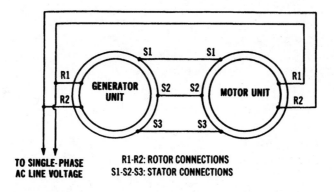

Figure 14-32. Circuit diagram of a basic synchro system

in a clockwise direction will cause a corresponding change in the motor unit. If calibrated dials were attached to the shaft of each unit, they would show the same angular displacement change.

Synchro units have unique construction features. The stator coils are wound inside a cylindrical laminated metal housing. The coils are uniformly placed in slots and connected to provide three poles spaced 120° apart. These coils serve as the secondary windings of a transformer. The rotor coil of synchro units is also wound on a laminated core. This type of construction causes north and south poles of the magnetic field to extend from the laminated area of the rotor. Insulated slip rings on the shaft are used to supply AC power to the rotor. The rotor coil responds as the primary winding of a transformer.

When AC is applied to the rotor coil of the synchro unit shown in Figure 14-31, it produces an alternating magnetic field. By transformer action, this field cuts across the stator coils and induces a voltage in each winding. The physical position of the rotor coil determines the amount of voltage induced in each stator coil. If the rotor coils are parallel with a stator coil, maximum voltage will be induced. The induced voltage will be of a minimum value when the rotor coils are at right angles to a stator-coil set.

The stator coils of the generator and motor of a synchro system are connected together, as indicated in the circuit diagram of Figure 14-32 Voltage induced in the stator coils of the generator, therefore, causes a resulting current flow in the stator coils of the motor. This, in turn, causes a corresponding magnetic field to be established in the stator of the motor. Line voltage applied to the rotor of the motor unit will cause it to align itself with the magnetic field of the stator coils.

Any change in rotor position of the generator unit is translated into an induced voltage and applied to the stator coils of the motor. Through this action, linear displacement changes can be effectively transmitted to the motor through three rather small stator coil wires. Systems of this type are becoming very important today in remote-control applications and in industrial automated-process control applications.

Servo System Operation

Servo systems are a specific type of rotating machine used typically for changing the mechanical position or speed of a device. Mechanical-position applications include numerical-control machinery and process-control-indicating equipment in industry. Speed applications are found in conveyor-belt control units, in spindle-speed control in machine-tool operations, and in disk or magnetic-tape drives for computers. As a general rule, a servo system is a rather complex unit that follows the commands of a closed-loop control path. Figure 14-33 shows the components of a typical servo system.

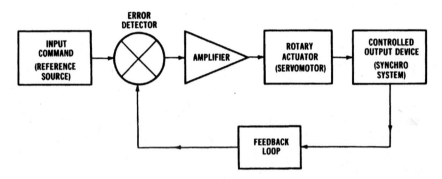

Figure 14-33. Block diagram of a typical servo system

The input of a servo system serves as the reference source, or as a set point to which the load element responds. When the input is changed in some way, a command is applied to the error detector. This device receives data from both the input source and from the controlled output device. If a correction is needed, with reference to the input command, it is amplified and applied to the actuator. The actuator is normally a servomotor that produces controlled shaft displacements. The controlled output device is usually a system that relays information back to the error detector for position comparison.

A servomotor is primarily responsible for producing mechanical changes from an electromagnetic actuating device. A device of this type is normally coupled to the work load by a gear train or some mechanical linkage. Both AC and DC servomotors are used to achieve this operation. As a general rule, a servomotor is unique, when compared with other electric motors, because a servomotor is a very special type of device that is used to achieve a precise degree of rotary motion. Servomotors, for example, are designed to do something other than change electrical energy into rotating mechanical energy. Motors of this type must first be able to respond accurately to signals developed by the amplifier of the system. Second, they must be capable of reversing direction quickly when a specific signal polarity is applied. Also, the amount of torque developed by a servomotor must be quite high. As a general rule, the torque developed is a function of the voltage and current source.

Of the two distinct types of servomotors in use today, an AC type of motor, called a synchronous motor, is commonly used in low-power applications. Excessive amounts of heat developed during starting conditions normally limit this motor to rather low-output-power applications. DC stepping motors are also used as servomotors.

ELECTRIC MOTOR APPLICATIONS

Certain factors must be considered when an electric motor is selected for a specific application. Among these considerations are (1) the source voltage and power capability available, (2) the effect of the power factor and efficiency of the motor on the overall system, (3) the effect of the starting current of the motor on the system, (4) the effect of the power system on the operation of the motor, (5) the type of mechanical load, and (6) the expected maintenance the motor will require.

Motor Performance

The major consumer of electrical power is the electric motor. It is estimated that electric motors account for 50 percent of the electrical power consumed in industrial usage, and that 35 percent of all the electrical power used is used by electrical motors. For these reasons, we must consider the efficient operation of motors to be a major part of our energy conservation efforts.

Both efficiency and the power factor must be considered in order to determine the effect of a motor, in terms of efficient power conversion. Remember the following relationships. First,

$$\text{Efficiency (\%)} = \frac{P_{out} \times 100}{P_{in}}$$

where:
P_{in} = the power input in horsepower, and
P_{out} = the power output in watts.

(To convert horsepower to watts, remember that 1 horsepower = 746 watts.) Then,

$$pf = \frac{P}{VA}$$

where:
    pf = the power factor of the circuit,
    P = the true power in watts, and
    VA = the apparent power in volt-amperes.

The maximum pf value is 1.0, or 100%, which would be obtained in a purely resistive circuit. This is referred to as unity power factor.

Effect of Load

Since electrical power will probably become more expensive and less abundant, the efficiency and power factor of electric motors will become increasingly important. The efficiency of a motor shows mathematically just how well a motor converts electrical energy into mechanical energy. A mechanical load placed on a motor affects its efficiency. Thus, it is particularly important for industrial users to load motors so that their maximum efficiency is maintained.

Power factor is also affected by the mechanical load placed on a motor. A higher power factor means that a motor requires less current to produce a given amount of torque or mechanical energy. Lower current levels mean that less energy is being wasted (converted to heat) in the equipment and circuits connected to the motor. Penalties are assessed on industrial users by the electrical utility companies for having low system power factors (usually less than 0.8 or 0.85 values). By operating at higher power factors, industrial users can save money on penalties, and can help, on a larger scale, with more efficient utilization of electrical power. Motor load affects the power factor to a much greater extent than it does the efficiency. Therefore, motor applications should be carefully studied to ensure that motors (particularly very large ones) are not overloaded or underloaded, so that the available electrical power will be used more effectively.

Effect of Voltage Variations

Voltage variation also has an effect on the power factor and efficiency. Even slight changes in voltage produce a distinct effect on the power factor. However, a less distinct effect results when the voltage causes a variation in the efficiency. Because proper power utilization is becoming more and more important, motor users should make sure that their motors do not operate at undervoltage or overvoltage conditions.

Considerations for Mechanical (Motor) Loads

There are three basic types of mechanical (motor) loads connected to electrical power systems. These are DC, single-phase AC, and three-phase AC systems. DC motors are ordinarily used for special applications, since they are more expensive than other types and require a DC power source. Typically, they are used for small, portable applications and powered by batteries, or for industrial/commercial applications with AC converted to DC by rectification systems. A major advantage of DC motors is their ease of speed control. The shunt-wound DC motor can be used for accurate speed control and good speed regulation. A disadvantage is the increased maintenance caused by the brushes and commutator of the machines. DC shunt-wound motors are used for variable speed drives on printing presses, rolling mills, elevators, hoists, and automated industrial machine tools.

Series-wound DC motors have a very high starting torque. Their speed regulation is not as good as that of shunt-wound DC motors. The series-wound motor also requires periodic maintenance because of the brush/commutator assembly. Typical applications of series-wound DC motors are automobile starters, traction motors for trains and electric buses, and mobile equipment operated by batteries. Compound-wound DC motors have very few applications today.

Single-phase AC motors are relatively inexpensive. Most types have good starting torque and are easily provided 120-volt and 240-volt electrical power. Disadvantages include maintenance problems due to centrifugal switches, pulsating torque, and rather noisy operation. They are used in fractional horsepower sizes (less than one horsepower) for residential, commercial, and industrial applications. Some integral horsepower sizes are available in capacitor-start types. Uses include machine cooling system blowers, and clothes dryer motors.

Specialized applications for single-phase motors include:

1. Shaded-pole motors used for portable fans, record players, dishwasher pumps, and electric typewriters. They are low-cost and small, but inefficient.
2. Single-phase synchronous motors used for clocks, appliance timers, and recording instruments (compact disk players). They operate at a constant speed.

3. Universal motors (AC/DC) motors used for many types of portable tools and appliances, such as electric drills, saws, office machines, mixers, blenders, sewing machines, and vacuum cleaners. They operate at speeds up to 20,000 r/min and have easy speed control. Remember that AC induction motors do not have speed control capability without the addition of expensive auxiliary equipment.

Three-phase AC motors of the induction type are very simple in construction, rugged, and reliable in operation. They are less expensive (per horsepower) than other motors. Applications of three-phase induction motors include industrial and commercial equipment and machine tools. Three-phase AC synchronous motors run at constant speeds and may be used for power factor correction of electrical power systems. However, they are expensive, require maintenance of brushes/slip rings, and need a separate DC power supply.

UNIT V
Electrical Power Control Systems

In this unit, *electrical power control* systems are studied. The fundamentals of electrical power control are discussed in Chapter 15. This chapter examines several of the fundamental types of *equipment* that are used for electrical power control. This equipment includes *electromechanical* contactors, relays and switches.

Chapter 16 examines the mechanical and electronic *power control* systems that are used with electrical machinery, and the *motor control* systems that are used extensively in day-to-day industrial, commercial, and residential applications. In addition, *programmable logic controllers (PLCs)* are introduced, since they now lay a significant role in machine control.

Chapter 17 discusses electronic control devices such as silicon controlled rectifiers (SCRs), triacs and diacs.

Figure V shows the *electrical power systems model* used in this book, and the major topics of Unit V—Electrical Power Control Systems.

UNIT OBJECTIVES

Upon completion of this unit, you should be able to:

1. Describe the function of standards organizations, such as NEMA, ANSI, and IEEE, in electrical power control.
2. Define important terms used with electrical power control systems.
3. Recognize and sketch symbols used for electrical power control systems.
4. Describe the following types of switches used for electrical power control:
   Toggle Switch
   Rocker Switch
   Pushbutton Switch
   Rotary Switch

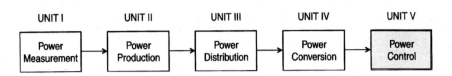

Figure V. Electrical power systems model

Power Control (Chapter 15)
Operational Power Control Systems (Chapter 16)
Control Devices (Chapter 17)

    Limit Switch
    Temperature Switch
    Float Switch
    Pressure Switch
    Foot Switch
    Drum Controller

5. Describe indicating lights used with electrical power control.
6. Recognize and describe various motor starting systems.
7. List the functions of motor starters.
8. Describe the method of assigning size ratings for motor contactors.
9. Describe bimetallic and melting alloy thermal overload devices and their current ratings.
10. Describe the classification system used for motor starters.
11. Describe the method of assigning sizes to manual starters.
12. Describe a combination starter.
13. List criteria used to select motor controllers.
14. List types of motor controller enclosures used by NEMA.
15. Explain the operation of a relay.
16. Explain the operation of a solenoid.
17. Describe an SCR, triac, and diac control devices that may be used for electrical power control.
18. Explain the use of programmable logic controllers and other computerized controllers used for electrical power control applications.
19. Explain the operation of the following power control circuits:
Start-Stop Control
Start-Stop Control from Multiple Locations

Start-Stop Control with Safe-Run Feature
Start-Stop Control with Run-Jog Feature
Start-Stop-Jog Control
Forward -Reverse-Stop Control
Full-Voltage Starting
Primary Resistance Starting
Primary Reactor Starting
Autotransformer Starting
Wye-Delta Starting
Part-Winding Starting
Forward-Reverse Starting
20. Describe dynamic braking for direct current (DC) and alternating current (AC) motors.
21. Describe universal motor speed control circuit operation.
22. Describe frequency conversion systems.
23. Describe silicon controlled rectifiers (SCRs), triacs and diacs.

Chapter 15

Power Control

The *control* of electrical power is a very important part of electrical power system operation. Control is the most complex part of electrical systems. It is, therefore, necessary to limit the discussion in this chapter primarily to some of the common types of electromechanical equipment used for electrical power control. Control equipment is used in conjunction with many types of electrical loads. Electrical motors (mechanical loads) use about 50 percent of all electrical *motor-control* equipment. In most cases, similar equipment is used to control electrical lighting and heating loads.

IMPORTANT TERMS

Chapter 15 deals with power control equipment. After studying this chapter, you should have an understanding of the following terms:

NEMA
ANSI
IEEE
Across-the-Line Starter Auxiliary Contact
Dynamic Braking
Normally Open Contact
Normally Closed Contact
Magnetic Contactor
Drum Controller
Jogging
Manual Starter
Overload Relay
Plugging
Normally Closed Pushbutton
Normally Open Pushbutton

Reduced-Voltage Starting
Control Relay
Safety Switch
Solenoid
Timing Relay
Switches
   SPST
   SPDT
   DPST
   DPDT
   4-Way
   3PST
   3PDT
Rotary Switch
Limit Switch
Temperature Switch
Float Switch
Pressure Switch
Foot Switch
Pilot Lights
   Full-Voltage
   Transformer-Operated
   Series Resistor
   Illuminated Pushbutton
Combination Starter
Controller Enclosures-NEMA

POWER CONTROL STANDARDS, SYMBOLS, AND DEFINITIONS

A great amount of fundamental knowledge is needed in order to fully understand electrical power control. Individuals who are concerned with electrical power control systems should have a good knowledge of the *standards, symbols,* and *definitions* that govern electrical power control.

NEMA Standards

The National Electrical Manufacturers' Association (NEMA) has developed standards for electrical control systems. NEMA standards are used extensively to obtain information about the construction and performance of various electrical power control equipment. These standards provide information concerning the voltage, frequency, power, and current ratings for various equipment.

ANSI and IEEE Standards

Two well-known organizations publish standards that are of importance to industry. These organizations are the American National Standards Institute (ANSI) and the Institute of Electrical and Electronic Engineers (IEEE). The standards published by these organizations are for the use of the manufacturing industries, as well as power consumers, and in some cases the general public. These standards are subject to review periodically; therefore, industrial users should keep up to date by obtaining the revisions.

These standards were developed from input from manufacturers, consumers, government agencies, and scientific, technical, and professional organizations. Often these published standards are used by industry as well as governmental agencies. It should be noted that the National Electrical Code, discussed previously, is an American National Standards Institute standard.

Definitions

There are several basic definitions that are used when dealing with electrical power control. These definitions are particularly important when interpreting control diagrams and standards. A listing of several of the important power control definitions follows. You should study these definitions.

Across-the-line-Control—A method of motor starting in which a motor is connected directly across the power lines when it is started.
Automatic Starter—A self-acting starter that is completely controlled by control switches, or some other sensing mechanism.
Auxiliary Contact—A contact that is part of a switching system. It is used in addition to the main contacts, and is operated by the main contacts.
Braking—A control method that is used to rapidly stop and hold a motor.
Circuit Breaker—An automatic device that opens under abnormally high current conditions, and can be manually or automatically reset.
Contact—A current-conducting part of a control device. It is used to open or close a circuit.
Contactor—A control device that is used to repeatedly open or close an

electric power circuit.

Controller—A device or group of devices that systematically controls the delivery of electric power to the load or loads connected to it.

Disconnect—A control device or group of devices that will open, so that electrical current in a circuit will be interrupted.

Drum Controller—A set of electrical contacts mounted on the surface of a rotating cylinder. It is usually used for controlling the on-off forward-reverse condition of a load.

Electronic Control—A usually solid-state device that performs part of the control function of a system.

Full Voltage Control—A control system that connects equipment directly across the power lines when the equipment is started.

Fuse—A circuit-protection device that disconnects a circuit when an overcurrent condition occurs. It is self-destructing and must be replaced.

Horsepower—The power output or mechanical work rating of an electrical motor.

Jogging—Momentary operation that causes a small movement of the load that is being controlled.

Magnetic Contactor—A contactor which is operated electromagnetically and usually controlled by pushbuttons activated by an operator.

Manual Controller—An electric control device that functions when operated by mechanical means, usually by an operator.

Master Switch—A switch that controls the power delivered to other parts of a system.

Motor—A device (mechanical load) for converting electrical power to mechanical power in the form of rotary motion.

Multispeed Starter—An electrical power-control device that provides for varying the speed of a motor.

Overload Relay—Overcurrent protection for a load. While it is in operation, it maintains the interruption of the load from the power supply until it is reset or replaced.

Overload Relay Reset—A pushbutton that is used to reset a thermal overload relay after the relay has been overloaded.

Pilot Device—A control device that directs the operation of another device or devices.

Plugging—A braking method that causes a motor to develop a retarding force in the reverse direction.

Pushbutton—A button-type control switch that is manually operated for actuating or disconnecting some load device.

Pushbutton Station—A housing for the pushbuttons that are used to control equipment.

Reduced-voltage Starter—A control device that applies a reduced voltage to a motor when it is started.

Relay—A control device that is operated by one electrical circuit to control a load that is part of another electrical circuit.

Remote Control—A system in which the control of an electrical load takes place from some distant location.

Safety Switch—An enclosed, manually operated, disconnecting switch used to turn a load off when necessary.

Solenoid—An electromagnetically actuated control device that is used to produce linear motion for performing various control functions.

Starter—An electric controller that is used to start, stop, and protect the motor that is connected to it.

Timer—A control device that provides variable time periods, so that a control function may be performed.

Symbols

One should have an understanding of the electrical symbols that are commonly used with power control systems. A few of these are somewhat different from basic electrical symbols. Some typical symbols are shown in Figure 15-1. You should especially observe the symbols that are used for the various types of switches and pushbuttons.

POWER CONTROL USING SWITCHES

An important, but often overlooked, part of electrical power control is the various types of switches used. This section will examine the many types of switches that are used to control electrical power. The primary function of a switch is to turn a circuit on or off; however, many more complex switching functions can be performed using switches. The emphasis in this section will be on switches that are used for motor control. Keep in mind that other load devices can also be controlled in a similar manner by switches.

Toggle Switches

Among the simplest types of switches are toggle switches. The symbols for several kinds of toggle switches are shown in Figure 15-2. You

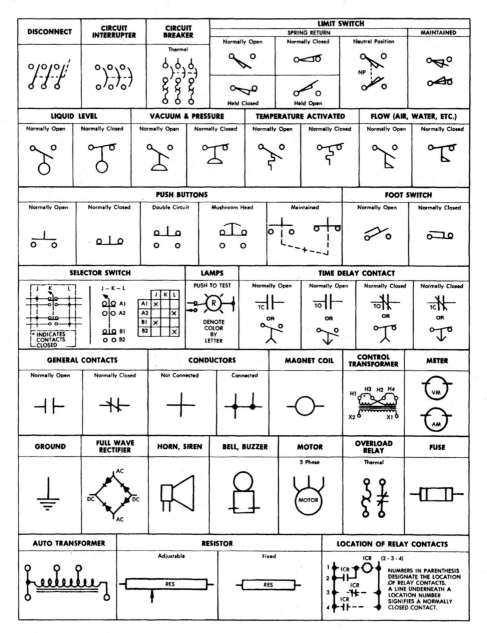

Figure 15-1. Common power control symbols (*Courtesy Furnas Electric Co.*)

should become familiar with the symbols that are used for various types of toggle switches and with the control functions that they can accomplish.

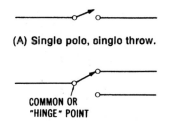

(A) Single pole, single throw.

(B) Three-way, or single pole, double throw.

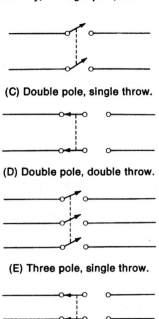

(C) Double pole, single throw.

(D) Double pole, double throw.

(E) Three pole, single throw.

(F) Three pole, double throw.

Figure 15-2. Types of toggle switches: (A) Single pole, single throw (SPST), (B) Three-way or single pole, double throw (SPDT), (C) Double pole, single throw (DPST), (D) Double pole, double throw (DPDT), (E) Three pole, single throw (3PST), (F) Three pole, double throw (3PDT)

Rocker Switches

Another type of switch is called a rocker switch. Rocker switches are used for on-off control. They may be either momentary-contact, for accomplishing temporary control, or sustained contact, for causing a load to remain in an on or off condition until the switch position is manually changed.

Pushbutton Switches

Pushbutton switches are commonly used. Many motor-control applications use pushbuttons as a means of starting, stopping, or reversing a motor; the pushbuttons are manually operated to close or open the control circuit of the motor. There are several types of pushbuttons used in the control of motors. Figure 15-3 diagrams some pushbutton styles. Pushbuttons are usually mounted in enclosures.

Ordinarily, pushbuttons are either the normally closed (NC) or normally open (NO) type. An NC pushbutton is closed until it is depressed manually.

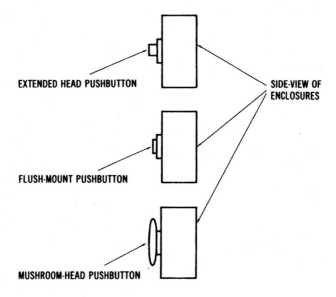

Figure 15-3. Pushbutton styles

It will open a circuit when it is depressed. The NO pushbutton is open until it is manually depressed, and then, once it is depressed, it will close a circuit. The "start" pushbutton of a motor control station is an NO type, while the "stop" switch is an NC type.

Rotary Switches

Another common type of switch is the rotary switch. Many different switching combinations can be wired using a rotary switch. The shaft of a rotary switch is attached to sets of moving contacts. When the rotary shaft is turned to different positions, these moving contacts touch different sets of stationary contacts, which are mounted on ceramic segments. The shaft can lock into place in any of several positions. A common type of rotary switch is shown in Figure 15-4. Rotary switches are usually controlled by manually turning the rotary shaft clockwise or counterclockwise. A knob is normally fastened to the end of the rotary shaft to permit easier turning of the shaft.

Limit Switches

Limit switches are made in a variety of sizes. Limit switches are merely on/off switches that use a mechanical movement to cause a change in the operation of the electrical control circuit of a motor or other load de-

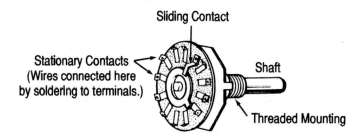

Figure 15-4. Rotary switch

vice. The electrical current developed as a result of the mechanical movement is used to limit movement of the machine, or to make some change in its operational sequence. Limit switches are often used in sequencing, routing, sorting, or counting operations in industry. Ordinarily, they are used in conjunction with hydraulic or pneumatic controls, electrical relays, or other motor-operated machinery, such as drill presses, lathes, or conveyor systems.

In its most basic form, a limit switch converts mechanical motion into an electrical control current. Part of the limit switch is the cam, an external part which is usually mounted on a machine. The cam applies force to the actuator of the limit switch. The actuator is the part of the limit switch that causes the internal NO or NC contacts to change state. The actuator operates by either linear or rotary motion of the cam, which applies force to the limit switch. Two other terms associated with limit switches are *pretravel* and *overtravel*. Pretravel is the distance that the actuator must move to change the normally open or normally closed state of the limit-switch contacts. Overtravel is the distance the actuator moves beyond the point at which the contacts change state. Both pretravel and overtravel settings are important in machine setups where limit switches are used.

Temperature Switches

Temperature switches are among the most common types of control devices used in industry. The control element of a temperature switch contains a specific amount of liquid. The liquid increases in volume when the temperature increases. Thus, changes in temperature can be used to change the position of a set of contacts within the temperature-switch enclosure. Temperature switches may be adjusted throughout a range of temperature settings.

Float Switches

Float switches are used when it is necessary to control the level of a liquid. The float switch has its operating lever connected to a rod and float assembly. The float assembly is placed into a tank of liquid, where the motion of the liquid controls the movement of the operating lever of the float switch. The float switch usually has a set of NO and NC contacts, which are controlled by the position of the operating lever. The contacts are connected to a pump-motor circuit. In operation, the NO contacts are connected in series with a pump-motor control circuit. When the liquid level is reduced, the float switch will be lowered to a point where the operating lever will be moved far enough that the contacts will be caused to change to a closed state. The closing of the contacts will cause the pump motor to turn on. More liquid will then be pumped into the tank until the liquid level has risen high enough to cause the float switch to turn the pump motor off.

Pressure Switches

Another type of electrical control device is called a pressure switch. A pressure switch has a set of electrical contacts that change states as the result of a variation in the pressure of air, hydraulic fluid, water, or some other medium. Some pressure switches are diaphragm operated. They rely upon the intake or expelling of a medium, such as air, which takes place in a diaphragm assembly within the pressure-switch enclosure. Another type of pressure switch uses a piston mechanism to initiate the action of opening or closing the switch contacts. In this type of switch, the movement of a piston is controlled by the pressure of the medium (air, water, et cetera).

Foot Switches

A foot switch is a switch that is controlled by a foot pedal. This type of switch is used for applications in which a machine operator has to use both hands during the operation of the machine. The foot switch provides an additional control position for the operation of a machine, for times when the hands cannot be used.

Drum-controller Switches

Drum controllers are special-purpose switches, which are ordinarily used to control large motors. They may be used with either single-phase or three-phase motors. The usual functions of a drum-controller switch are start/stop control or the forward/reverse/stop control of electrical motors. Contacts are moved as the handle of the controller is turned to provide machine control.

Pilot Lights for Switches

Pilot lights usually operate in conjunction with switches. Motor-control devices often require that some visual indication of the operating condition of the motor be provided. Pilot lights of various types are used to provide such a visual indication. They usually indicate either an on or off condition. For instance, a pilot light could be wired in parallel with a motor to indicate when it is on. Some types of pilot lights are:

1. *Full-voltage Across-the-line Lights*—These are relatively inexpensive lights, but they do not last as long as other types.
2. *Transformer-operated Lights*—These use a relatively low voltage to activate the lamp, but they require the expense of an additional transformer, which ordinarily reduces the operating voltage to 6 volts.
3. *Resistor-type Lights*—These lights use a series resistor to reduce the voltage across the lamp.
4. *Illuminated Pushbutton*—In these, the functions of the pushbutton and the pilot-light device are combined into one item of equipment, which reduces the mounting-space requirement.

CONTROL EQUIPMENT FOR ELECTRIC MOTORS

There are several types of electromechanical equipment used for the control of electric-motor power. The selection of power-control equipment will affect the efficiency of the power-system operation and the performance of the machinery. It is very important to use the proper type of equipment for each power-control application. This section will concentrate on the types of equipment used for motor control.

Motor-starting Control

A motor-starting device is a type of power control used to accelerate a motor from a "stopped" condition to its normal operating speed. There are many variations in motor starter design, the simplest being a manually operated on/off switch connected in series with one or more power lines. This type of starter is used only for smaller motors that do not draw an excessive amount of current.

Another type of motor starter is the magnetic starter, which relies upon an electromagnetic effect to open or close the power-source circuit of the motor. Often, motor starters are grouped together for the control of adjacent equipment in an industrial plant. Such groupings of motor starters and associated control equipment are called control centers. Control centers provide a relatively easy access to the power distribution system, since they are relatively compact, and the control equipment is not scattered throughout a large area. Motor-starting systems will be discussed in greater detail in Chapter 16.

Function of Motor Starters

Various types of motor starters are used for control of motor power. The functions of a starter vary in complexity; however, motor starters usually perform one or more of the following functions:
1. On and off control.
2. Acceleration.
3. Overload protection.
4. Reversing the direction of rotation.

Some starters control a motor by being connected directly across the power input lines. Other starters reduce the level of input voltage that is applied to the motor when it is started, so as to reduce the value of the starting current. Ordinarily, motor overload protection is contained in the same enclosure as the magnetic contactor.

Sizes of Motor Contactors

The contactors used with motor starters are rated according to their current capacity. The National Electrical Manufacturers Association (NEMA) has developed standard sizes for magnetic contactors according to their current capacity. Table 15-1 lists the NEMA standard sizes for magnetic contactors. By looking at this chart, you can see that a NEMA size 1 contactor has a 30-ampere current capacity if it is open (not mounted in a metal enclosure), and a 27-ampere capacity if it is enclosed. The corresponding maximum horsepower ratings of loads for each of the NEMA contactor sizes are also shown in Table 15-1.

Manual Starters

Some motors use manual starters to control their operation. This type

Table 15-1. Sizes of Magnetic Contactors

			Maximum Horsepower Rating of Load					
	Ampere Rating		Single-Phase			Three-Phase		
NEMA Size	Open	Enclosed	115 V	230 V	115 V	200 V	230 V	460 V
00	10	9	0.33	1	0.75	1.5	1.5	2
0	20	18	1	2	2	3	3	5
1	30	27	2	3	3	7.5	7.5	10
2	50	45	3	7.5	7.5	10	15	25
3	100	90	7.5	15	15	25	30	50
4	150	135	—	—	25	40	50	100
5	300	270	—	—	—	75	100	200
6	600	540	—	—	—	150	200	400
7	900	810	—	—	—	—	300	600
8	1350	1215	—	—	—	—	450	900
9	2500	2250	—	—	—	—	800	1600

of starter provides starting, stopping, and overload protection similar to that of a magnetic contactor. However, manual starters must be mounted near the motor that is being controlled. Remote-control operation is not possible, as it would be with a magnetic contactor. This is due to the small control current that is required by the magnetic contactor. Magnetic contactors also provide a low-voltage protection by the dropout of the contacts when a low-voltage level occurs. Manual starters remain closed until they are manually turned off. They are usually limited to small sizes.

Motor Overload Protection

Both manual and magnetic starters can have overload protection contained in their enclosures. It is common practice to place either bimetallic or melting-alloy overload relays in series with the motor branch circuit power lines. These devices are commonly called heaters. An overload protective relay is selected according to the current rating of the motor circuit to which it is connected. An identification number is used on overload protective devices. This number is used to determine the current that will cause the overload device to "trip" or open the branch circuit. Some typical overload protection or heater tables are given in Tables 15-2 and 15-3.

Table 15-2 is used for melting-alloy-type devices. The Table Number (26142) is selected according to the type and size of controller, the size of motor, and the type of power distribution (single-phase or three-phase). For example, a heater with an H37 code number is used with a motor that has a full-load current of 18.6 to 21.1 amperes.

Table 15-3 is used with bimetallic overload relays. They are selected in the same manner as melting-alloy relays. However, this table shows the trip amperes of the heater. A heater with a *K53* code number has a 13.9 trip-ampere rating. Thus, a current in excess of 13.9 amperes would cause the heater element to trip. This would open the motor branch circuit by removing the power from the motor.

Table 15-2. Melting Allow Devices

Table 26142					
Full Load Motor Amps.	Heater Code	Full Load Motor Amps.	Heater Code	Full Load Motor Amps.	Heater Code
12.6 - 14.5	H33	21.2 - 22.2	H38	34.6 - 38.9	H43
14.6 - 16.0	H34	22.3 - 23.6	H39	39.0 - 44.7	H44
16.1 - 16.9	H35	23.7 - 26.2	H40	44.8 - 48.8	H45
17.0 - 18.5	H36	26.3 - 30.1	H41	48.9 - 54.2	H46
18.6 - 21.1	H37	30.2 - 34.5	H42	54.3 - 60.0	H48

Table 15-3. Bimetallic Devices

Table 62K					
Trip Amperes	Heater Code	Trip Amperes	Heater Code	Trip Amperes	Heater Code
1.95	K21	5.59	K36	17.0	K56
2.12	K22	6.39	K37	18.3	K57
2.31	K23	6.88	K39	19.8	K58
2.49	K24	7.78	K41	21.0	K60
2.75	K26	8.42	K42	22.5	K61
3.05	K27	9.54	K43	24.1	K62
3.29	K28	10.1	K49	25.7	K63
3.69	K29	11.5	K50	28.3	K64
4.01	K31	12.6	K52	31.1	K67
4.32	K32	13.9	K53	34.6	K68
4.77	K33	15.1	K54		
5.14	K34	16.0	K55		

Classes of Motor Starters

The types of motor starters that are commercially available are divided into five classes. These classes, which were established by NEMA, are:

1. Class A—Alternating current (AC), manual or magnetic, air-break or oil-immersed starters that operate on 600 volts or less.
2. Class B—Direct current (DC), manual or magnetic, air-break starters that operate on 600 volts or less.
3. Class C—AC, intermediate voltage starters.
4. Class D—DC, intermediate voltage starters.
5. Class E—AC, magnetic starters that operate on 2200 volts to 4600 volts. Class E1 uses contacts, and Class E2 uses fuses.

Sizes of Manual Starters

A uniform method has also been established for sizing manual motor starters. Some examples of the sizes of full-voltage manual starters are:

1. Size M-O—For single-phase 115-volt motors up to 1 horsepower.
2. Size M-1—For single-phase 115-volt motors up to 2 horsepower.
3. Size M-1P—For single-phase 115-volt motors up to 3 horsepower.

These sizes are summarized in Table 15-4.

Table 15-4. Sizes of Manual Starters

	Maximum Horsepower Rating of Load				
Starter Size	Single-Phase 115 V	230 V	Three-Phase 200-230 V	460-575 V	Ampere Rating
M-O	1	2	3	5	20
M-1	2	3	7.5	20	30
M-1P	3	5	—	—	30

Combination Starters

A popular type of motor starter used in control applications is the combination starter. These starters incorporate protective devices such as fused-disconnect switches, air-type circuit breakers, or a system of fuses and circuit breakers mounted in a common enclosure. They are used on systems of 600 volts or less.

Criteria for Selecting Motor Controllers

There are several important criteria that should be considered when selecting electric motor controllers. Among these are:

1. The type of motor-AC or DC, induction or wound rotor.
2. The motor ratings-voltage, current, duty cycle, and service factor.
3. Motor operating conditions-ambient temperature and type of atmosphere.
4. Utility company regulations-power factor, demand factor, load requirements, and the local codes.
5. Type of mechanical load connected to motor-torque requirement.

In order to become more familiar with the criteria listed above, you must be able to interpret the data on a motor nameplate. The information contained on a typical nameplate is summarized as follows:

Manufacturing Co.—The company that built the motor.
Motor Type—A specific type of motor, that is: split-phase AC, universal, three-phase induction, et cetera.
Identification Number—Number assigned by the manufacturer.
Model Number—Number assigned by the manufacturer.
Frame Type—Frame size defined by NEMA.
Number of Phases (AC)—Single-phase or three-phase.
Horsepower—The amount produced at rated speed.
Cycles (AC)—Frequency the motor should be used with (usually 60 Hz).
Speed (r/min)—The amount at rated hp, voltage, and frequency.
Voltage Rating-Operating voltage of motor.
Current Rating (amperes)—Current drawn at rated load, voltage, and frequency.
Thermal Protection—The type of overload protection used.
Temperature Rating (°C)—Amount of temperature that the motor will rise over ambient temperature, when operated.
Time Rating—Time the motor can be operated without overheating (usually continuous).
Amps—Current drawn at rated load, voltage, and frequency.

Motor-controller Enclosures

The purpose of a motor-controller enclosure is obvious. The operator is protected against accidental contact with high voltages that could cause death or shock. In some cases, however, the enclosures are used to protect the control equipment from its operating environment, which may contain water, heavy dust, or combustible materials. The categories of motor-controller enclosures have been standardized by NEMA. The following list, courtesy of Furnas Electric Company, summarizes various classifications.

NEMA 1 General-Purpose enclosures protect personnel from accidental electrical contact with the enclosed apparatus. These enclosures satisfy indoor applications in normal atmospheres that are free of excessive moisture, dust, and explosive materials.

NEMA 3 (3R) Weatherproof enclosures protect the control from weather hazards. These enclosures are suitable for applications on ship docks, canal locks, and construction work, and for application in subways and tunnels. The door seals with a rubber gasket. They are furnished with conduit hub and pole mounting bracket.

NEMA 4 Weathertight enclosures are suitable for application outdoors on ship docks and in dairies, breweries, et cetera. This type meets standard hose test requirements. They are sealed with a rubber gasket in the door. NEMA 4X Corrosion Resistant fiberglass enclosures are virtually maintenance free. These U.L.-listed units are adaptable to any conduit system by means of readily available metal, fiberglass, or PVC conduit hubs, and are suitable for NEMA 3, 3R, 3S, 4, and 12 applications because they are dust-tight, rain-tight, water-tight, and oil-tight.

NEMA 12 Industrial-Use enclosures also satisfy dust-tight applications. These enclosures exclude dust, lint, fibers, and oil or coolant seepage. The hinged cover seals with a rubber gasket.

NEMA 13 Oil-tight pushbutton enclosures protect against dust, seepage, external condensation, oil, water, and coolant spray.

NEMA 7. For atmospheres containing hazardous gas, NEMA 7 enclosures satisfy Class 1, Group C or D applications, as outlined in Section 500 of the *National Electrical Code Standard of the National Board of Fire Underwriters for Electrical Wiring and Apparatus*. These cast aluminum enclosures are designed for use in atmospheres containing ethylether vapors, ethylene, cyclopropane, gasoline, hexane, naptha, benzine, butane, propane, alcohol, acetone, benzol, lacquer solvent vapors, or natural gas. Machined surfaces between the cover and the base provide the seal.

NEMA 9, For atmospheres containing explosive dust, NEMA 9 enclosures satisfy Class II, Group E, F, and G applications, as outlined by the *National Electri-*

cal Code for metal dust, carbon black, coal, coke, flour, starch, or grain dusts. The enclosure is cast aluminum with machined surfaces between cover and base to provide seal.

NEMA TYPES 1B1, 1B2, IB3 Flush Types provide behind-the-panel mounting into machine bases, columns, or plaster walls to conserve space and to provide a more pleasant appearance. NEMA 1B1 mounts into an enclosed machine cavity, NEMA 1B2 includes its own enclosure behind the panel to exclude shavings and chips that might fall from above. NEMA IB3 for plaster walls includes an adjustment to compensate for wall irregularities.

OTHER ELECTROMECHANICAL POWER CONTROL EQUIPMENT

There are so many types of electromechanical power control equipment used today that it is almost impossible to discuss each type. However, some of the very important types will be discussed in the following paragraphs.

Relays

Relays represent one of the most widely used control devices available today. The electromagnet of a relay contains a stationary core. Mounted close to one end of the core is a movable piece of magnetic material called the armature. When the coil is activated electrically, it produces a magnetic field in the metal core. The armature is then attracted to the core, which in turn produces a mechanical motion. When the coil is de-energized, the armature is returned to its original position by spring action.

The armature of a relay is generally designed so that electrical contact points respond to its movement. Activation of the relay coil will cause the contact points to "make" or "break," according to the design of the relay. A relay could be described as an electromagnetic switching mechanism. There are an almost endless number of special-purpose relays and switch combinations used for electrical power control. Figure 15-5 shows a simplified diagram of the construction of a relay that is used to control a motor.

Relays use a small amount of current to create an electromagnetic field that is strong enough to attract the armature. When the armature is attracted, it either opens or closes the contacts. The contacts then either turn on or turn off circuits that are using large amounts of current. The minimal current that must flow through the relay coil, in order to create a magnetic field strong enough to "attract" the armature, is known as the

Power Control

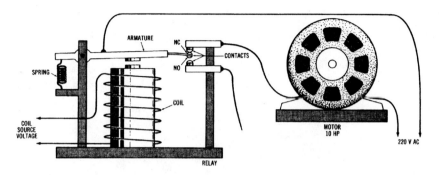

Figure 15-5. Simplified diagram of the construction of a relay that is used to control a motor

"pickup" or "make" current. The current through the relay coil that allows the magnetic field to become weak enough to release the armature is known as the "break" or "dropout" current.

There are two types of contacts used in conjunction with most relays—normally open (NO) and normally closed (NC). The NO contacts remain open when the relay coil is de-energized, and are closed only when the relay is energized. The NC contacts remain closed when the relay is de-energized, and are open only when the coil is energized.

Solenoids

A solenoid, shown in Figure 15-6 is an electromagnetic coil with a movable core that is constructed of a magnetic material. The core, or plunger, is sometimes attached to an external spring. This spring causes the plunger to remain in a fixed position until moved by the electromagnetic field that is created by current through the coil. This external spring also causes the core or plunger to return to its original position when the coil is de-energized.

Solenoids are used for a variety of control applications. Many gas and fuel oil furnaces use solenoid valves to automatically turn the fuel supply on or off upon demand. Most dishwashers use one or more solenoids to control the flow of water.

Specialized Relays

There are many types of relays used for electrical power control. General-purpose relays are used for low-power applications. They are rel-

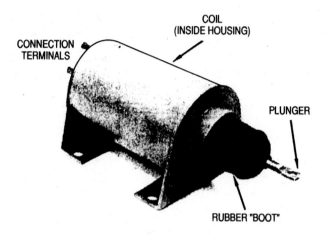

Figure 15-6. A solenoid

atively inexpensive and small in size. Many small, general-purpose relays are mounted in octal base (8-pin) plug-in sockets. Latching relays are another type of relay. They are almost identical to the relays discussed previously, but they have a latching mechanism that holds the contacts in position after the power has been removed from the coil. A latching relay usually has a special type of unlatching coil connected in series with a pushbutton stop switch. Solid-state relays are used when improved reliability, or a rapid rate of operation, is necessary. Electromagnetic relays will wear out after prolonged use, and have to be replaced periodically. Solid-state relays, like other solid state devices, have a long life expectancy. They are not sensitive to shock, vibration, dust, moisture, or corrosion. Timing relays are used to turn a load device on or off after a specific period of time.

The operation of a pneumatic timing relay is dependent upon the movement of air within a chamber. Air movement is controlled by an adjustable orifice that controls the rate of air movement through the chamber. The air-flow rate determines the rate of movement of a diaphragm or piston assembly. This assembly is connected to the contacts of the relay. Therefore, the orifice adjustment controls the air-flow rate, which determines the time from the activation of the relay until a load connected to it is turned on or off. There are other types of timing relays, such as solid-state, thermal, oil-filled, dashpot, and motor-driven timers. Timing relays are useful for sequencing operations that require a time delay between operations. A typical application would be as follows: (1) a "start" push-button is pressed, (2) a timing relay is activated, (3) after a 5-second time delay, a motor is turned on.

ELECTRONIC POWER CONTROL

Solid-state power controllers are capable of replacing electromagnetic circuit breakers and relays used for electrical power control. The introduction of computerized control for industrial equipment has brought about a new technology of machine control. Many industrial machines, such as automated manufacturing and robotic equipment, are now controlled by computerized circuits. An understanding of the basic principles of electrical control is, however, still very important for technicians.

Chapter 16

Operational Power Control Systems

Chapter 15 dealt with the equipment used for electrical power control. Now, the *operational power control systems* will be discussed in Chapter 16. Some specific applications of the equipment and devices that were studied in Chapter 15 will be investigated in this chapter.

IMPORTANT TERMS

Chapter 16 deals with operational power control systems. After studying this chapter, you should have an understanding of the following terms:

Start-Stop Pushbutton
Normally Open (NO) Pushbutton
Normally Closed (NC) Pushbutton
Overload Protection
Jogging
Limit Switch
High-Low Speed Selection
Magnetic Contactor
Full-Voltage Starting
Primary Resistance Starting
Primary Reactor Starting
Wye-Delta Starting
Autotransformer Starting
Part-Winding Starting
Direct Current (DC) Starting Systems
Forward-Reverse-Stop Control
    Three-phase Alternating Current (AC) Motors

Single-phase AC Motors
DC Motors
Motor Starting Protection
Dynamic Braking
Speed Control Circuits
Frequency Conversion Systems
Programmable Logic Controller

BASIC CONTROL SYSTEMS

Electrical power control systems are used with many types of loads. The most common electrical loads are motors, so our discussion will deal mainly with electric *motor control*. However, many of the basic control systems are also used to control lighting and heating loads. Generally, the controls for lighting and heating loads are less complex.

Several *power control circuits* are summarized in Figures 16-1 through 16-9. Figure 16-1 is a *start-stop pushbutton* control circuit with *overload protection* (OL). Notice that the *"start"* pushbutton is normally open (NO), and the *"stop"* pushbutton is normally closed (NC). *Single-phase lines* L1 and L2 are connected across the control circuit. When the start pushbutton is pushed, a *momentary* contact is made between points 2 and 3. This causes the NO contact (M) to close. A complete circuit between L1 and L2 results, which causes the electromagnetic *coil* M to be energized. When the NC stop pushbutton is pressed, the circuit between L1 and L2 will open. This causes contact M to open and turn the circuit off.

The circuit of Figure 16-2 is the same type of control as the circuit given in Figure 16-1. In the circuit of Figure 16-2, the *start-stop control* of a load can be accomplished from *three* separate locations. Notice that the start pushbuttons are connected in *parallel,* and the stop pushbuttons are connected in *series*. The control of one load, from as many locations as is desired, can be accomplished with this type of control circuit.

The next circuit (Figure 16-3) is the same as the circuit in Figure 16-1, except that a "safe-run" switch is provided. The "safe" position assures that the start pushbutton will not activate the load. A "start-safe" switch circuit often contains a key, which the machine operator uses to turn the control circuit on or off.

Figure 16-4 is also like the circuit of Figure 16-1, but with a "jog-run" switch added in series with the NO contact (M). In the "run" position, the circuit will operate just like the circuit of Figure 16-1. The "jog" position is used so that a complete circuit between L1 and L2 will be achieved and sustained only while the start pushbutton is pressed. With the selector switch in the "jog" position, a motor can be rotated a small amount at a time, for positioning purposes. *Jogging* or *inching* is defined as the momentary operation of a motor to provide small movements of its shaft.

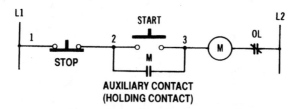

Figure 16-1. A start-stop pushbutton control circuit with overload protection (one-line diagram)

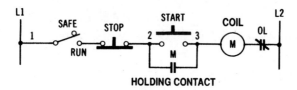

Figure 16-2. A start-stop control circuit with low-voltage protection and control from three locations

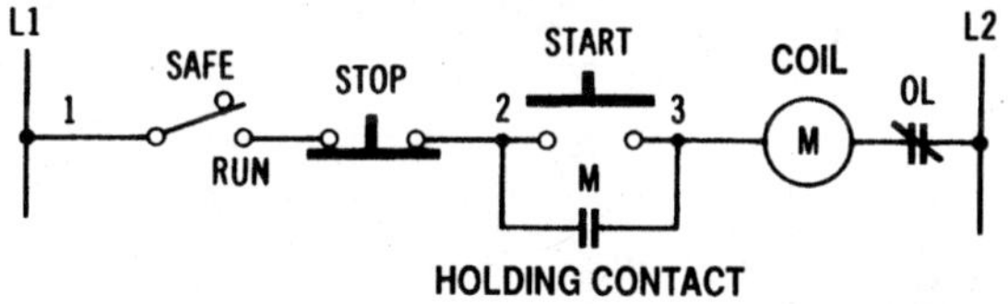

Figure 16-3. A start-stop control circuit with a safe-run selector switch

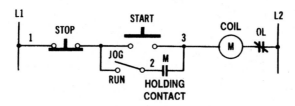

Figure 16-4. A start-stop pushbutton control circuit with a jog-run selector switch

Figure 16-5 shows a circuit that is another method of motor-jogging control. This circuit has a separate pushbutton (which relies upon an NO contact [CR] to operate) for jogging. Two control relays are used with this circuit.

The circuit in Figure 16-6 is a forward-reverse pushbutton control circuit with both forward and reverse limit switches (normally closed switches). When the "forward" pushbutton is pressed, the load will operate until the "forward" limit switch is actuated. The load will then be turned off, since the circuit from L1 to L2 will be opened. The reverse circuit operates in a similar manner. Two control relays are needed for forward-reverse operation.

The circuit of Figure 16-7 is the same as the circuit in Figure 16-6, except for the pushbutton arrangement. The forward and reverse pushbuttons are arranged in sets. Pressing the "forward" pushbutton automatically opens the reverse circuit, and pressing the "reverse" pushbutton automatically opens the forward circuit. Limit switches are also used with this circuit. Their function is the same as in the circuit shown in Figure 16-6. In the circuit in Figure 16-7, when the "forward" pushbutton is pressed, the

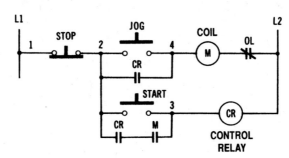

Figure 16-5. A pushbutton control circuit for start-stop-jogging

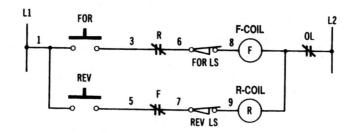

Figure 16-6. A forward-reverse pushbutton control circuit with forward and reverse limit switches

Operational Power Control Systems   431

top pushbutton will momentarily close, and the lower pushbutton will momentarily open. When points 2 and 3 are connected, current will flow from L1 to L2 through coil F. When coil F is energized, NO contact F will close, and the NC contact F will open. The "forward" coil will then remain energized. The reverse pushbuttons cause similar actions of the reversing circuit. Two control relays are also required in this case.

The circuit of Figure 16-8 is similar in function to the circuit in Figure 16-7. The pushbutton arrangement of this circuit is simpler. When the NO forward pushbutton is pressed, current will flow through coil F. When the forward control relay is energized, NO contact F will close, and NC contact F will open. This action will cause a motor to operate in the forward direction. When the NC stop pushbutton is pressed, the current through coil F is interrupted. When the NO reverse pushbutton is pressed, current will flow through coil R. When the reverse coil is energized, NO contact F will close, and NC contact F will open. This action will cause a motor to operate in the reverse direction, until the stop pushbutton is pressed again.

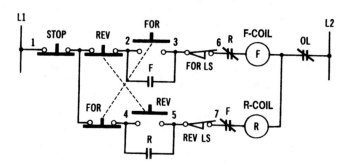

Figure 16-7. An instant forward-reverse-stop pushbutton control circuit

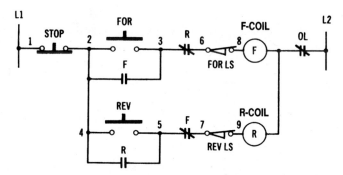

Figure 16-8. A forward-reverse-stop pushbutton control circuit

Figure 16-9 shows a circuit that is another method of forward-reverse-stop control. This control circuit has the added feature of a high- and low-speed selector switch for either the forward or reverse direction. The selector switch is placed in series with the windings of the motor. When the selector is changed from the HIGH position to the LOW position, a modification in the windings of the motor is made.

There are many other pushbutton combinations that can be used with control relays to accomplish motor control, or control of other types of loads. The circuits discussed in this section represent some basic power control functions, such as start-stop, forward-reverse-stop, jogging, and multiple-speed control.

MOTOR-STARTING SYSTEMS

The equipment used with motor-starting systems was studied in Chapter 15. Some specific systems used for starting electric motors are discussed in this section.

Function of Magnetic Contactors

Most motor-starting systems utilize one or more magnetic contactors. A schematic diagram of a magnetic contactor circuit used for control-

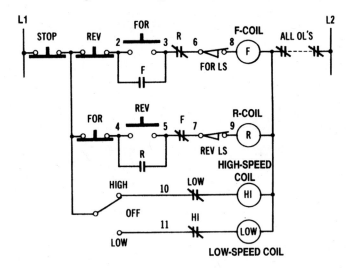

Figure 16-9. Pushbutton control circuit with a high- and low-speed selector switch

ling a single-phase motor is shown in Figure 16-10. Note that a magnetic contactor relies upon an electromagnetic coil, which energizes when current passes through it. The activated coil performs the function of closing a set of normally open contacts. These contacts are connected in series with the power input to the motor that is being controlled.

In Figure 16-10, the START pushbutton switch is an NO switch. When the start switch is pressed, current will flow through the coil of the magnetic contactor. This action energizes the coil, and the solenoid of the contactor is drawn inward to close the contacts, which are in series with the power lines. Once these contacts are closed, current will continue to flow through the electromagnetic coil through the holding contacts. Current will continue to flow until the STOP pushbutton switch is pressed. The stop switch is an NC switch. When it is pressed, the circuit to the electromagnetic coil is broken. At this time, current no longer flows through the coil. The contacts now release and cause the flow of current through the power lines to be interrupted. Thus, the motor will be turned off. Magnetic contactor circuits are sometimes referred to as *across-the-line* starters. The relay principle is utilized, since a small current through the coil controls a larger current through a motor.

Types of Starting Systems

Motor starting, particularly for large motors, can play an important role in the efficient operation of an electrical power system. There are several systems used to start electric motors. The motor-starting equipment that is used is placed between the electrical power source and the motor. Electric motors draw a larger current from the power lines during starting

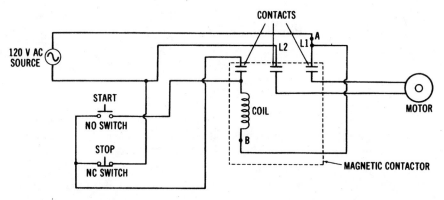

Figure 16-10. Magnetic contactor motor control circuit (Schematic diagram)

than during normal operation. Motor-starting equipment should attempt to reduce starting currents to a level that can be handled by the electrical power system where they are being used.

Full-voltage Starting—One method of starting electric motors is called *full-voltage starting*. This method is the least expensive and the simplest to install. Since full power-supply voltage is applied to the motor initially, maximum starting torque and minimum acceleration time result. However, the power system must be able to handle the starting current drawn by the motor.

Full-voltage starting is illustrated by the diagram in Figure 16-11. In this power control circuit, a start-stop pushbutton station is used to control a three-phase motor. When the NO start pushbutton is pressed, current will flow through the relay coil (M), causing the NO contacts to close. The line contacts allow full voltage to be applied to the motor when they are closed. When the start pushbutton is released, the relay coil remains energized, because of the holding contact. This contact provides a current path from L1 through the NC stop pushbutton, through the holding contact, through the coil (M), through a thermal overload relay, and back to L2. When the stop pushbutton is pressed, this circuit is opened, causing the coil to be de-energized.

Primary Resistance Starting—Another motor starting method is called *primary resistance starting*. This method uses resistors in series with the power lines to reduce the motor-starting current. Ordinarily, the resistance connected into the power lines may be reduced in steps until full voltage is applied to the motor. Thus, starting current is reduced according to the value of series resistance in the power lines. Since starting torque is directly proportional to the current flow, starting torque is reduced according to the magnitude of current flow.

OPERATIONAL POWER CONTROL SYSTEMS

Figure 16-12 shows the primary resistance starting method used to control a three-phase motor. When the start pushbutton is pressed, coils S and TR are energized. Initially, the start contacts (S) will close, applying voltage through the primary resistors to the motor. These resistors reduce the value of starting current. Once the time delay period of timing relay TR has elapsed, contact TR will close. The run contacts (R) will then close and

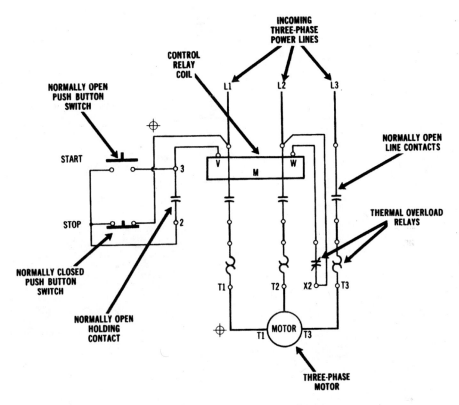

Figure 16-11. Full-voltage starting circuit for a three-phase motor (*Courtesy Furnas Electric Co.*)

apply full voltage to the motor. Notice that a step-down transformer is used for applying voltage to the control portion of the circuit. This is a commonly used technique for reducing the voltage applied to the relay coils.

Primary Reactor Starting—Another method, similar to primary resistance starting, is called the *primary reactor starting* method. Reactors (coils) are used in place of resistors, since they consume smaller amounts of power from the AC source. Usually, this method is more appropriate for large motors that are rated at 600 volts.

Autotransformer Starting—Autotransformer starting is another method used to start electric motors. This method employs one or more autotransformers to control the voltage that is applied to a motor. The autotransformers used are ordinarily tapped to provide a range of starting-cur-

436   Electrical Power Systems Technology

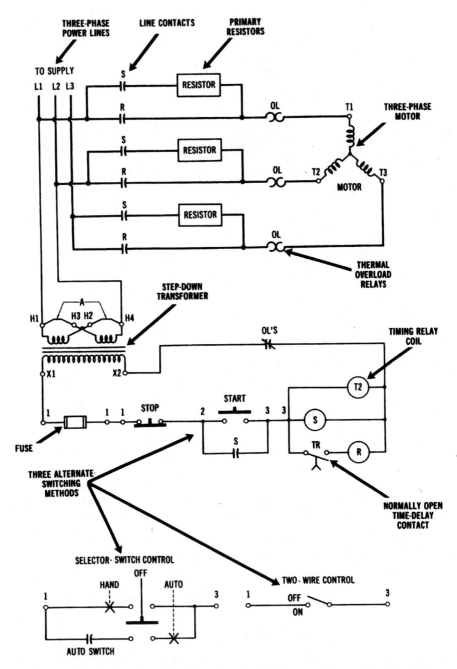

Figure 16-12. Primary resistance starter circuit (*Courtesy Furnas Electric Co.*)

rent control. When the motor has accelerated to near its normal operating speed, the autotransformer windings are removed from the circuit. A major disadvantage of this method is the expense of the autotransformers.

An autotransformer starting circuit is shown in Figure 16-13. This is an expensive type of control that uses three autotransformers and four relays. When the start pushbutton is pressed, current will flow through coils 1S, 2S, and TR. The 1S and 2S contacts will then close. Voltage will be applied through the autotransformer windings to the three-phase motor. One NC and one NO contact are controlled by timing relay TR. When the specified time period has elapsed, the NC TR contact will open, and the NO TR contact will close. Coil R will then energize, causing the NO R contacts to close and apply full voltage to the motor. NC R contacts are connected in series, with coils 1S, 2S, and TR to open their circuits when coil R is energized. When the stop pushbutton is pressed, the current to coil R will be interrupted, thus opening the power line connections to the motor.

Notice that the 65 percent taps of the autotransformer are used in Figure 16-13. There are also taps for 50 percent, 80 percent, and 100 percent, to provide more flexibility in reducing the motor-starting current.

Wye-delta Starting—It is possible to start three-phase motors more economically by using the *wye-delta-starting* method. Since, in a wye configuration, line current is equal to the phase current divided by 1.73 (or 3), it is possible to reduce the starting current by using a wye connection rather than a delta connection. This method, shown in Figure 16-14, employs a switching arrangement that places the motor windings in a wye configuration during starting, and in a delta arrangement for running. In this way, starting current is reduced. Although starting torque is reduced, running torque is still high, since full voltage appears across each winding when the motor is connected in a delta configuration.

In Figure 16-14, when the start pushbutton is pressed, coil S is energized. The normally open S contacts will close. This action connects the motor windings in a wye (or star) configuration and also activates timing relay TR and coil 1M. The NO 1M contacts then close to apply voltage to the wye-connected motor windings. After the time delay period has elapsed, the TR contacts will change state. Coil S will de-energize, and coil 2M will energize. The S contacts, which hold the motor windings in a wye arrangement, will then open. The 2M contacts will close and cause the motor windings to be connected in a delta configuration. The motor will then continue to run with the motor connected in a delta arrangement.

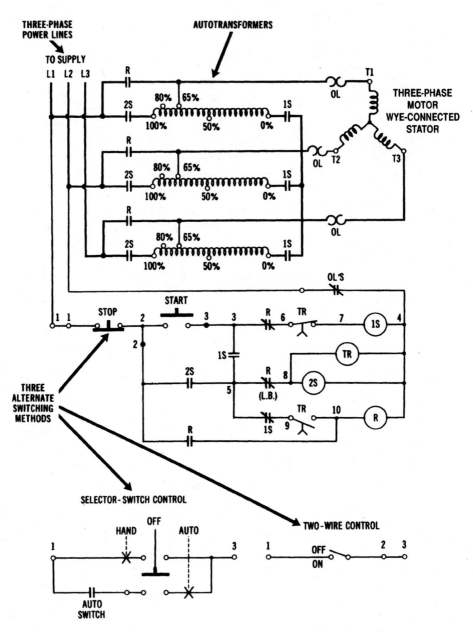

Figure 16-13. Autotransformer starter circuit used with a three-phase motor (*Courtesy Furnas Electric Co.*)

Operational Power Control Systems 439

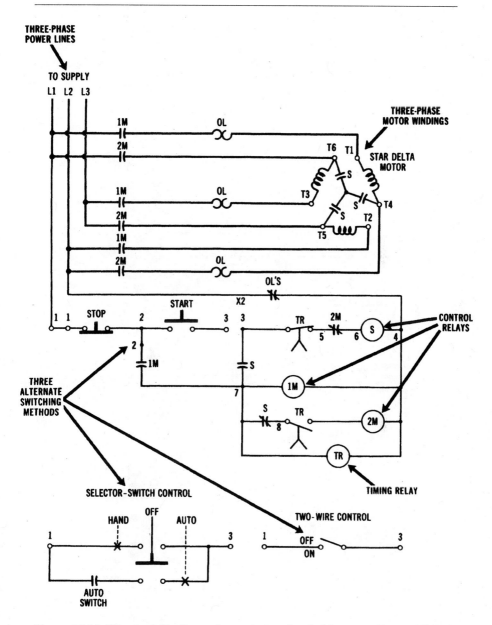

Figure 16-14. Wye-to-delta three-phase starter circuit (*Courtesy Furnas Electric Co.*)

Part-winding Starting—Figure 16-15 shows the *part-winding starting* method, which is usually simpler and less expensive than other starting methods. However, motors must be specifically designed to operate in this manner. During starting, the power-line voltage is applied across only

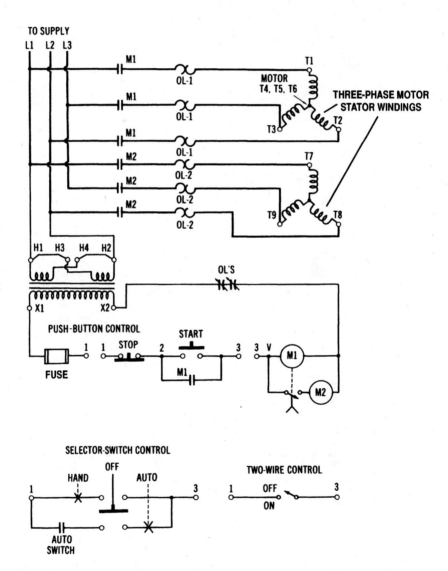

Figure 16-15. Part-winding circuit for three-phase motor starting (*Courtesy Furnas Electric Co.*)

part of the parallel-connected motor windings, thus reducing starting current. Once the motor has started, the line voltage is placed across all of the motor windings. This method is undesirable for many heavy-load applications, because of the reduction of starting torque.

In Figure 16-15, when the start pushbutton is pressed, current will flow through coil M1 of the time-delay relay. This will cause the NO contacts of M1 to close, and a three-phase voltage will be applied to windings T1, T2, and T3. After the time-delay period has elapsed, the NO contact located below coil M1 in Figure 16-15 will close. This action energizes coil M2 and causes its NO contacts to close. The M2 contacts then connect the T7, T8, and T9 windings in parallel with the T1, T2, and T3 windings. When the stop pushbutton is pressed, coils M1 and M2 will be de-energized.

Direct Current Starting Systems—Since DC motors have no counterelectromotive force (cemf) when they are not rotating (see Chapter 14), they have tremendously high starting currents. Therefore, they must use some type of control system to reduce the initial starting current. Ordinarily, a series resistance is used. This resistance can be manually or automatically reduced until a full voltage is applied. The four types of control systems commonly used with DC motors are (1) current limit, (2) definite time, (3) counter-emf, and (4) variable voltage. The *current-limit* method allows the starting current to be reduced to a specified level, and then advanced to the next resistance step. The *definite-time* method causes the motor to increase speed in timed intervals, with no regard to the amount of armature current or to the speed of the motor. The *counter-emf* method samples the amount of cemf generated by the armature of the motor to reduce the series resistance accordingly. This method can be used effectively, since cemf is proportional to both the speed and the armature current of a DC motor. The *variable-voltage* method employs a variable DC power source to apply a reduced voltage to the motor initially, and then to gradually increase the voltage. No series resistances are needed when the latter method is used.

SPECIALIZED CONTROL SYSTEMS

Electrical power control is usually desired for some specific application. In this section, we will discuss some common types of specialized, electrical power control systems that are used today.

Forward and Reverse Operation of Motors

Most types of electrical motors can be made to rotate in either forward or reverse direction by some simple modifications of their connections. Ordinarily, motors require two magnetic contactors to accomplish forward and reverse operation. These contactors are used in conjunction with a set of three pushbutton switches—FORWARD, REVERSE, and STOP. When the FORWARD pushbutton switch is depressed, the forward contactor will be energized. It is deactivated when the STOP pushbutton switch is depressed. A similar procedure takes place during reverse operation.

DC Motor Reversing

DC motors can have their direction of rotation reversed by changing either the armature connections, or the field connections to the power source. In Figure 16-16, a DC shunt-motor-control circuit is shown. When the forward pushbutton is pressed, the F coil will be energized, causing the F contacts to close. The armature circuit is then completed from L1 through the lower F contact, up through the armature, through the upper F contact, and back to L2. Pressing the stop pushbutton will de-energize the F coil.

The direction of rotation of the motor is reversed when the reverse pushbutton is pressed. This is due to the change of the current direction through the armature. Pressing the reverse pushbutton energizes the R coil and closes the R contacts. The armature current path is then from L1 through the lower R contact, down through the armature, through the upper R contact, and back to L2. Pressing the stop button will de-energize the R coil.

Single-phase Induction Motor Reversing

Single-phase AC induction motors that have start and run windings can have their direction of rotation reversed by the circuit in Figure 16-18. If we modify the diagram by replacing the shunt field coils with the run windings, and the armature with the start windings, directional reversal of a single-phase AC motor can be accomplished. Single-phase induction motors are reversed by changing the connections of either the start windings or the run windings, but not of both at the same time.

Three-phase Induction-Motor Reversing

Three-phase motors can have their direction of rotation reversed by simply changing the connections of any two power input lines. This

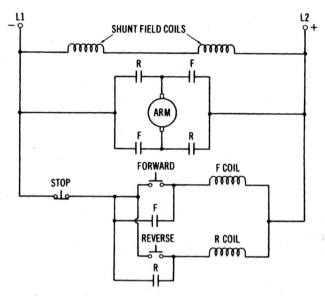

Figure 16-16. Control circuit for the forward and reverse operation of a DC shunt motor

changes the phase sequence applied to the motor. A control circuit for three-phase induction-motor reversing is shown in Figure 16-17.

When the forward pushbutton is pressed, the forward coil will energize and close the F contacts. The three-phase voltage is applied from L1 to T1, L2 to T2, and L3 to T3, to cause the motor to operate. The stop pushbutton de-energizes the forward coil. When the reverse pushbutton is pressed, the reverse coil is energized, and the R contacts will close. The voltage is then applied from L1 to T3, L2 to T2, and L3 to T1. This action reverses the L1 and L3 connections to the motor, and causes the motor to rotate in the reverse direction.

Starting Protection

Overload protection was discussed previously. Protection of expensive electric motors is necessary to extend the lifetime of these machines. The cost of overload protection is small compared to the cost of replacing large electric motors. Motors present unique problems for protection since their starting currents are much higher than their running currents during normal operation. Solid-state overload protection and microprocessor-based monitoring systems are now available to provide motor protection.

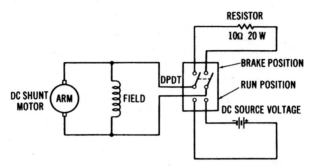

Figure 16-17. Control circuit for the forward and reverse operation of a three-phase AC induction motor

Dynamic Braking

When a motor is turned off, its shaft will continue to rotate for a short period of time. This continued rotation is undesirable for many applications. Dynamic braking is a method used to bring a motor to a quick stop whenever power is turned off. Motors with wound armatures utilize a resistance connected across the armature as a dynamic braking method. When power is turned off, the resistance is connected across the armature. This causes the armature to act as a loaded generator, making the motor slow down immediately. This dynamic braking method is shown in Figure 16-18.

AC induction motors can be slowed down rapidly by placing a DC voltage across the winding of the motor. This DC voltage sets up a constant magnetic field, which causes the rotor to slow down rapidly. A circuit for the dynamic braking of a single-phase AC induction motor is shown in Figure 16-19.

Universal Motor Speed Control

An important advantage of a universal AC/DC motor is the ease of speed control. The universal motor has a brush/commutator assembly, with the armature circuit connected in series with the field windings. By varying may be varied from zero to maximum.

The circuit used for this purpose, shown in Figure 16-20 uses a gate-controlled triac. The triac is a semiconductor device whose conduction may be varied by a trigger voltage applied to its gate. A silicon-controlled rectifier (SCR) could also be used as a speed-control device for a universal motor. Speed-control circuits like this one are used for many applications,

Operational Power Control Systems 445

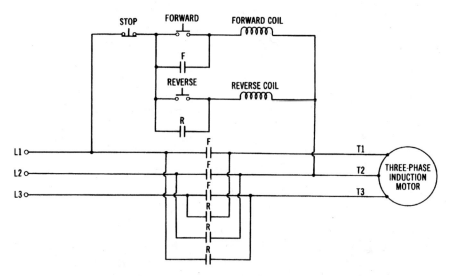

Figure 16-18. Dynamic braking circuit for a DC shunt motor

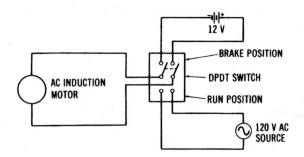

Figure 16-19. Dynamic braking circuit for a single-phase AC induction motor

such as electric drills, sewing machines, electric mixers, and industrial applications.

FREQUENCY-CONVERSION SYSTEMS

The power system frequency used in the United State is 60 hertz, or 60 cycles per second. However, there are specific applications that require other frequencies in order to operate properly. Mechanical frequency converters may be used to change an incoming frequency into some other fre-

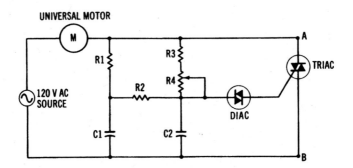

Figure 16-20. Speed control circuit for a universal motor

quency. Frequency converters are motor/generator sets that are connected together, or solid-state variable frequency drives.

For example, a frequency of 60 Hz could be applied to a synchronous motor that rotates at a specific speed. A generator connected to its shaft could have the necessary number of poles to cause it to produce a frequency of 25 Hz. Recall that frequency is determined by the following relationship:

$$\text{Freq. (Hz)} = \frac{\text{Speed of Rotation (rpm)} \times \text{No. of poles}}{120}$$

A frequency-conversion system is shown in Figure 16-21. Synchronous units, such as the one shown, are used wherever precise frequency control is required. It is also possible to design units that are driven by induction motors, if some frequency variation can be tolerated. Variable frequency drives are now being used to control many types of industrial machinery. They are used because they are not as expensive as variable speed controllers have been in the past. Solid-state variable frequency drives are used to control overhead cranes, hoists, and many other types of industrial equipment that operate from AC power lines. Variable speed can be accomplished less expensively, in most cases, with solid-state drives than with DC motors or electromechanical variable-frequency drives. Solid-state drives can be used with AC induction motors to change their speed by varying the input frequency to the motor. There are fewer maintenance problems with AC induction motors than with DC motors. In addition, fewer electromechanical parts are involved in the control operation; thus, equipment has a longer life expectancy.

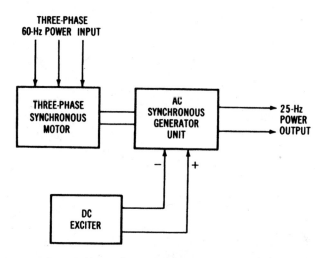

Figure 16-21. A 60 hertz to 25 hertz frequency conversion system

PROGRAMMABLE LOGIC CONTROLLERS (PLCS)

For a number of years, industrial control was achieved by electromechanical devices, such as relays, solenoid valves, motors, linear actuators, and timers. These devices were used to control large production machines where only switching operations were necessary. Most controllers were used simply to turn the load device on or off. In addition, some basic logic functions could also be achieved. Production line sequencing operations were achieved by motor-driven drum controllers with timers. As a rule, nearly all electromagnetic controllers were hardwired into the system, and responded as permanent fixtures. Modification of the system was rather difficult to accomplish and somewhat expensive. In industries where production changes were frequent, this type of control was rather costly. It was, however, the best way-and in many cases the only way-that control could be effectively achieved with any degree of success.

In the late 1960s, solid-state devices and digital electronics began to appear in controllers. These innovations were primarily aimed at replacing the older electromechanical control devices. The transition to solid-state electronics has, however, been much more significant than expected. The use of solid-state electronic devices, digital logic integrated circuits (ICs), and microprocessors has led to the development of programmable logic controllers (PLCs). These devices have capabilities that far exceed the older relay controllers. Programmable logic controllers are extremely flexible, have reduced downtime when making changeovers, occupy very little space, and have improved operational efficiency.

A programmable logic controller is very similar to a small computer. In fact, most programmable logic controllers are classified as dedicated computers. This type of unit is usually designed to perform a number of specific control functions in the operation of a machine or industrial process. The degree of sophistication or "power" of a programmable logic controller is dependent on its application. Many PLCs respond like a computer terminal and interface with a mainframe computer. Other units are completely independent and respond only to those things that are needed to control a specific machine's operation.

Programmable Logic Controller Components

A programmable logic controller is basically a software-based equivalent of the older electromagnetic relay control panel. Essentially, the PLC is a flexible system that can be easily modified and still be used as a general-purpose control device. Most PLCs can be programmed to control a variety of machine functions at one time. When a production change is necessitated, the program can be altered by a keyboard to make the system conform to the needed changes. The new control procedure may be entirely different from the original.

Mini-Programmable Logic Controller Systems

Recent improvements in large-scale IC and power transistor manufacturing technology are responsible for the development of mini-PLCs. These systems can be used economically to control simple machine operations and numerous manufacturing processes. A number of companies are now producing mini-PLCs.

Mini-PLCs are classified as systems that can economically replace as few as four relays in a control application. They are capable of providing timer and counter functions, as well as relay logic, and are small enough to fit into a standard rack assembly. Most systems of this type have fewer than 32 I/O ports or modules. Some units can be expanded to drive up to 400 I/O devices. Typically, the I/O of this type of system responds to digital signals. Some units are capable of responding to analog information. This makes it possible for the system to respond to temperature, pressure, flow, level, light, weight, and practically any analog control application.

Mini-PLCs, in general, can achieve the same control functions as the larger programmable controllers, only on a smaller basis. Mini-PLCs are usually smaller, less expensive, simpler to use, and rather efficient, compared with mainframe programmable controllers. Mini-units are now beginning to make their way into the PLC field. In the future, these systems will obviously playa greater role in the control of industrial systems.

PROGRAMMING THE PLC

Instructions for the operation of a programmable logic controller are given to the PLC through pushbuttons, a keyboard programmer, magnetic disks, or cassette tape. Each PLC has a special set of instructions and procedures that makes it functional. How the PLC performs is based on the design of its programming procedure. In general, PLCs can be programmed by relay ladder diagrams or logic diagrams. These procedures can be expressed as language words or as symbolic expressions on a CRT display. One manufacturer describes these methods of programming as *assembly language* and *relay language*. Assembly language is generally used by the microprocessor of the system. Relay language is a symbolic logic system that employs the relay ladder diagram as a method of programming. This method of programming relies heavily on relay symbols instead of words and letter designations.

Assembly Language

Assembly language is a basic instructional set that is specific to the type of the microprocessor used in the construction of the PLC. For one type of microprocessor, the assembly language program is a combination of mnemonics and labels that the programmer uses to solve a control problem. If the system is programmed directly with binary numbers, the programmer will probably use a lookup table to write the program in assembly language. It is usually easier to keep track of loops and variables in an assembly language. It is also easier for others to look at the program when they are trying to see how it works. Ultimately, the programmer must type the coded information into the computer in binary form. This generally means that an assembly language program needs to be translated into binary data before it can be entered. This can be done by laboriously writing down each mnemonic and label, line by line, and then entering the translation into memory. A convenient alternative to this procedure is to have a program that takes lines of assembly language as its input and does the translation for the programmer automatically. A program called "assembler" functionally performs this operation. It actually puts together the operational codes and the addresses that take the place of mnemonics and labels. When the procedure is complete, the assembler has a duplicate of the assembly language in the machine language of the microprocessor. These data can then be entered by the programmer manually, but they are generally applied directly to the input, since it is in memory when the translation is completed.

Relay Language or Logic

The processor of a programmable logic controller dictates the language and programming procedure to be followed by the system. Essentially, it is capable of doing arithmetic and logic functions. It can also store and handle data, and continuously monitor the status of its input and output signals. The resulting output being controlled is based on the response of the signal information being handled by the system. The processor is generally programmed by a keyboard panel program panel, or CRT terminal.

In a relay language system, the basic element of programming is the relay contact. This contact may be NO or NC. Typically, each contact has a four-digit reference number. This number is used to identify specific contacts being used in the system. The contact is then connected in either series or parallel to form a horizontal rung of a relay ladder diagram.

Once a relay program has been entered into the PLC, it may be monitored on the CRT and modified by a keystroke. Monitoring the operation of a program is achieved by illuminating the current path by making it brighter than the remainder of the circuit components. Modification of the diagram can be achieved by simply placing the cursor of the CRT on the device to be altered, and making the change with a keystroke. Cursor control is achieved by manipulation of the four arrow keys on the right of the keyboard. Contact status can be changed or bypassed, and different outputs can be turned on or off, by this procedure. Each input or output has a four-digit number that can be altered by the numerical data entry part of the keyboard.

All the control components of a PLC are identified by a numbering system. As a rule, each manufacturer has a unique set of component numbers for its system. One manufacturer has a four-digit numbering system for referencing components. The numbers are divided into discrete component references and register references. A discrete component is used to achieve on and off control operations. Limit switches, pushbuttons, relay contacts, motor starters, relay coils, solenoid valves, and solid-state devices are examples of discrete component references. Registers are used to store some form of numerical data or information. Timing counts, number counts, and arithmetic data may be stored in register devices. All component references and register references are identified by a numbering system. Each manufacturer has a particular way of identifying system components.

Programming Basics

Programmable logic controllers are provided with the capability to program or simulate the function of relays, timers, and counters. Programming is achieved on a format of up to 10 elements in each horizontal row or rung of a relay diagram, and up to seven of these rungs may be connected to form a network. A network can be as simple as a single rung, or can be a combination of several rungs, as long as there is some interconnection between the elements of each rung. The left rail of the ladder can be the common connecting element. Each network can have up to seven coils connected in any order to the right rail of the ladder. These coils can be assigned any valid number for identification. The coil number can only by used once in the operational sequence. The quantity of discrete devices and registers available for use depends on the power or capacity of the system.

When programming a relay ladder diagram into a PLC, the discrete devices and registers are placed in component format. Each component in this case is assigned a four-digit identification number. The specific reference number depends on the memory size of the system. In a low-capacity system, number assignments could be 0001 to 0064 for output coils, and 0258 to 0320 for internal coils. A system with a larger capacity might use number assignments of 0001 to 0256 for output coils, and 0258 to 0512 for internal coils. Any coil output or internal coil can only be used once in the system. References to contacts controlled by a specific coil can be used as many times as needed to complete the control operation. Output coils that are not used to drive a specific load can be used internally in the programming procedure.

When the response of a particular input module is programmed, it may be identified as a relay contact. In this regard, the symbol may be a NC contact or a NO contact. The coil or actuating member of the contact takes on the same numbering assignment as the coil. The coil, however, is identified as a circle on the diagram, and the contacts are identified by the standard contact symbol. The contacts can be programmed to achieve either the NO or NC condition, according to their intended functions.

Chapter 17

Control Devices

One of the most efficient methods of electrical power control that is circuit switching. When a switch is turned on, power is consumed by the load device. When a switch in any circuit is turned off, no power is consumed. This switching method of power control is shown in Figure 17-1 using a lamp as the load device. Since the switch is low resistant, it consumes little power. When the switch is open, it consumes no power. By switching the circuit on and off rapidly, the average current flow can be reduced. The brightness of the lamp will also be reduced. In effect, lamp brightness is controlled by the switching speed. Power in this method of control is not consumed by the control device. This type of power control is both efficient and effective.

The switching method of electrical power control cannot be achieved effectively with a manual switch. The mechanical action of the switch will not permit it to be turned on and off quickly. A switch would soon wear out if used in this manner. Electronic switching can be used to achieve the same result. Switching action of this type can be accomplished in a circuit without a noticeable flicker of a lamp. Electronic control devices such as silicon controlled rectifiers and triacs are used with the switching method for controlling power. The operation of these devices is described in this chapter.

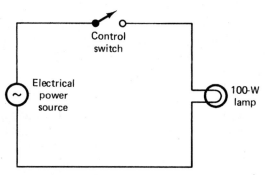

Figure 17-1. Switch-controlled lamp circuit.

IMPORTANT TERMS

Chapter 17 describes control devices that are commonly used in electrical power control systems. After studying this chapter, you should have an understanding of the following terms:

Silicon Controlled Rectifier (SCR)
Circuit Switching
I – V Characteristics
DC Power Control
AC Power Control
Triac
Static Switching
Start-Stop Control
Diac

SILICON-CONTROLLED RECTIFIERS

A silicon-controlled rectifier or SCR is probably the most popular electronic power control device today. The SCR is used primarily as a switching device. Power control is achieved by switching the SCR on and off during one alter-nation of the ac source voltage. For 60 Hz ac the SCR would be switched on and off 60 times per second. Control of electrical power is achieved by altering or delaying the turn-on time of an alternation.

An SCR, as the name implies, is a solid-state rectifier device. It conducts current in only one direction. It is similar in size to a comparable silicon power diode. SCRs are usually small, rather inexpensive, waste little power, and require practically no maintenance. The SCR is available today in a full range of types and sizes to meet nearly any power control application. Presently, they are available in current ratings from less than 1 A to over 1400 A. Voltage values range from 15 to 2600 V.

SCR Construction

An SCR is a solid-state device made of four alternate layers of P- and N-type silicon. Three P-N junctions are formed by the structure. Each SCR has three leads or terminals. The anode and cathode terminals are similar to those of a regular silicon diode. The third lead is called the *gate*. This lead determines when the device switches from its off to on state. An SCR will usually not go into conduction by simply for-ward biasing the anode and cathode. The gate must be forward biased at the same time. When these conditions occur, the SCR becomes conductive. The internal resistance of a conductive SCR is less than 1 Ω. Its reverse or off-state resistance is generally in excess of 1 MΩ. This allows the device to be similar to a mechanical switch.

A schematic symbol and the crystal structure of an SCR are shown in Figure 17-2. Note that the device has a *PNPN* structure from anode to cathode. Three distinct *P-N* junctions are formed. When the anode is made positive and the cathode negative, junctions 1 and 3 are forward biased. J_2 is reverse biased. Reversing the polarity of the source alters this condition. J_1 and J_3 would be reverse biased and J_2 would be forward biased and would not permit conduction. Conduction will occur only when the anode, cathode, and gate are all forward biased at the same time.

Some representative SCRs are shown in Figure 17-3. A few of the more popular packages are shown here. As a general rule, the anode is connected to the largest electrode if there is a difference in their physical size. The gate is usually smaller than the other electrodes. Only a small gate current is needed to achieve control. In some packages, the SCR symbol is used for lead identification.

To turn off a conductive SCR, it is necessary to momentarily remove or reduce the anode-cathode voltage. The device will then remain in this state until the anode, cathode, and gate are forward biased again. With

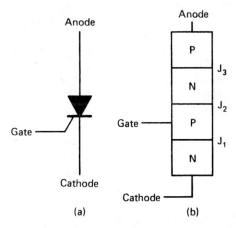

Figure 17-2. SCR crystal: (a) symbol; (b) structure

ac applied to an SCR, it will automatically turn off during one alternation of the input. Control is achieved by altering the turn-on time during the conductive or "on" alternation.

SCR *I-V* Characteristics

The current-voltage characteristics of an SCR tell much about its operation. The *I-V* characteristic curve of Figure 17-4 shows that an SCR has two conduction states. Quadrant I shows conduction in the forward direction, which shows how conduction occurs when the forward breakover voltage (V_{BO}) is exceeded. Note that the curve returns to approximately zero after the V_{BO} has been exceeded. When conduction occurs, the internal resistance of the SCR drops to a minute value similar to that of a forward-biased silicon diode. The conduction current (I_{AK}) must be limited by an external resistor. This current, however, must be great enough to maintain conduction when it starts. The holding current or I_H level must

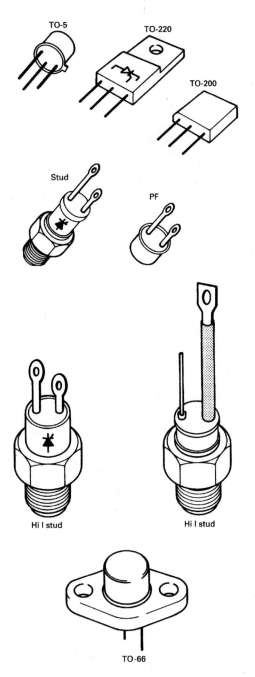

Figure 17-3. Representative SCR packages

Control Devices

be exceeded for this to take place. Note that the I_H level is just above the knee of the I_{AK} curve after it returns to the center.

Quadrant III of the I-V characteristic curve shows the reverse breakdown condition of operation. This characteristic of an SCR is similar to that of a silicon diode. Conduction occurs when the peak reverse voltage (PRV) value is reached. Normally, an SCR would be permanently damaged if the PRV is exceeded. Today, SCRs have PRV ratings of 25 to 2000 V.

For an SCR to be used as a power control device, the forward V_{BO} must be altered. Changes in gate current will cause a decrease in the V_{BO}. This occurs when the gate is for-ward biased. An increase in I_G will cause a large reduction in the forward V_{BO}. An enlargement of quadrant I of the I-V characteristics is shown in Figure 8-9, which also shows how different values of I_G change the V_{BO}. With 0 I_G it takes a V_{BO} o of 400 V to produce conduction. An increase in I_G reduces this quite significantly. With 7 mA of I_G the SCR conducts as a forward-biased silicon diode. Lesser values of

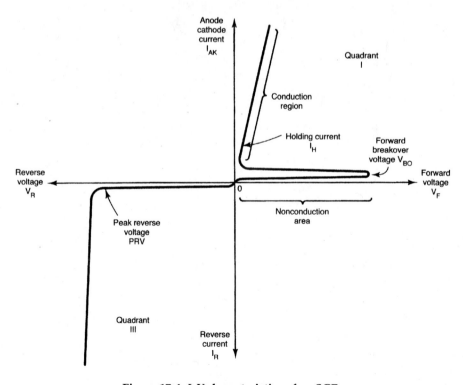

Figure 17-4. I-V characteristics of an SCR

I_G will cause an increase in the V_{BO} needed to produce conduction.

The gate current characteristic of an SCR shows an important electrical operating condition. For any value of I_G there is a specific V_{BO} that must be reached before conduction can occur, which means that an SCR can be turned on when a proper combination of I_G and V_{BO} is achieved. This characteristic is used to control conduction when the SCR is used as a power control device.

DC Power Control with SCRs

When an SCR is used as a power control device, it responds primarily as a switch. When the applied source voltage is be-low the forward breakdown voltage, control is achieved by increasing the gate current. Gate current is usually made large enough to ensure that the SCR will turn on at the proper time. Gate current is generally applied for only a short time. In many applications this may be in the form of a short-duration pulse. Continuous I_G is not needed to trigger an SCR into conduction. After conduction occurs, the SCR will not turn off until the I_{AK} drops to zero.

Figure 17-5 shows an SCR used as a dc power control switch. In this type of circuit, a rather high load current is controlled by a small gate current. Note that the electrical power source (V_S) is controlled by the SCR. The polarity of V_S must forward bias the SCR, which is achieved by mak-

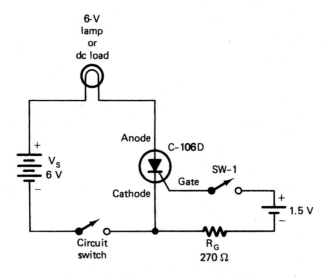

Figure 17-5. Dc power control switch

Control Devices

ing the anode positive and the cathode negative.

When the circuit switch is turned on initially, the load is not energized. In this situation the V_{BO} is in excess of the V_S voltage. Power control is achieved by turning on SW-1 which forward biases the gate. If a suitable value of I_G occurs, it will lower the V_{BO} and turn on the SCR. The I_G can be removed and the SCR will remain in conduction. To turn the circuit off, momentarily open the circuit switch. With the circuit switch on again, the SCR will remain in the off state. It will go into conduction again by closing SW-1.

Dc power control applications of the SCR require two switches to achieve control, but this application of the SCR is not practical. The circuit switch would need to be capable of handling the load current. The gate switch could be rated at an extremely small value. If several switches were needed to control the load from different locations, this circuit would be more practical. More practical dc power circuits can be achieved by adding a number of additional components. Figure 17-6 shows a dc power control circuit with one SCR being controlled by a second SCR. SCR_1 would control the dc load current. SCR_2 controls the conduction of SCR_1. In this circuit, switching of a high-current load is achieved with two small, low-current switches. SCR_1 would be rated to handle the load current. SCR_2 could have a rather small current-handling capacity. Control of this type could probably be achieved for less than a circuit employing a large electrical contactor switch.

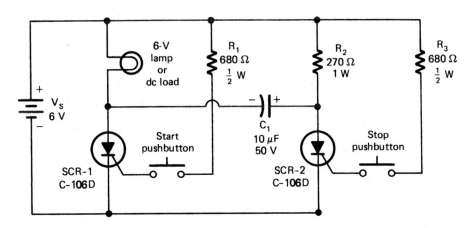

Figure 17-6. Dc power control circuit

Operation of the dc control circuit of Figure 17-6 is based on the conduction of SCR_1 and SCR_2. To turn on the load, the "start" pushbutton is momentarily closed, and thus forward biases the gate of SCR_1. The V_{BO} is reduced and SCR1 goes into conduction. The load current latches SCR_1 in its conduction state. This action also causes C_1 to charge to the indicated polarity. The load will remain energized as long as power is supplied to the circuit.

Turn-off of SCR_1 is achieved by pushing the stop button, which momentarily applies I_G to SCR_2 and causes it to be conductive. The charge on C_1 is momentarily applied to the anode and cathode of SCR_1, which reduces the I_{AK} of SCR_1 and causes it to turn off. The circuit will remain in the off state until it is energized by the start button. An SCR power circuit of this type can be controlled with two small pushbuttons. As a rule, control of this type would be more reliable and less expensive than a dc electrical contactor circuit.

Ac Power Control with SCRs

Ac electrical power control applications of an SCR are common. As a general rule, control is easy to achieve. The SCR automatically turns off during one alternation of the ac input and thus eliminates the turn-off problem with the dc circuit. The load of an ac circuit will see current only for one alternation of the input cycle. In effect, an SCR power control circuit has half-wave output. The conduction time of an alternation can be varied with an SCR circuit. We can have variable output through this method of control.

A simple SCR power control switch is shown in Figure 17-7. Connected in this manner, conduction of ac will only occur when the anode is positive and the cathode negative. Conduction will not occur until SW-1 is closed. When this takes place, there is gate current. The value of I_G lowers the V_{BO} to where the SCR becomes conductive. R_G of the gate circuit limits the peak value of I_G. Diode (D_1) prevents reverse voltage from being applied between the gate and cathode of the SCR. With SW-1 closed, the gate will be forward biased for only one alternation, which is the same alternation that forward biases the anode and cathode. With a suitable value of I_G and correct anode-cathode voltage (V_{AK}), the SCR will become conductive.

The ac power control switch of Figure 17-7 is designed primarily to take the place of a mechanical switch. With a circuit of this type, it is possible to control a rather large amount of electrical power with a rather small switch. Control of this type is reliable. The switch does not have contacts

Control Devices 461

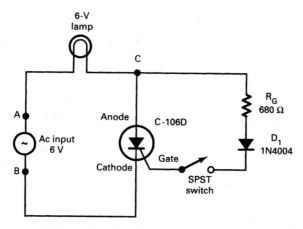

Figure 17-7. SCR power control switch

that spark and arc when changes in load current occur. Control of this type, however, is only an on-off function. SCRs are widely used to control the amount of electrical power supplied to a load device. Circuits of this type respond well to 60 Hz ac.

TRIAC POWER CONTROL

Ac power control can be achieved with a device that switches on and off during each alternation. Control of this type is accomplished with a special solid-state device known as a *triac*. This device is described as a three-terminal ac switch. Gate current is used to control the conduction time of either alternation of the ac waveform. In a sense, the triac is the equivalent of two reverse-connected SCRs feeding a common load device.

A triac is classified as a gate-controlled ac switch. For the positive alternation it responds as a *PNPN* device. An alternate crystal structure of the *NPNP* type is used for the negative alternation. Each crystal structure is triggered into conduction by the same gate connection. The gate has a dual-polarity triggering capability.

Triac Construction

A triac is a solid-state device made of two different four-layer crystal structures connected between two terminals. We do not generally use the terms *anode* and *cathode* to describe these terminals. For one alternation they would be the anode and cathode. For the other alternation they would respond as the cathode and anode. It is common practice therefore to use the terms *main 1* and *main 2* or *terminal 1* and *terminal 2* to describe these leads. The third connection is the gate. This lead determines when the device switches from its off to its on state. The gate G will normally go into conduction when it is forward biased, and is usually based on the polarity of terminal 1. If T_1 is negative, G must be positive. When T_1 is positive, the gate must be negative. This means that ac volt-age must be applied to the gate to cause conduction during each alternation of the T_1-T_2 voltage. The schematic symbol and the crystal structure of a triac are shown in Figure 17-8. Notice the junction of the crystal structure simplification. Looking from T_1 to T_2, the structure involves crystals N_1, P_1, N_2, and P_2. The gate is used to bias P_1. This is primarily the same as an SCR with T_1 serving as the cathode and T_2 the anode.

Looking at the crystal structure from T_2 to T_1, it is N_3, P_2, N_2, and P_1. The gate is used to bias N_4 for control in this direction, which is similar to the structure of an SCR in this direction. Notice that T_1, T_2, and G

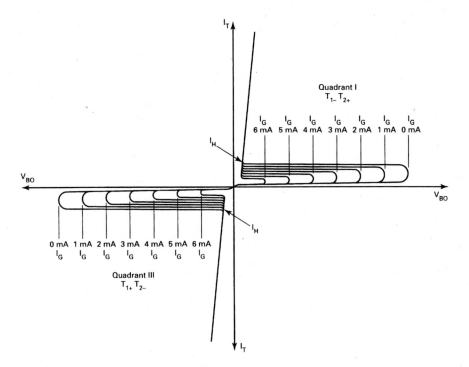

Figure 17-8. *I-V* characteristics of a triac

are all connected to two pieces of crystal. Conduction will take place only through the crystal polarity that is forward biased. When T_1 is negative, for example, N_1 is forward biased and P_1 is reverse biased. Terminal selection by bias polarity is the same for all three terminals.

The schematic symbol of the triac is representative of reverse-connected diodes. The gate is connected to the same end as T_1, which is an important consideration when connecting the triac into a circuit. The gate is normally forward biased with respect to T_1.

When ac is applied to a triac, conduction can occur for each alternation, but T_1 and T_2 must be properly biased with respect to the gate. Forward conduction occurs when T_1 is negative, the gate (G) is positive, and T_2 is positive. Reverse conduction occurs when T_1 is positive, G is negative, and T_2 is negative. Conduction in either direction is similar to that of the SCR.

Triac I-V Characteristics

The *I-V* characteristic of a triac shows how it responds to forward and reverse voltages. Figure 17-8 is a typical triac *I-V* characteristic. Note that conduction occurs in quadrants I and III. The conduction in each quadrant is primarily the same. With 0 IG, the breakover voltage is usually quite high. When breakover occurs, the curve quickly returns to the center. This shows a drop in the internal resistance of the device when conduction occurs. Conduction current must be limited by an external resistor. The holding current or I_H of a triac occurs just above the knee of the I_T curve. I_H must be attained or the device will not latch during a specific alternation.

Quadrant III is normally the same as quadrant I and thus ensures that operation will be the same for each alter-nation. Because the triac is conductive during quadrant III, it does not have a peak reverse voltage rating. It does, however, have a maximum reverse conduction current value the same as the maximum forward conduction value. The conduction characteristics of quadrant III are mirror images of quadrant I.

Triac Applications

Triacs are used primarily to achieve ac power control. In this application the triac responds primarily as a switch. Through normal switching action, it is possible to control the ac energy source for a portion of each alternation. If conduction occurs for both alternations of a complete sine wave, 100% of the power is delivered to the load device. Conduction for half of each alternation permits 50% control. If conduction is for one-fourth of each alternation, the load receives less than 25% of its normal power. It is possible through this device to control conduction for the entire sine wave, which means that a triac is capable of controlling from 0% to 100% of the electrical power supplied to a load device. Control of this type is efficient as practically no power is consumed by the triac while performing its control function.

Static Switching. The use of a triac as a static switch is primarily an on-off function. Control of this type has a number of advantages over mechanical load switching. A high-current energy source can be controlled with a very small switch. No contact bounce occurs with solid-state switching which generally reduces arcing and switch contact destruction. Control of this type is rather easy to achieve. Only a small number of parts are needed for a triac switch.

Two rather simple triac switching applications are shown in Figure 17-9. The circuit in Figure 17-9(a) shows the load being controlled by an SPST switch. When the switch is closed, ac is applied to the gate. Resistor R_1 limits the gate current to a reasonable operating value. With ac applied to the gate, conduction occurs for the entire sine wave. The gate of this circuit requires only a few milliamperes of current to turn on the triac. Practically any small switch could be used to control a rather large load current.

The circuit of Figure 17-9(b) is considered to be a three-position switch. In position 1, the gate is open and the power is off. In position 2, gate current flows for only one alternation. The load receives power during one alternation, which is the half-power operating position. In position 3, gate cur-rent flows for both alternations. The load receives full ac power in this position.

Start-stop Triac Control

Some electrical power circuits are controlled by two push-buttons or start-stop switches. Control of this type begins by momentarily pushing the start button. Operation then continues after releasing the depressed button. To turn off the circuit, a stop button is momentarily pushed. The circuit then resets itself in preparation for the next starting operation. Control of this type is widely used in motor control applications and for lighting circuits. A triac can be adapted for this type of power control.

A start-stop triac control circuit is shown in Figure 17-10. When electrical power is first applied to this circuit, the triac is in its nonconductive

Control Devices 465

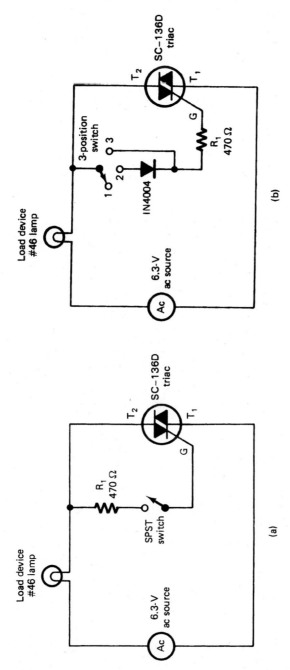

Figure 17-9. Triac switching circuits: (a) static triac switch; (b) three-position static switch

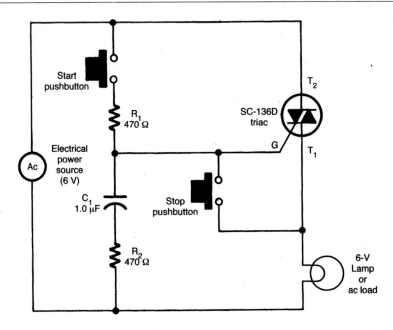

Figure 17-10. Start-stop control

state. The load does not receive any power for operation. All the supply voltage appears across the triac because of its high resistance. No voltage appears across the RC circuit, the gate, or the load device initially.

To energize the load device, the start pushbutton is momentarily pressed. C_1 charges immediately through R_1 and R_2, which in turn causes I_G to flow into the gate. The V_{BO} of the triac is lowered and it goes into conduction. Voltage now appears across the load, R_2-C_1, and the gate. The charging current of $R2$-C_1 and the gate continue and are at peak value when the source voltage alternation changes. The gate then retriggers the triac for the next alternation. C_1 is recharged through the gate and R_2. The next alternation change causes I_G to again flow for retriggering of the triac. The load receives full power from the source. The process will continue into conduction as long as power is supplied by the source. To turn off the circuit, the stop button is momentarily depressed. This action immediately bypasses the gate current around the triac. With no gate current, the triac will not latch during the alternation change. As a result, C1 cannot be recharged. The triac will then remain off for each succeeding alternation change. Conduction can be restored only by pressing the start button. This circuit is a triac equivalent of the ac motor electrical contactor.

Triac Variable Power Control

The triac is widely used as a variable ac power control device. Control of this type is generally called *full-wave control*. Full wave refers to the fact that both alternations of a sine wave are being controlled. Variable control of this type is achieved by delaying the start of each alternation. This process is similar to that of the SCR. The primary difference is that triac conduction applies to the entire sine wave. For this to be accomplished, ac must be applied to both the gate and the conduction terminals.

Variable ac power control can be achieved rather easily when the source is low voltage. Figure 17-11 shows a simple low-voltage variable lamp control circuit. Note that the gate current of this circuit is controlled by a potentiometer. Connected in this manner, adjustment of R_1 determines the value of gate current for each alternation.

Conduction of a triac is controlled by the polarity of T_1 and T_2 with respect to the gate voltage. For the positive alternation, assume that point A is positive and B is negative, thus causing a $+T_2$, a $-T_1$, and a $+I_G$. The value of the circuit gate current is determined by the resistance of R_1 and R_2. For a high-resistance setting of R_1, the triac may not go into conduction at all. For a smaller resistance value, conduction can be delayed in varying amounts. Generally, conduction delay will occur only during the first 90°

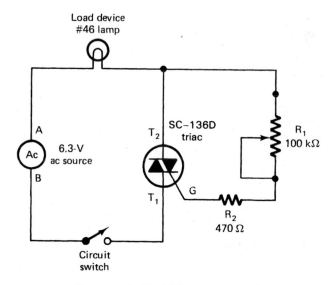

Figure 17-11. Variable lamp control

of the alternation. If conduction does not occur by this time, it will be off for the last 90° of the alternation.

Variable control of the same type also occurs during the negative alternation. For this alternation, point A is negative and point B is positive, thus causing a $-T_2$, a $+T_1$, and a $-I_G$. Gate current will flow and cause conduction during this alternation. The resistance setting of R_1 influences I_G in the same manner as it did for the positive alternation. Both alternations will therefore be controlled equally. Variable control of this type applies to only 50% of the source volt-age. If conduction does not occur in the first 90% of an alternation, no control will be achieved.

DIAC POWER CONTROL

A diac is a special diode that can be triggered into conduction by voltage. This device is classified as a bidirectional trigger diode, meaning that it can be triggered into conduction in either direction. The word *diac* is derived from "*di*ode for *ac*." This device is used primarily to control the gate current of a triac. It will go into conduction during either the positive or the negative alternation. Conduction is achieved by simply exceeding the breakover voltage.

Figure 17-12 shows the crystal structure, schematic symbol, and *I-V* characteristics of a diac. Note that the crystal is similar to that of a transistor without a base. The $N1$ and N_2 crystals are primarily the same in all respects. A diac will therefore go into conduction at precisely the same negative or positive voltage value. Conduction occurs only when in-put voltage exceeds the breakover voltage. A rather limited number of diacs are available today. The one shown here has a minimum V_{BO} of 28 V. This particular device is a standard trigger for triac control. Note that the voltage across the diac decreases in value after it has been triggered.

ELECTRONIC CONTROL CONSIDERATIONS

Electronic power control with an SCR or a triac is efficient when used properly. These devices are, however, attached directly to the ac power line. Severe damage to a load device and a potential electrical hazard may occur if this method of control is connected improperly. As a general rule, SCR and triac control should not be attempted for ac-only equipment such as fluorescent lamps, TV receivers, induction motors, and other transformer-operated devices.

SCR and triac control can be used effectively to control resistive loads such as incandescent lamps, soldering irons or pencils, heating devices and electric blankets. It is also safe to use this type of control for universal motors. This type of motor is commonly used in portable power tools and some small appliances such as portable drills, saws, sanders, electric knives, and mixers. When in doubt about a particular device, check the manufacturer's instruction manual. It is also important that the wattage rating of the load not exceed the wattage of the control device. Wattage ratings are nearly always marked on the device.

Appendix A
Trigonometric Functions

Trigonometry is a very valuable form of mathematics for anyone who studies electricity/electronics. Trigonometry deals with angles and triangles, particularly the right triangle, which has one angle of 90°. An electronic example of a right triangle is shown in Figure A-1. This example illustrates how resistance, reactance, and impedance are related in AC circuits. We know that resistance (R) and reactance (X) are 90° apart, so their angle of intersection forms a right angle. We can use the law of right triangles, known as the *Pythagorean theorem*, to solve for the value of any side. This theorem states that *in any right triangle, the square of the hypotenuse is equal to the sum of the squares of the other two sides*. With reference to Figure A-1, we can express the Pythagorean theorem mathematically as:

$$Z^2 = R^2 + X^2$$

or,

$$Z = \sqrt{R^2 + X^2}$$

By using trigonometric relationships, we can solve problems dealing with phase angles, power factor, and reactive power in AC circuits. The three most used trigonometric functions are the *sine*, the *cosine*, and the *tangent*. These functions show the ratios of the sides of the triangle, which determine the size of the angles. Figure A-2 illustrates how these ratios are expressed mathematically, and their values can be found.

Trigonometric ratios hold true for angles of any size; however, we use angles in the first quadrant of a standard graph (0° to 90°) as a reference, and in order to solve for angles greater than 90° (second-, third-, and fourth-quadrant angles), we can convert them to first-quadrant angles (see Figure A-3). All first-quadrant angles have positive functions, while angles in the second, third, and fourth quadrants have two negative functions and one positive function.

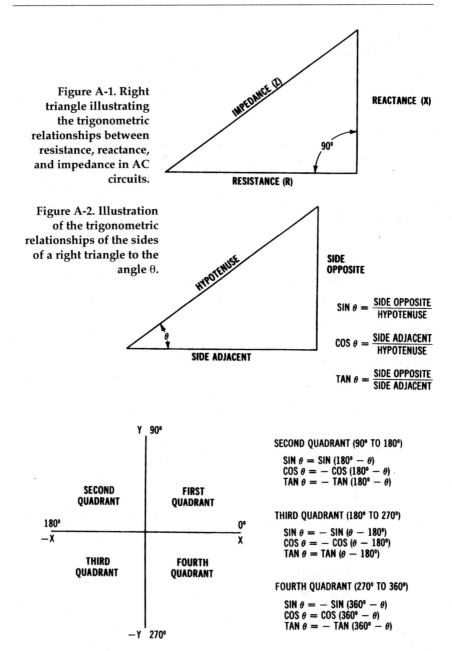

Figure A-1. Right triangle illustrating the trigonometric relationships between resistance, reactance, and impedance in AC circuits.

Figure A-2. Illustration of the trigonometric relationships of the sides of a right triangle to the angle θ.

$$\text{SIN } \theta = \frac{\text{SIDE OPPOSITE}}{\text{HYPOTENUSE}}$$

$$\text{COS } \theta = \frac{\text{SIDE ADJACENT}}{\text{HYPOTENUSE}}$$

$$\text{TAN } \theta = \frac{\text{SIDE OPPOSITE}}{\text{SIDE ADJACENT}}$$

SECOND QUADRANT (90° TO 180°)

SIN θ = SIN (180° − θ)
COS θ = − COS (180° − θ)
TAN θ = − TAN (180° − θ)

THIRD QUADRANT (180° TO 270°)

SIN θ = − SIN (θ − 180°)
COS θ = − COS (θ − 180°)
TAN θ = TAN (θ − 180°)

FOURTH QUADRANT (270° TO 360°)

SIN θ = − SIN (360° − θ)
COS θ = COS (360° − θ)
TAN θ = − TAN (360° − θ)

Figure A-3. Standard graph illustrating how trigonometric functions for angles greater than 90° (second-, third-, and fourth-quadrant angles) are derived by converting to first-quadrant angles.

Appendix B
The Elements

Element	Symbol	Atomic No.	Atomic Weight	Element	Symbol	Atomic No.	Atomic Weight	Element	Symbol	Atomic No.	Atomic Weight
Actinium	Ac	89	*227	Hafnium	Hf	72	178.6	Praseodymium	Pr	59	140.92
Aluminum	Al	13	26.97	Hahnium	Ha	105	*262	Promethium	Pm	61	*145
Americium	Am	95	*243	Helium	He	2	4.003	Protactinium	Pa	91	*231
Antimony	Sb	51	121.76	Holmium	Ho	67	164.94	Radium	Ra	88	226.05
Argon	A	18	39.944	Hydrogen	H	1	1.0080	Radon	Rn	86	222
Arsenic	As	33	74.91	Indium	In	49	114.76	Rhenium	Re	75	186.31
Astatine	At	85	*210	Iodine	I	53	126.91	Rhodium	Rh	45	102.91
Barium	Ba	56	137.36	Iridium	Ir	77	192.2	Rubidium	Rb	37	85.48
Berkelium	Bk	97	*247	Iron	Fe	26	55.85	Ruthenium	Ru	44	101.1
Beryllium	Be	4	9.013	Krypton	Kr	36	83.8	Rutherfordium or Kurchatonium	Rf or Ku	104	*260
Bismuth	Bi	83	209.00	Lanthanum	La	57	138.92	Samarium	Sm	62	150.43
Boron	B	5	10.82	Lawrencium	Lw	103	*257	Scandium	Sc	21	44.96
Bromine	Br	35	79.916	Lead	Pb	82	207.21	Selenium	Se	34	78.96
Cadmium	Cd	48	112.41	Lithium	Li	3	6.940	Silicon	Si	14	28.09
Calcium	Ca	20	40.08	Lutetium	Lu	71	174.99				
Californium	Cf	98	*251	Magnesium	Mg	12	24.32	Silver	Ag	47	107.880
Carbon	C	6	12.01	Manganese	Mn	25	54.94	Sodium	Na	11	22.997
Cerium	Ce	58	140.13	Mendelevium	Mv	101	*256	Strontium	Sr	38	87.63
Cesium	Cs	55	132.91	Mercury	Hg	80	200.61	Sulfur	S	16	32.066
Chlorine	Cl	17	35.457	Molybdenum	Mo	42	95.95	Tantalum	Ta	73	180.95
Chromium	Cr	24	52.01	Neodymium	Nd	60	144.27	Technetium	Tc	43	*97
Cobalt	Co	27	58.94	Neon	Ne	10	20.183	Tellurium	Te	52	127.61
Copper	Cu	29	63.54	Neptunium	Np	93	*237	Terbium	Tb	65	158.93
Curium	Cm	96	247	Nickel	Ni	28	58.69	Thallium	Tl	81	204.39
Dysprosium	Dy	66	162.46	Niobium	Nb	41	92.91	Thorium	Th	90	232.12
Einsteinium	E	99	*254	Nitrogen	N	7	14.008	Thulium	Tm	69	168.94
Erbium	Er	68	167.2	Nobelium	No	102	253	Tin	Sn	50	118.70
Europium	Eu	63	152.0	Osmium	Os	76	190.2	Titanium	Ti	22	47.90
Fermium	Fm	100	*255	Oxygen	O	8	16.000	Tungsten	W	74	183.92
Fluorine	F	9	19.00	Palladium	Pd	46	106.7	Uranium	U	92	238.07
Francium	Fr	87	*233	Phosphorus	P	15	30.975	Vanadium	V	23	50.95
Gadolinium	Gd	64	156.9	Platinum	Pt	78	195.23	Xenon	Xe	54	131.3
Gallium	Ga	31	69.72	Plutonium	Pu	94	244	Ytterbium	Yb	70	173.04
Germanium	Ge	32	72.60	Polonium	Po	84	210	Yttrium	Y	39	88.92
Gold	Au	79	197.0	Potassium	K	19	39.100	Zinc	Zn	30	65.38
								Zirconium	Zr	40	91.22

* Mass number of the longest-lived of the known available forms of the element.

Appendix C

Metric Conversions

This appendix lists some typical means of converting quantities in the English system into values in the metric system (and vice versa). Tables C-1 through C-6 list the typical multiplying factor for various units. However, temperature conversions require specific formulas. These formulas follow:

1. To convert degrees Fahrenheit to degrees Celsius, use one of the following formulas:

$$°C = \frac{5}{9} (°F - 32°)$$

or

$$°C = \frac{°F - 32°}{1.8}$$

2. To change degrees Celsius to degrees Fahrenheit, use one of the following formulas:

$$°F = \frac{9}{5} (°C) + 32°$$

or

$$°F = 1.8 (°C) + 32°$$

3. Degrees Celsius may be converted to Kelvins by using:

$$°C = K - 273.16°$$

4. Kelvins can be converted to degrees Celsius by simply changing the formula in Step 3 to:

$$K = °C + 273.16°$$

5. If you wish to change degrees Fahrenheit to degrees Rankine, use this formula:

$$°F = °R + 495.7°$$

6. Then, by changing the preceding formula, degrees Rankine can be converted to degrees Fahrenheit.

$$°R = °F + 459.7°$$

By using a combination of the above listed formulas, temperature values given for any scale can be changed to a temperature on one of the three other scales. Figure C-1 gives a comparison of the various temperature scales.

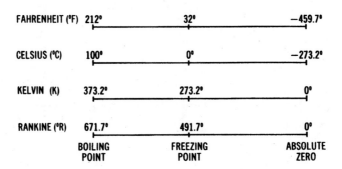

Figure C-1. Comparison of temperature scales.

Table C-1. Length

Known Quantity	Multiply by	Quantity To Find
inches (in)	2.54	centimeters (cm)
feet (ft)	30	centimeters (cm)
yards (yd)	0.9	meters (m)
miles (mi)	1.6	kilometers (km)
millimeters (mm)	0.04	inches (in)
centimeters (cm)	0.4	inches (in)
meters (m)	3.3	feet (ft)
meters (m)	1.1	yards (yd)
kilometers (km)	0.6	miles (mi)
centimeters (cm)	10	millimeters (mm)
decimeters (dm)	10	centimeters (cm)
decimeters (dm)	100	millimeters (mm)
meters (m)	10	decimeters (dm)
meters (m)	1000	millimeters (mm)
dekameters (dam)	10	meters (m)
hectometers (hm)	10	dekameters (dam)
hectometers (hm)	100	meters (m)
kilometers (km)	10	hectometers (hm)
kilometers (km)	1000	meters (m)

Table C-2. area

Known Quantity	Multiply by	Quantity To Find
square inches (in^2)	6.5	square centimeters (cm^2)
square feet (ft^2)	0.09	square meters (m^2)
square yards (yd^2)	0.8	square meters (m^2)
square miles (mi^2)	2.6	square kilometers (km^2)
acres	0.4	hectares (ha)
square centimeters (cm^2)	0.16	square inches (in^2)
square meters (m^2)	1.2	square yards (yd^2)
square kilometers (km^2)	0.4	square miles (mi^2)
hectares (ha)	2.5	acres
square centimeters (cm^2)	100	square millimeters (mm^2)
square meters (m^2)	10,000	square centimeters (cm^2)
square meters (m^2)	1,000,000	square millimeters (mm^2)
ares (a)	100	square meters (m^2)
hectares (ha)	100	ares (a)
hectares (ha)	10,000	square meters (m^2)
square kilometers (km^2)	100	hectares (ha)
square kilometers (km^2)	1,000,000	square meters (m^2)

Table C-3. Mass

Known Quantity	Multiply by	Quantity To Find
ounces (oz)	28	grams (g)
pounds (lb)	0.45	kilograms (kg)
tons	0.9	tonnes (t)
grams (g)	0.035	ounces (oz)
kilograms (kg)	2.2	pounds (lb)
tonnes (t)	100	kilograms (kg)
tonnes (t)	1.1	tons
centigrams (cg)	10	milligrams (mg)
decigrams (dg)	10	centigrams (cg)
decigrams (dg)	100	milligrams (mg)
grams (g)	10	decigrams (dg)
grams (g)	1000	milligrams (mg)
dekagram (dag)	10	grams (g)
hectogram (hg)	10	dekagrams (dag)
hectogram (hg)	100	grams (g)
kilograms (kg)	10	hectograms (hg)
kilograms (kg)	1000	grams (g)
metric tons (t)	1000	kilograms (kg)

Table C-4. Volume

Known Quantity	Multiply by	Quantity To Find
milliliters (mL)	0.03	fluid ounces (fl oz)
liters (L)	2.1	pints (pt)
liters (L)	1.06	quarts (qt)
liters (L)	0.26	gallons (gal)
gallons (gal)	3.8	liters (L)
quarts (qt)	0.95	liters (L)
pints (pt)	0.47	liters (L)
cups (c)	0.24	liters (L)
fluid ounces (fl oz)	30	milliliters (mL)
teaspoons (tsp)	5	milliliters (mL)
tablespoons (tbsp)	15	milliliters (mL)
liters (L)	1000	milliliters (mL)

Table C-5. Cubic Measure

Known Quantity	Multiply by	Quantity To Find
cubic meters (m³)	35	cubic feet (ft³)
cubic meters (m³)	1.3	cubic yards (yd³)
cubic yards (yd³)	0.76	cubic meters (m³)
cubic feet (ft³)	0.028	cubic meters (m³)
cubic centimeters (cm³)	1000	cubic millimeters (mm³)
cubic decimeters (dm³)	1000	cubic centimeters (cm³)
cubic decimeters (dm³)	1,000,000	cubic millimeters (mm³)
cubic meters (m³)	1000	cubic decimeters (dm³)
cubic meters (m³)	1	steres
cubic feet (ft³)	1728	cubic inches (in³)
cubic feet (ft³)	28.32	liters (L)
cubic inches (in³)	16.39	cubic centimeters (cm³)
cubic meters (m³)	264	gallons (gal)
cubic yards (yd³)	27	cubic feet (ft³)
cubic yards (yd³)	202	gallons (gal)
gallons (gal)	231	cubic inches (in³)

Table C-6. Electrical

Known Quantity	Multiply by	Quantity To Find
Btu per minute	0.024	horsepower (hp)
Btu per minute	17.57	watts (W)
horsepower (hp)	33,000	foot-pounds per min (ft-lb/min)
horsepower (hp)	746	watts (W)
kilowatts (kW)	57	Btu per minute
kilowatts (kW)	1.34	horsepower (hp)
watts (W)	44.3	foot-pounds per min (ft-lb/min)

Index

A
active power 50
AC generators 142
AC power control 460
AC ripple 194
admittance 46
air conditioning 319
alkaline cells 163
alternating current 25
   (AC) circuits 30
alternators 142
American wire gage (AWG) 225,
   226, 230, 286
ampacity 230, 234
ampere-hour rating 166
analog instruments 60
ANSI 407
apparent power (VA) 48, 50
assembly language 449
autotransformer 222
autotransformer starter/starting
   435

B
base units 7, 8
batteries 160
biomass 135
boilers 90, 91, 94, 95, 96
branch circuit 249, 252, 274, 275,
   278, 339, 342, 343
bridge rectifier 185
brushless DC motors 363

C
candlepower 329
capacitance 37
capacitive AC circuits 37
capacitive circuit 38
capacitive heating 310
capacitive reactance (XC) 40
capacitor 37
   motors 375, 379
capacitor-run motor 376
capacitor-start motor 376
   induction 376
capacitor filter 193, 198
carbon-zinc batteries 163
cathode ray tube 63
cemf 32
chart recorders 65
circuit breakers 241, 242, 246
clamp-on meters 76
coal-fired 98
   systems 88
coal gasification 133, 134
combination starters 419
comparative instruments 62
compound-wound DC generator
   178, 179
compound-wound DC motor 360
condensers 97
conductance 46
conductors 225, 227-230, 234, 243,
   249, 271, 273, 276, 281-283, 285-
   286
connection 55
control 19, 21
cosine 51
counterelectromotive force 354
cryogenic cable 211
current transformers 223

481

D

d'Arsonval 61
DC generator 169, 170, 172
DC motors 353, 357, 358, 360, 361, 362, 398
DC power control 458, 459
DC stepping motor 364, 365, 366, 367
delta configuration 53, 150
delta connection 54, 55
delta system 56, 263
demand factor 85, 294, 295
derived units 7, 8
diac 468
dielectric 37
direct conversion 119
disconnect switches 242
distribution 20
 systems 211, 213
drum-controller switches 414
dynamic braking 444, 445
dynamotor 362, 363

E

eddy currents 309
efficiency 90, 155, 217, 396
electrical metallic tubing (EMT) 287
electrical systems 21, 30
 power 24, 25
electric arc welding 311
electric heating 314, 317, 320
electromagnetic induction 138, 140
electromagnetic spectrum 328
elements 473
enclosures 421
energy 22, 33, 39, 66
 source 18, 21
English system of units 6

F

Faraday's Law 138, 141
farad (F) 39
feeders 249, 252
feeder circuits 280, 281, 283
filter circuits 191
float switches 414
fluorescent lighting 332, 333, 334
foot switches 414
fossil fuel 87, 97
 systems 86
frequency 71, 72, 153
 converters 445, 446
fuel cells 119, 131, 132
full-voltage starting 434
full-wave rectifier 183, 184, 185, 189
fuse 241, 244, 245, 246

G

geothermal 118, 122, 123
ground-fault indicators 74, 75
ground-fault interrupters (GFIs) 264, 265, 266, 267, 268
grounding 262, 263, 284, 285

H

half-wave rectifier 186
harmonics 153
heaters 417
heating loads 319
heat pump 318, 322, 323, 325
heat transfer 96
henry (H) 37
high-voltage direct current (HVDC) 210
horsepower 355, 356
HVAC systems 319
HVDC 211
hydraulic turbines 104

Index

hydroelectric 101, 102, 104, 105
   systems 100

I
IEEE 407
impedance 45
   (Z) 44, 46
impedance triangles 46, 47, 48
incandescent lighting 331, 337
indicator 19, 22
inductance 36
   (L) 36
induction heating 308
induction motor 369
induction welding 312
inductive circuit 37
   AC 32
inductive reactance (XL) 37
inductive susceptance 46
insulation 235, 237, 315
insulators 243
internal resistance 159, 161, 162
International System of Units 6

J
jogging 429

K
kilowatt-hour 28

L
lagging 42
lead-acid cell 165
leading 42
left-hand rule 140
lighting fixtures 344
lighting loads 347
lighting system 346
lightning arresters 242

limit switches 412
line currents 54
line voltages (VL) 54
load 19, 22, 25, 294, 308, 397, 398

M
magnetic contactors 417, 432, 433
magnetohydrodynamic (MHD) 118, 126
manual starters 416, 419
maximum power transfer 29
measurement systems 60
megohmmeters 75
mercury cells 163
metal-clad cable 286
metric measurement system 5
metric system 475
microfarad (μF) 39
motor 350, 351
   contactors 416
   overload protection 417
   reversing 442
   starters 415, 416, 419

N
National Electrical Code (NEC) 230, 269, 339
negative power 34, 35
NEMA standards 406
nickel-cadmium cell 168
nickel-iron cell 167
nickel-oxide cell 167
nonmetallic-sheathed cable (NMC) 284
nuclear-fusion 119, 128
nuclear cells 163, 164
nuclear fission 106, 107, 108, 109
nuclear fusion 118, 127, 130
nuclear reactors 110, 111

numerical readouts 64

O
oil-fired 98
oil shale 134
oscilloscope 63
overload relays 418

P
panelboards 250, 251
parallel AC circuits 43, 46, 48
parallel RC circuit 46
parallel RLC circuit 43, 45, 46
part-winding starting 440
peak demand 116
permanent-magnet DC generator 172
permanent-magnet DC motor 358
permanent capacitor motor 377
phase angle 34, 51
   (θ) 50
phase current 54
phase sequence 72
phase voltages (VP) 54
phase windings 53
pi-type filter 198
pilot lights 415
power 23, 26, 27, 28, 48, 54
   control 428, 441, 453
   demand 70, 71, 115
   distribution 209, 214, 249, 250, 256, 257, 262, 269, 278, 279
   factor 49, 52, 69, 296, 302, 397
      correction 297, 299
      meter 69
   measurement 24
   of 10 13
   outlets 253
   plants 84

   source 253
   transformers 223
   triangle 50, 51, 52
prefixes 9, 10, 11
pressure switches 414
primary cells 159, 163
primary reactor starting 435
primary resistance starting 434
programmable logic controllers (PLCs) 447
pulsating direct current 171, 182, 191
pumped-storage 106
purely capacitive circuit 39
purely inductive 34
pushbutton switches 411

R
R-L circuit 36
raceways 249, 253
RC circuit 40, 41
RC filter 198
reactance (XT) 46
reactive power 50, 297, 298
rectifiers 181
relay 246, 247, 422, 423
   language 450
repulsion motor 379, 380, 381
resistance heating 308, 321
resistance welding 311
resistive AC circuits 32
resistivity 229
right triangle 50, 52, 471, 472
rigid conduit 286
ripple factor 192
RL circuit 34
rocker switches 411
rotary converter 190
rotary switches 412

Index

rotating armature 146
rotating field 147

S

scientific notation 11
SCR 455, 456, 457, 458, 459, 460
    contactors 314
secondary cells 165, 166, 168
self-excited DC generators 174
separately excited DC generator 172, 173
series-wound DC generator 174, 175
series-wound DC motors 359, 398
series AC circuit 42, 44, 47
series RLC circuit 43
service entrance 252, 258
servomechanisms 391
servomotor 395
servo systems 394, 395
shaded-pole motor 379
    induction 378
shunt-wound DC generator 176, 177, 178
shunt-wound DC motor 359, 360
shunt regulator 200, 201
silicon-controlled rectifier (SCRs) 313, 454
sine 51
sine wave 144, 145
single-phase 256
    AC generators 145
    AC motors 368, 398
    alternator 143
solar 121
    energy 120, 121
    power 118
solenoid 423, 424
specific gravity 166

speed control 444
speed regulation 357
split-phase induction motor 372, 373
split-ring commutator 170
standards 6
static switch 464, 465
steam turbine 92, 93, 94, 148
street lighting 336
substation 240
switchgear 251
synchronous motor 382, 383, 384
synchroscope 72
synchro system 391, 392, 393, 394
system 18
    concept 17

T

telemetering 77, 78
temperature switches 413
thermal overload relays 247
thermal resistance (R) 316
three-phase 299
    AC alternator 53
    AC circuits 52
    AC generator 147, 148, 149
    AC motors 384, 399
    induction motor 384, 385
    power 56, 57, 151, 259, 260
    power analyzers 69
    rectifiers 186
    synchronous motor 384, 386, 387, 388, 389, 390
    systems 304
    transformers 259
tidal energy 119
tidal power 133
timing relay 424
toggle switches 409, 411

torque 353
total power (PT) 56
transformer 214-222, 225
transmission lines 208, 210
transmission path 19, 21
triac 461, 463, 464, 466, 467
trigonometric ratios 50
trigonometry 471
true power 49, 50

U

uninterruptible power supply (UPS) 253
units of measurement 6
unity 69
universal motors 368, 399
uranium-235 109, 110

V

vapor lighting 334
vectors 40, 42, 43
vector diagram 40, 41
visible light 328, 330
volt-amperes reactive (vars) 50
volt-ohm-milliammeter (vom) 60
voltage drop 270, 272, 274, 275, 277, 282, 283
voltage regulation 154, 199, 200
voltage regulators 202, 243
voltage variation 397

W

watt 28
watt-hour meter 66, 68
watt-second 28
wattmeter 66
Wheatstone bridge 62, 63
wind systems 118, 124, 125
work 22, 23
wound-rotor induction motor 391
wye 55
    configuration 53, 150
    delta starting 437
    system 54, 56, 57, 263, 302, 303, 304

Z

zener diodes 200